EXPLORATION IN DEVELOPMENT ISSUES

To my wife and children

Exploration in Development Issues

Selected Articles of Nurul Islam

NURUL ISLAM
International Food Policy Research Institute
Washington DC
USA

LONDON AND NEW YORK

First published 2003 by Ashgate Publishing

Reissued 2018 by Routledge
2 Park Square, Milton Park, Abingdon, Oxon OX14 4RN
711 Third Avenue, New York, NY 10017, USA

Routledge is an imprint of the Taylor & Francis Group, an informa business

A Library of Congress record exists under LC control number: 2002028376

ISBN 13: 978-1-138-74165-2 (hbk)
ISBN 13: 978-1-138-74162-1 (pbk)
ISBN 13: 978-1-315-18274-2 (ebk)

Contents

PART IV FOOD SECURITY: AGRICULTURAL PROGRESS: PAST PERFORMANCE

PART V FOOD SECURITY: POLICY REFORMS IN AGRICULTURE

PART VI FOOD SECURITY: POVERTY AND UNDERNUTRITION

List of Figures

List of Tables

Introduction: an Overview

The articles collected in this volume have been written at different times over a period of almost four decades. They cover three main areas – development strategy, food security and trade and aid – corresponding roughly to three successive phases of my professional career, starting as a professor in Dhaka University, then as the head of a research organization in development economics in Pakistan and Bangladesh, subsequently as the head of economic and social policy department of the Food and Agriculture Organization of the United Nations (FAO), and as an adviser at the International Food Policy Research Institute (IFPRI). The articles on food security were written mainly in the last phase. The articles have been classified into three areas as indicated above, even though a certain degree of overlapping could not be avoided, and have been presented in three different sections. In addition, the articles have not been arranged strictly in order of the dates of their publication; in each section they have been subclassified on the basis of closely linked issues or themes, irrespective of the exact dates of their publication.

Over this long period, research and thinking among the development specialists and/or economists on these issues have evolved and undergone changes and refinement in many ways. New approaches or methodologies in analyzing these or similar issues have emerged; new insights have been gained from the lessons of experience. Issues which were of primary concern or were in the forefront of analytical or policy debate in some period or decade were relegated to relative unimportance in the next, only to be revived again in the subsequent period.

I have done my research and writing predominantly on economic policy issues, be it trade and aid, development strategy, or food security. The articles are overwhelmingly policy-oriented; they tend to echo or represent the climate of debate, thinking and research in the different periods when they were written. The selection of articles was made in such a way that they fulfill, as far as possible, one or several of the following criteria: 1) they represent my main thinking and writing on that particular issue; 2) they were prominent in debate and discussion among academic economists/policy analysts during the time when they were written; 3) they retain some relevance and/or are of major or continuing interest in the current academic/policy discussions.

Development Policy

This section presents articles that deal with a few specific aspects of development policy. The first subsection deals with two concepts in development economics – i.e., surplus labour and balanced growth – that were once high on the agenda in the early years when problems of underdevelopment began to be looked at differently from the growth in rich countries and there was an active search for a new framework of analysis relevant to the developing countries. These new concepts went through elaboration and critical scrutiny and empirical verification; the elements of these early concepts were modified, expanded, and incorporated into a wider framework over time. A few of these concepts and the subsequent development of more rigorous analytical approaches found expression in varying degrees in the attempts to formulate comprehensive long-term (mostly five year) development plans for individual developing countries. This is illustrated in the second subsection, dealing with planning models and implementation. The last subsection takes up a few important lessons of development experience, many of which continue to be of considerable interest, debates, and discussion among both development analysts and practitioners. One set of issues represents the important aspects of political economy of development that go beyond consideration of economic growth or increase in per capita income to those of income distribution and poverty. Another set is related to the current debate regarding the pros and cons of open versus closed economy, as well as the speed and sequence of adjustment to and integration with the process of globalization, which in a limited way had its beginning already in the 1970s. The high point in this debate as it pertained to the developing countries was the spate of research and writings on the lessons of experience of the so-called East Asian miracle.

Concepts in Development Economics

Beginning mid-1950s on the publication of the pioneering works by R. Nurske and Arthur Lewis (1954, 1958) development economists came to consider the existence of surplus labour, especially in the rural areas, as the most significant characteristic of developing economies. Analytically, the prevalence of the surplus labour was ascribed to the rigidity of production structure, involving an almost fixed proportion between labour and capital that ruled out a substitution of labour for capital in response to a low relative price of labour due to the presence of a large unemployed rural labour force, not to speak in many cases, rigidity of relative factor prices as well. Fei and Ranis

(1964) subsequently explored in great detail the multifaceted implications of the phenomenon of surplus labour defined as a large unemployed labour force in relation to the existing capital stock.

A number of questions came up as the concept was subjected to further elaboration and was sought to be quantified. Is this concept relevant to a situation of self-employment or that of wage employment? Does the surplus labour express itself in open unemployment or so-called 'disguised' or invisible unemployment? Could labour force be withdrawn from rural economy or agriculture without reducing output? How is this concept related to underemployment? Is underemployment or surplus labour another way of characterizing low-wage-low productivity employment and therefore an aspect of poverty? Were not the poor too poor to remain unemployed? How could underemployed or underdeveloped rural labour be more productively mobilized, either in agriculture or in related nonfarm activities? The chapter 'Development of a Labour Surplus Economy' explores both these and other issues that occupied the interest of policy makers, national and international, for many years. It argues that the essence of rural surplus labour is the availability of labour force for urban industry at constant real wage, i.e. a perfectly elastic supply. An increasing population combined with limited, rural employment opportunities results in a highly elastic supply of labour. However, an elastic supply of labour force does not necessarily require a so-called 'unlimited' supply of labour; increased agricultural productivity reduces labour requirements per unit of output and thus releases labour for industry at unchanged real wage. Moreover, increase in agricultural productivity that produces an elastic supply of wage goods (mainly food) helps in keeping down the cost of labour and in maintaining profitability of the urban industries. The chapter explores, in the context of a two-sector model (agriculture and industry), the implications of a highly elastic supply of rural labour for agricultural progress, industrial development, intersectoral flows of resources and terms of trade, etc.

The chapter 'Concepts and Measurement of Underemployment in Developing Countries' sharpens the concepts of unemployment and underemployment further for the purposes of quantification, illustrated with empirical data from Bangladesh. The extent of 'visible' rural unemployment (defined as the number of labour days without work in a year) during 1960s was found to vary between 15 per cent and 25 per cent of total annual man-days. This was estimated on the basis of the interviews of sample rural households, who were not, however, asked whether they were available for work outside the rural areas. Furthermore, there were considerable seasonal variations, with a shortage of agricultural labour in peak seasons. To remove permanently a segment of the agricultural labour force to urban areas would have required a reorganization of agricultural operations, including measures to meet excess labour demand during peak seasons. 'Disguised' unemployment measured as employment at wage rate below the marginal product of labour was difficult to measure; there was some evidence that average income of a self-employed worker employed on his own land was frequently less than the wage earned by the landless ordinary labour. Measurement of income of the self-employed person was not satisfactory; neither the preference of a worker between self-employment and wage employment was considered. The subject of rural unemployment as well as of low productivity rural employment remains relevant even now. Current discussions on policies to alleviate rural poverty are clearly linked with the issues or concerns raised in these two chapters.

In the history of economic thought on development in poor countries, the doctrine of balanced growth widely acclaimed during 1950s and 1960s occupies a prominent place. The essence of the doctrine was that self-sustaining process of industrialization required a simultaneous expansion of mutually supporting industrial activities that provided markets for each other's output. This policy helped circumvent the most important constraint, i.e. a narrow domestic market in a low-income developing country. This concept spawned considerable debate and analysis as to extent to which a big push of this kind was needed for successful industrial development. Drawing upon contemporary debates and discussion, the chapter 'External Economies and the Doctrine of Balanced Growth' points out the additional supplementary links in the chain of interindustry relationships/interactions of which only one aspect – no doubt a very crucial aspect – i.e., demand interrelationships, was emphasized by this approach. Moreover, demand for the output of an industry is derived from income generated in not only another industry but also in agriculture, apart from export markets. Moreover, all sectors (including industry) compete for productive factors such as labour and capital. This concept was subsequently elaborated and expanded in the formulation of a variety of input/output and linear programming models. The latter incorporated not only the whole range of intersector/interindustrial relationships, including input/output relationships, but also external demand and final consumer demand-all competing for the use of primary factors of production.

Planning Models and Implementation

These latter-day programming approaches are well illustrated in the chapter

'The Relevance of Development Models to Economic Planning'. It analyzes the characteristics, merits and shortcomings of types of models used (aggregative, input/output, and optimizing models, etc.) in development planning, their data requirements and feasibility of applying them in developing countries. The chapter represents the climate of thinking and research in those early days when interest in development modeling exercises was at its height.

In fact, the decades of 1950s and 1960s constituted the heyday of planning models and development plans not only among the economic theorists but also among the policy makers in developing countries, frequently encouraged by the international donor community. They were considered essential for presenting a framework of programmes and projects (with associated policies and institutions) on the basis of which development assistance could be provided.

As the developing countries embarked on the application of these techniques, the practical problems of plan implementation came to the front. The chapter 'Development Planning and Plan Implementation: Lessons of Experience' examines a number of issues that arose in the preparation and implementation of a medium-term plan in Bangladesh in early 1970s. The limited availability and quality of socioeconomic data, inadequate knowledge about costs and benefits of alternative production and investment possibilities, uncertainty of future demand and supply possibilities, variability in the agricultural sector in export markets and in foreign aid flow, all complicated the task of long- or medium-term planning. The chapter analyzes the compromises that had to be made under conditions of uncertainty and inadequate data to provide over time for flexibility and adjustment in the plan by way of 'learning by doing'. The task was further complicated by the need to incorporate income distribution and employment in the plan objectives; no less challenging was the task of integrating investment in social sectors – i.e. education, manpower training and health, etc. – with that in economic sectors. The plan implementation required that appropriate and enabling policies (both macro and sectoral) were formulated and institutions, both private and public sectors, including the local governments were devised. There was a need for conciliation between different interest groups (rural and urban, industrial and agricultural groups), the roles of the bureaucrats and the politicians, the latter faced with decisions on trade-off between economic and noneconomic objectives.

The concept of planning has fallen into disrepute in recent years; by the 1990s the reaction against the concept of planning in view of its association with state control and regulation went so far as to make the UN Committee of

Development Planning, created in the 1960s under the guidance of Professor J. Tinbergen (one of the pioneers of planning models), change its name to UN Committee of Development Policies in 1998, under the author's chairmanship. However, issues raised and lessons learned during the earlier periods remain valid today for exercises in policy making in a developing country. Whatever may be the relative role of private/public sector, the policies are needed to manage the market, so to speak. Also, so long as public sector expenditures remain an important component of total national expenditures, choices between alternative uses of scarce public resources has to be made and the role of techniques, methods, and procedures for achieving consistency and efficiency remains important. The policy makers today face similar uncertainties and challenges as the planners of yesterday.

Lessons of Past Experience

Import substituting, domestic market-oriented industrialization was the dominant strategy in many developing countries in 1960s and 1970s. Export pessimism and the tendency for a long-run decline in the agricultural terms of trade constituted the main rationales for such a strategy. The chapter 'National Import Substitution and Inward Looking Strategies: Policies of Less-Developed Countries' provides a detailed historical perspective on the strategy of import substitution, its evolution, causes, consequences and the subsequent modification of or retreat from an unqualified acceptance of this strategy. The instruments used for the implementation of the strategy were direct control over trade and foreign exchange transactions through quotas and administrative allocations. These policy instruments required a degree of information and knowledge on the part of the administrators that was seldom available and hence decisions were highly ad hoc and arbitrary. Little room was left for price and markets in the choice of techniques or patterns of production; the levels of protection were not only very high but also were greatly diverse as between sectors and commodities. Emphasis on import substituting industrialization discouraged agriculture, the major source of income and employment that in turn limited the domestic market and contributed to the creation of considerable excess capacity in industry. At the same time, the policy discriminated against exports, since incentives were heavily skewed in favor of the import competing sector. Part III of this book provides further empirical evidence on the effects of this policy. Over time, inefficient and growth inhibiting consequences of this policy became clear to the policy makers. It was realized that the dichotomy between import substitution and export promotion was unnecessary; that import substitution was frequently carried out too inefficiently and for a longer period than was justifiable to encourage 'infant' industry and 'learning by doing'. While the domestic industrialization in the early stages necessarily involved the substitution of imports, the view that the strategy should focus on sectors and industries where a country enjoys long-run comparative advantage started to gain wide acceptance. There was no need to discriminate against exports; the structure of incentives should as far as possible be neutral between import and export sectors. In course of time, the import substitution industries when grounded on comparative advantage successfully entered the export market.

The foregoing issues remain relevant even today as the developing countries pursue a policy of structural adjustment and economic reforms. During the 1990s, the majority were engaged in dismantling the old structure of incentives, retreating from direct controls over and regulation of allocation of resources and relying increasingly on price incentives via competitive exchange rates and more uniform and lower tariffs. An important question that has arisen in respect of the ongoing policy of liberalization and deregulation is the speed and sequencing of the process. In recent debates on globalization, faced with pressures from international financial and trade organizations, many developing countries consider that the pace is too fast and exceeds the capacity of the domestic economy and institutions to adjust to the new market-biased rules. The chapter 'Learning from East Asia: Lessons for South Asia – an Epilogue' discusses a few of the above issues and much more, regarding the East Asian experience – the characteristics and causes of the so-called Asian Miracle – a phenomenon that seems to have stimulated numerous reexaminations (sometimes widely differing) of their experience in the aftermath of the Asian financial crisis of late 1990s. The 'beauty' of the East Asian experience seems to be in the eyes of the beholder; both 'interventionists' and 'free marketeers' draw evidence for their particular points of view. However, there seems to be a broad consensus that macroeconomic stability, high rates of savings and investment, investment in human capital, quality of governance, and orientation of leadership were crucial factors in contributing to high growth and poverty reduction. Also important was the initial redistribution of land, i.e. the most important asset in a poor country.

The controversy relates to the relative role of the strategic intervention by the government through subsidies, tax policies, and trade controls, etc. in achieving the structural transformation needed for high growth. How critical, for example, was the export orientation of the industrial policy in promoting competitive efficiency and growth? The strategic intervention by the state in the markets was time bound, transparent and reversible. Incentives were performance based. While incentives for import substitution were greater than export incentives in the early stages, they were not as high nor so dispersed as to cause significant misallocation of resources. At a later stage, in some cases, incentives for export promotion might have been stronger than those for import substitution. This chapter examines these questions in detail.

Policies, institutions, development-oriented governance and human capital formation – all contributed to growth and poverty alleviation. Debate on the relative importance of these factors, which also differed as between individual countries, continues unabated. The vision and commitment of the political leadership in East Asia to development was partly due to the historical circumstances, both external and international. The reasons why bureaucracy was relatively efficient and development oriented, or why rentseeking could be kept within limits, are yet to be fully understood. Nor it is clear to what extent the specific sociocultural and historical circumstances might have contributed and, if so, how to make them applicable for other countries. The chapter derives a few lessons for South Asia; it also compares with the high performing Southeast Asian countries, even though the latter did not follow the East Asian model in all respects.

Irrespective of whether the policies followed by the East Asian countries in their periods of high growth were effective or efficient, the developing countries today face new constraints on their freedom of action, so that they cannot replicate those policies even if they want to? The process of globalization of markets in goods, services, capital, technology and ideas as well as the new rules and obligations under the various international economic institutions, restrict their freedom of action. Moreover, there is a growing international recognition that matters previously considered as the exclusive domain of national governments – such as social protection, labour standards, human rights, democracy, environmental standards, competition and investment policies, etc. – are all legitimate subjects for international negotiations to be partly governed by international rules and regulations. This was certainly not the case when the East Asian countries were pursuing their high growth policies.

The pendulum in development thinking, after swinging from one extreme to the other, tends to settle in the middle, at least for some time. For example, the mainstream economists seem to be veering round to a more balanced view of the role of the state and the market away from an unqualified reliance on the unregulated play of market forces that seemed to be dominant during 1980s and 1990s. Public action is considered necessary to deal with the problems posed by missing markets, market failures and risks, uncertainties, or instabilities that are further aggravated by inadequate or asymmetric of information; however, the optimum extent or the best forms of such public action are far from universally agreed upon.

Development experience in East Asia and elsewhere in high performing countries confirmed that governance on a wide sense was of crucial importance in promoting and sustaining development. Governance, in this context, includes not only rule of law, protection of property rights, peace and security, but also management and organization on the one hand, and political decision-making processes on the other. The chapter 'Lessons of Experience: Development Policy and Practice' discusses a range of issues in this regard. The structure of administrative system and the incentives, training and skills that determine the efficiency and motivation of the administrative personnel are important. On the matter of technical competence, skill and training there is much scope for transfers from advanced countries through technical cooperation and assistance. The essential requirement of the political process must be the provision of mechanisms and institutions for reconciling the conflicting demands or interests of various groups. Development involves changes in the relative position of different groups; some lose and some gain. Even when all gain, they seldom gain in equal measures. It is necessary to ensure that the distribution of gains and losses – accompanied, wherever feasible, by compensation when needed – is accepted by society at large.

The debate as to whether democracy or authoritarianism is more growth-promoting is unresolved. Evidence so far indicates that economic development can be promoted by either type of political system; however, as income rises and material standards reach higher levels, demand for political freedom seems to emerge. In the long run, demand for political freedom and rights becomes powerful enough to destabilize an authoritarian government. Authoritarian governments promise and often produce good results in the short run; the democratic route is slow, disorganized and time-consuming, as it requires consensus-building through compromises and negotiations among divergent interest groups. The process of development is not always smooth or stable; nor is it necessarily or universally peaceful. There are occasional disruptions and dislocations; the chances are that they are less traumatic and destabilizing in an open democratic, pluralistic and decentralized system than in an authoritarian one. Compromises are more likely in a democratic system with a high probability that disruptions are minimized.

The relationship between growth, inequality of income and poverty alleviation has been a matter of perennial interest among economic analysts and policy makers. Experience showed that growth did not necessarily lead to a reduction of income inequalities or alleviation of poverty. The concern with income distribution and equity as distinct from absolute poverty suffered a relative decline in the 1970s and 1980s and drew renewed attention in the 1990s. The chapter 'Reflections on Development Perspectives since the 1960s' examines the interrelationship between growth, equity and poverty, and discusses, among others, the issues of whether growth lead to inequality; whether attempts to redistribute assets or achieve a greater equity hamper growth; whether growth lead to a fall in poverty irrespective of the degree of inequality associated with or generated by the growth process; whether the proposition that growth 'first' and equity and poverty alleviation 'later' work in practice; and what is the most appropriate pattern of growth, including sectoral priorities, that promotes equitable growth?

Most high performing countries suffered increasing inequality at a time of rapid growth. In some countries, inequality persisted even at a relatively advanced stage of development. As distinct from income inequality, poverty, however, in most cases, diminished in response to rapid growth. Also, when public action was taken to redistribute assets (such as land) in the beginning of the development process, growth was not only more equitable but also led to a faster reduction of poverty. Slow down in population growth and expansion of nonfarm employment opportunities, both rural and urban, helped promote poverty reduction.

Human development implies more than an increase in material wealth and a rise in per capita income; it also includes wide access to health and education, especially on the part of the poor, low infant/child mortality, long life expectancy, low illiteracy rates and high school enrolment, etc. A high degree of access to health and education does not merely depend on a high level of income, either aggregate or per capita; it also critically depends on high pubic expenditures on the provision of health and educational facilities. Among the high income countries, some provided a wide access to adequate health and education facilities, while many others did not do so.

The chapter 'Economic Regress: Concepts and Features' elaborates these issues in terms of the relation between income growth, poverty, or human deprivation as indicated by high infant mortality or low life expectancy. It discusses *not* how per capita income growth is related to the indices of human deprivation, *but* takes up the reverse issue, i.e. how economic regress (as against economic progress), defined in terms of a decline in absolute income or negative rate of growth in per capita income, is related to changes in infant mortality. This chapter examines the evidence produced by A.K. Sen that, in many cases, in spite of a decline in per capita income, there was an improvement or decline in infant mortality rate and that there was no correlation between the worst performers in respect of the per capita income and the worst performers in respect of the decline in infant mortality or life expectancy. The determining variables in either achieving a high level of human development in a relatively poor country and in preventing a decline in human development in a situation of economic regression is the relative magnitude of public expenditures on health and education.

Food Security

The concern with food and agriculture has been uppermost in development literature right from the beginning. After all, this sector provided the largest share of income and employment in the developing countries; the large majority of the poor in these countries lived in the rural areas and on agriculture; it provided the basic sustenance and source of nutrition for the population as a whole. To provide an adequate and stable food supply in the developing countries and to ensure that all the people, especially the poor, get access to the basic food they require were major policy issues. All these wide-ranging issues continue to be of high priority at the present time. They are succinctly adumbrated in the concept of food security. The chapters in this section deal with food security in this broad sense. They cover: a) future food demand and supply prospects; b) policy reforms in agriculture; c) access of households to food and related problems of undernourishment and poverty; and d) variability in food supplies and prices, their implications, and measures to deal with them.

Agricultural Progress: Past Performance and Future Prospects

The significant acceleration of growth in agriculture, especially in Asia during 1960s and 1970s, attracted considerable attention. Many analyses focused on lessons that could be derived from it for others as well as for sustained growth in Asia in the future years. The chapter 'The Green Revolution in Asia: A Few Aspects and Some Lessons' deals with the now widely known factors that contributed to the cereal-based Green Revolution in Asia, with particular emphasis on the role of irrigation, new seed varieties, and fertilizer – all interacting synergistically – and supplemented by investment in research, institutions and infrastructure. The most important lesson was the crucial importance of technology – mechanical, biological, and chemical innovations that generate high growth in productivity. Added to this was a favorable policy environment of high incentives such as input subsidies and output price support policies. Also, ensured and adequate supply of inputs was more important than price subsidies; equally important were education and training, extension, on the one hand, and physical infrastructure for wide diffusion and efficient utilization of new technology, on the other.

To accelerate or maintain a high growth rate in the future, a sustained increase in agricultural research, both national and international, is needed to generate new technology as well as to preserve and manage land and water. Water is going to be in the future in greater scarcity than land. As the chapter on the 'Issue of Sustainability in Third World Food Production' discusses, the challenge for the future. It emphasizes the role of high productivity, environment-friendly technology, to mitigate or avoid the adverse environmental consequences of high input technology; at the same time technology appropriate for low-productivity regions with inferior soils and water is to be designed. Policies (such as fiscal and credit, etc.) as well as institutions (such as property rights) that tend to encourage a wasteful and excessive use of inputs, land, and water are to be abandoned.

Looking towards the next two decades or so, the two chapters 'Population and Food in the Early Twenty-first Century' and 'Asia's Food and Agriculture: Future Perspectives and Policy Influences' analyze the future food supply and demand prospects. The implications of alternative assumptions about income and consumption growth, land and water constraints, and technological possibilities are explained. The chapters were written in response to the concerns expressed in early 1990s, with rising food prices and declining world food stocks, regarding the prospects of increasing food supply.

The projections confirm that increase in food supply will be able to match the increase in demand. There is expected to be a significant change in the composition of food demand with a rising proportion of food demand devoted to livestock and horticultural products. A substantial increase in indirect demand for cereals to be used as feedgrains is projected. Possible climate change such as global warming will tend to aggravate uncertainties in weather and hence the stability of agricultural production.

The developing countries as a whole would be net food importers and their import requirements would be met by the export surpluses of the developed world. Two regions, Asia and Sub-Saharan Africa, are likely to face significant food gaps and would in general remain poorer than the rest of the developing world. In Asia, South Asia would face a larger food gap than the rest of Asia, i.e. East and Southeast Asia. The second chapter also explores quantitatively the possible impact of food supply and demand projections on the prospects of poverty in Asia. The impact of growth in agriculture and associated growth in overall per capita income on poverty alleviation is analyzed; and investment in health, education, sanitation for the reduction of poverty. It examines how important is the growth in nonfarm income.

The basic component of food security is the access to food for all, especially the poor. This, in turn, is related to poverty and undernutrition. This is acute in South Asia and Sub-Saharan Africa. How did the Green Revolution affect inequality and poverty? The lessons that were derived from the Asian Green Revolution in 1960s were found subsequently to be relevant in Asia and elsewhere. Neither the farm size nor the system of tenure was an important factor in the diffusion of the Green Revolution. The small farmers, after a time lag, obtained new technology; however, their access to inputs and services was more limited, since they had limited financial resources; also they paid higher prices. Hence, even though their income increased, the percentage increase in their income was less than that for the large farmers. Since new technology was evolved only for a few food crops, those producing other crops, both food and nonfood, did not benefit, thus causing a disparity between the incomes of the two groups of fanners. Similarly, there was an increasing disparity between more productive/fertile regions, to which the new technology was more appropriate and less productive regions.

A small farmer-oriented strategy which ensures adequate supply of inputs accompanied by access to credit and education and training was twice blessed. It not only improved income and employment of the poor farmers but also promoted a higher growth rate since their productivity, i.e. yield per hectare, was higher than that of the large farmers. The chapter 'Agricultural Growth, Technical Programmes, and Rural Poverty' examines in some detail these interrelated issues. This chapter thus links up with the earlier chapter on 'The Green Revolution in Asia, etc.'. It elaborates the various linkages between technical progress, growth and poverty, and examines how and to what extent the pace and pattern of growth can be designed to have a positive impact on poverty reduction. For example, it discusses how the direct impact of agricultural growth on poverty depends on the extent to which the choice of technology, access of the poor farmers to credit, the competitiveness or otherwise of the markets for inputs and outputs are, etc. increases the employment or income of the poor. The indirect impact of agricultural growth on poverty depends on what happens to food prices as well as on the nature and magnitude of intersectoral production and consumption linkages and their impact on income and employment of the poor in the rest of the economy. Increased production or productivity of the small farmers was frequently not large enough to lift them above the poverty line; hence was the need for direct measures to alleviate poverty either through income transfers or price subsidies or through the direct provision of employment, in especially designed public employment schemes.

Policy Reforms in Agriculture

Since the achievement of adequate food supply is essential for providing food security, much attention has been given to the optimum or efficient policy instruments needed for expanding food supply. During 1970s and 1980s a number of policies (macro and sectoral) that were pursued during 1960s and earlier – and are explained in the chapter on 'The Green Revolution in Asia, etc.' – were subject to re-examination and evaluation. Two sets of issues have dominated the debate; one related to the appropriate set of incentives, including output and input pricing policies, on the one hand; and, on the other hand, nonprice factors such as public investment in research, education and infrastructure, etc. Their interrelations and role in promoting output growth, technological progress and efficiency of resource use have attracted much attention in past years and continues, even now, to attract considerable research and analysis. The second set of issues is concerned with institutions including systems of land ownership and land tenure, training and extension services, credit and marketing, etc. During 1980s and 1990s, policy analysis shifted heavily towards the structure of incentives in agriculture, in particular input and output pricing policies; the second set of issues suffered a relative decline in emphasis with an indication only in very recent years of a re-emergence of interest in them.

The two chapters 'Government: Issues in Policy Reforms' and 'Tensions Between Economics and Politics in Dealing with Agriculture: A Comment' cover both sets of issues, including an analysis of various policy reforms undertaken during 1980s and 1990s and their consequences. A policy of low food prices in favor of consumers partly offset by low subsidized input prices was found to be neither efficient nor equitable. Similar was the case with the state marketing and distribution as well as the state's control over the farmers' resource allocation decisions. Both sets of policies were gradually revised during 1980s and 1990s, but there were a few unintended adverse effects of policy reforms. Rapid changes in favor of free markets without appropriate institutional framework resulted, for example, in a substitution of private monopoly for public monopoly or a total disruption in the supply when state marketing agencies were dismantled without private enterprises being ready to fill the gap. The changes in trade and exchange rate policies that discriminated against agriculture, without corresponding changes in the domestic markets and infrastructure, did not benefit the farmers. In a few middle-income developing countries, agriculture, instead of being discriminated against, became the beneficiary of protectionism, supported by politically influential farmers' lobby.

The above two chapters seek to emphasize firstly the role of nonprice factors (such as technology, infrastructure, institutions) and secondly the appropriate sequencing of policy reforms. The much-debated question of the impact of output price policies on supply response, technology generation and income distribution and poverty is examined. Also, the two chapters – especially the second – discuss the politics of policy reforms, including the pressure of rural and urban interest groups. The role of human capital in agricultural progress in the promotion of equity and poverty reduction is emphasized. The importance of this factor, human capital development, has stood the test of time; in fact, in most recent times, it has gained primacy in policy debates. The chapter further examines to what extent price incentives for farmers may depress the real income of the poor food consumers and how the adverse impact on poverty can be mitigated by such measures as direct income transfers, subsidies or public employment programmes for the poor.

While on the subject of efficiency of resource allocation in agriculture, the choice between food crops and cash crops or nonfood crops has raised some controversy. Is food security of nations or individuals compromised if nonfood crops such as export crops or agricultural raw materials are produced in lieu of cereals and noncereal food crops? Should the choice between alternative cropping patterns be guided by relative rates of returns, especially when the possibilities of exchange through trade exist? In other words, should not the comparative cost considerations determine the choice? The 'food-first' school, for example, believes that the production of nonfood crops for either domestic or world markets mitigates against food security at the level of both households and countries. Commercialization, it is argued, exposes individuals/nations to the vagaries of market and endangers an assured access to food. It is argued that food supplies and prices are volatile, whereas cash or export, if available for exchange against food crops, and also crops, suffer not only from high/short-run price fluctuations in export markets but also from a long-run decline in the terms of trade.

Research and experience have demonstrated that this is a very simplistic approach. The chapter 'Commercialization of Agriculture and Food Security: Development Strategy and Trade Policy' elaborates both analytically as well as on the basis of available quantitative information, both historical and current, the range of issues or considerations bearing on the choice between food and nonfood crops or between food crops for consumption or production for the market. In reality, this is a very complex issue depending on a variety of factors. It discusses how and to what extent the choice based on the relative costs and returns may be modified by such considerations as the degree of relative output and price volatility, risk aversion of farmers, relative ease or cost of access to markets, availability of marketing infrastructure and future market prospects. The implications of choice between food and export crops for employment, poverty, and nutrition are also examined.

Poverty and Undernutrition

Undernutrition, hunger and poverty that determine access to food are interrelated phenomena. The chapter 'Undernutrition and Poverty: Magnitude, Pattern, and Measures for Poverty Alleviation' provides a broad survey of several aspects of this triangular relationship in developing countries, including the magnitude as well as the pattern/characteristics of undernutrition and factors contributing to the failure of entitlements or effective demand. The broad range of measures for alleviating hunger and undernutrition, including subsidized food distribution system and public employment programmes, etc. are discussed. The merits and shortcomings of these measures are illustrated with reference to the conditions of South Asia, where experience with direct poverty alleviating measures such as subsidized public distribution, public works programmes and credit for the rural poor has been very extensive. The lessons learned from the experience are also elaborated; this discussion provides a link to the discussions in the chapter on 'World Food Security: National and International Means for Stabilization of Supplies'.

It has been argued that countries that suffer from endemic hunger and poverty have in the past often succeeded in preventing the occurrence of famines, while those making progress in reducing endemic hunger or providing widespread access to health and education fail to prevent famine. Second, among countries with higher per capita income, some have provided a wide access to health and education, while others have not done so.

The chapter 'Hunger, Famine, and Poverty: A Few Considerations of Political Economy' critically examines these two prepositions and related issues drawing upon empirical evidence. A positive relationship between democracy, free press and prevention of famine has been much emphasized by the Nobel Laureate, Professor A.K. Sen, who also argues that while democratic governments with free press and vocal opposition are committed to the prevention of famines, the alleviation of long-term deprivation and poverty do not seem to engage serious commitment on their part. This chapter elaborates the implications of this hypothesis and discusses a few circumstances in which this hypothesis does not seem to hold.

Instability of Food Supplies and Prices

Food security requires that access to adequate food supplies is available at all times at prices that are within the reach of the poor. Fluctuations in food supplies and prices from season to season or from year to year disrupt access to food and hence food security, causing considerable deprivation, and even famine, at times of severe scarcity and high prices. The chapter on 'World Food Security: National and International Means for Stabilization of Supplies' provides a very comprehensive treatment of all the interrelated issues. It deals with the nature, magnitude and sources of supply instability, including weather variation, market imperfections, interrelated instability in cereals and feed sectors and price instability, as well as consumption stability. Food price instability has always been a major concern among policy makers in all countries, developing and developed. Governments frequently intervene in the food markets to stabilize prices. Also at the international level during 1960s and 1970s, food price instability in world markets was a subject of international negotiations with an attempt to devise or explore cooperative arrangements among nations to stabilize the world market.

A complete withdrawal from public intervention in markets with a view to stabilizing food price has been found to be politically unfeasible until now. The search for appropriate and least market-'distorting' public intervention schemes, through public stocks or for an appropriate combination of trade policy and domestic public stocks, continues. At the same time, the role of private sector marketing and distribution is sought to be expanded. The chapter describes the debate among the mainstream economists as how to use foreign trade as the principal mechanism for stabilizing prices in the domestic market and to minimize the role of public stocks that often tend to be very costly and inefficient. On the international level, there were attempts to: a) negotiate multilateral agreements to build up international food stocks and/or to coordinate national food stocks; and b) to provide financial assistance to developing countries to meet high food import costs at times of scarcity. The chapter traces the history of negotiations to build up the world food security reserves that failed; the IMF facility that was introduced to meet sudden upsurge in food import costs during 1980s was later modified or weakened substantially. These episodes in a way signify two recent trends: a) a decline of international economic cooperation in general; and b) a retreat from state intervention in markets and prices.

The issues of price instability in the world food markets and of its impact on the domestic economy have assumed renewed importance in the context of the WTO negotiations. While agricultural trade liberalization and a consequent reduction/elimination of domestic support programmes (that is dominant in several developed, food-exporting countries) if it is on a significant scale, may help reduce price instability in world markets, at the same time it would cause a fall in food surpluses, and a corresponding decline in world food stocks, thus reducing their stabilizing role in world markets. This may lead to both a rise in food prices and an increase in price instability. Moreover, climate change with its side effects leading to a higher variability of weather is likely to aggravate supply variations. The chapter 'World Food Prices in the Post Uruguay Round Situation: Prospects and Policy Issues' analyzes how and to what extent in the new post-WTO trading environment individual countries can use external trade policy, i.e. taxes and subsidies on trade and domestic stocks, to reduce the degree of domestic price instability. It explains that as the new round of WTO negotiation on food and agriculture gets going, the developing countries will have a vital interest in placing the question of the appropriate scope for trade intervention for food security high on the agenda.

Trade and Aid

Early in this volume, in Part I dealing with development policy, the rationale of the import substituting and industrialization policy, the economic instruments/tools used for its implementation and its consequences for growth and efficiency are discussed. During the 1960s, trade and exchange rate policies in a number of developing countries were subject to considerable scrutiny and examination by academic economists, both in developed and developing countries. There was a series of quantitative studies in this field of research relating to Pakistan that frequently applied what was considered at that time relatively new concepts, as well as techniques or methods of analysis for: a) measuring import protection and export subsidy; and b) analyzing their impact on growth and efficiency.

National Trade Policy

A direct estimation of comparative costs of different industries was an essential component and in fact provided the foundation for the study of effective or nominal rate of protection. The direct comparison of import prices and costs of production of specific industrial outputs/activities was difficult to come by, as it required actual firm or industrial level of data on costs and prices. The chapter 'Factor Proportions and Industrial Efficiency in Pakistan' contributed to this body of research and provided a first set of quantitative estimates in this respect. It provides an estimate, on the basis of a direct comparison between ex-factory prices and CIF import prices, of comparative costs of individual manufacturing industries and seeks to relate the interindustry differences in cost ratios to such possible explanatory factors such as excess capacity, monopoly profits in a protected domestic industry and factor proportions, etc. The costs of large number of industries in Pakistan were almost 100 per cent higher than the landed cost of competing imports and the protection was provided by a combination of high tariffs and quantitative restrictions. An overvalued exchange rate kept the prices of imported inputs low and they were allocated among the selected industries by a highly differentiated system of quotas. Over time, the cost ratios did not decline, even though in some cases very high protection created large excess profits; in other cases, access to 'undervalued' imports of frequently indivisible, large capital equipment, on the one hand, and scarcity of imported raw materials, on the other, combined with small domestic market, led to excess capacity and waste of scarce capital resources. The adverse effects on exports due to: a) high cost inputs provided by protected industries; and b) higher returns in import-competing sectors than in export sectors were partially offset by an array of subsidies to export sectors. The chapter 'Export Incentives and Export Incentives in Pakistan' for the first time, provided a quantitative estimate of nominal and effective export subsidy (in the same way 'nominal and effective rates of protection' were distinguished) received by different industries, from a wide variety of incentives such as financial concessions, tax rebates, exemptions and directed, cheap credit, etc. By comparing the effective rates of import protection with those of export subsidy for individual industries, the relative attractiveness of the domestic market was compared with that of the export market. It also examined how far, if any, interindustry differences in export performance were related to differential export subsidy.

In recent years, a number of new issues and developments in world trade have assumed increasing importance in trade analysis and policy. They have implications for national trade policies as well as for international negotiations for trade liberalization. Important among such issues are the following:

1) rapidly expanding trade in intermediate inputs, including a considerable interfirm trade among corporations across nations;
2) the role of strategic trade intervention in search of a large market share, when the world market for a commodity is oligopolistic or monopolitistic;
3) the role of 'industrial policy' for picking winners for preferential treatment and trade protection;
4) the impact of multinational corporations that procure inputs from sources distributed worldwide as well as sell outputs in the world market.

The chapter 'New Protectionism and the Nature of World Trade' discusses a few of the issues and their implications for trade policy.

Trade in intermediate inputs, especially if it is between different branches of the same corporation does not share the same characteristics as trade in consumption goods, or trade among independent/unrelated enterprises. Intrafirm trade across national borders has implications for pricing of imports or exports, which are no longer determined in a competitive market, and which could incorporate elements of subsidy and protectionism through, for example, the mechanisms of transfer pricing. Similarly, a multinational corporation, dependent on imports from multiple sources, on the other hand, and worldwide export markets, on the other, would be in favor of trade liberalization, unless of course they sell their output mainly or only in the location in which it is produced so that tariffs on imported inputs can be offset by tariffs on competing imported outputs.

The emergence of few firms in a particular commodity or a narrow sector, i.e. oligopoly or cartels, makes the assumption of competitive behavior unrealistic; in that situation, government interventions do not necessarily militate against the long-run efficient use of the world's resources. Under the circumstances, comparative advantage leading to a large market share in world trade is not entirely a matter of factor endowment; it is also to large extent a result of 'acquired advantage' that accrues from being first in the field and/ or from technological pre-eminence acquired through research and development efforts. Many of these issues are currently high on the agenda in the WTO forums, as evidenced in the debate on competition policy, TRIP (trade-related intellectual property rights) and TRIM (trade-related investment measures), etc.

World Trade Issues

Developing countries have a vital interest in agricultural trade liberalization, as exporters of agricultural raw materials and tropical products as well as importers of food. Trade restrictions in agriculture, combined with domestic subsidies or price support programmes in developed, food-exporting countries in particular, had significant restrictive effects on trade that were much higher than those in manufactured goods. These policies in several instances, while manufacturing high domestic prices, depressed prices in world markets, through their programmes of surplus disposal. The chapter 'Progress of the GATT Negotiations in Agriculture and Developing Countries: Options and Strategies', written ahead of the conclusion of the Uruguay Round in 1994, elaborates a number of issues of importance to the developing countries and discusses the directions in which the process of trade liberalization could enhance their gains. Their exports of nontraditional commodities such as livestock products and horticultural products were subject not only to tariffs and quotas/variable levies, but also to sanitary and phytosanitary regulations of a very detailed nature, highly differentiated among importing countries and subject to frequent changes. Therefore, they had a very special interest in the setting of internationally agreed rules based on science (SPs) and, to as extent, possibly harmonized among importing countries. There was also the question of a possible rise or a higher instability in food prices, if reforms in food-exporting developed countries were significant and measures to deal with the emerging situation were to be designed.

The WTO agreement that was eventually negotiated did incorporate a few of these concerns, including assistance to the developing countries to improve domestic capacity to deal with SPs regulations or food aid or financial compensation to the food-importing developing countries, etc. These provisions were mainly in the nature of exhortations or recommendations rather than commitments or undertakings incorporated in the WTO Treaty. Some of the issues were treated inadequately or halfheartedly. Similarly, the 'escalation' of tariffs on agricultural commodities related to the degree of pressing was barely changed. The issues examined in this chapter constitute a part of the 'unfinished' agenda that needs to be dealt with in the next round WTO negotiations in agriculture to start in 2000.

The interrelations or interfaces between trade and environment have become a very controversial issue, especially due to the growing importance and influence of the environmental NGOs. Following the United Nations Rio Conference on Environment and Development, this subject has penetrated the various areas of international economic negotiations. The convention on biodiversity does have implications for trade in respect of crop materials, genetic resources, and forest products, etc. The recently concluded agreement on biosafety deals with trade in genetically-modified foods. Following the WTO agreement, there is a WTO Committee on Trade and Environment dealing with the environmental consequences of the trade agreements as well as the trade implications of the environmental agreements. This issue, as evidenced during December 1999 the Seattle Ministerial Meeting on WTO negotiations, would occupy the attention of the international community to an increasing extent.

As pointed out earlier, the setting of SPs standards for environmental reasons (danger to human, plant life, or natural environment, etc.) as provided in the WTO agreement will involve the careful application of science as well as consensus building among the negotiating parties, requiring a high degree of transparency and participation in appropriate forums by the consumers and other interest groups. There is always an element of uncertainty in this area, as science does not always provide conclusive answers and differences among scientists persist on many occasions. This raises the question of 'precautionary principles' play a role on setting such standards as evidenced in the food trade dispute between the European Union and the United States in respect of genetically modified foods.

The chapter 'Is Agricultural Trade Liberalization Bad for the Environment' provides an analysis of the interrelations between trade and environment in the context of domestic policies such as tax, credit and pricing policies as well as institutions. It discusses how trade liberalization will affect instability of agricultural prices, agricultural investment, use of chemical inputs, and pressure for agricultural intensification on marginal lands, etc. and hence what will be their attendant environmental implications. The reforms of agricultural support policies in developed countries, for example, may reduce the intensity of use of agricultural resources and thus diminish the risk of environmental degradation. It analyzes why it is neither efficient nor always possible to deal with environmental consequences with trade policy instruments. Supplementary measures are needed to preserve the environment, while avoiding as far as possible trade distorting measures.

Issues in Development Assistance

The developing countries are now urged, as well as required, to rely increasingly on the inflow of foreign private capital, including portfolio investment, equity and bond, as well as foreign direct investment. This is partly because of a declining trend in overall public development assistance in relation to requirements; partly there is a growing belief that private capital is more efficient as it is driven by productivity and profit considerations.

The chapter 'Recent Trends in the Theory of International Investment' was written in early 1960s in the context of the prevailing pattern of international capital flow with a focus on developing countries. Hence, it deals with: a) borrowing through bonds from international financial markets; and b) direct foreign investment. The other forms of portfolio investment in the developing countries (such as foreign investment in the domestic stock markets, private corporations or governments floating equities in the developed country stock exchanges or direct borrowing by banking or nonbanking corporations from the developed-country banks) were not very popular or prevalent in those days. However, it is important to stress that several issues raised in this chapter are relevant in slightly modified forms to the debates on the present-day capital inflows to developing countries. The examples include the implications of foreign capital inflow (when they result in a transfer of external financial resources through the domestic financial system) for the demand for nontradable resources and inflation in the domestic economy (an aspect of the now familiar phenomenon of the so-called Dutch disease). What is today discussed in terms of 'aid effectiveness' was formulated in the early days in terms of what was called the 'absorptive capacity' of the recipient country. The chapter analyzes the considerations that may determine the choice between portfolio investment and foreign direct investment. What is, however, not analyzed is the short-term capital movement; the focus in the chapter was exclusively on the long-term capital flow. The volatility of short-term capital inflows in all the different forms, including speculative inflows that are invested in real estate or stock markets, as well as their unproductive or fraudulent use by unregulated or poorly regulated domestic financial system, etc. are among the issues that are currently very high on the agenda. The chapter distinguishes the implications of foreign capital investment in the import substituting sector, export sector or in the purely domestic sector (now called nontradable sector) and their interrelationships. This is relevant to the present-day debate as to whether foreign direct investment should be required to directly generate exports.

International development assistance which reached its peak during the 1950s and 1960s became increasingly subject to critical appraisal towards the end of 1960s, leading to the establishment of an international commission consisting of several wise men called the Commission on International Development, or the Pearson Commission, to examine the past performance and future prospects and to suggest a future strategy for aid. The debates and discussions during the 1960s and 1970s regarding foreign aid are reminiscent of the current debates. For example, the 'political right' complained about the waste and inefficiency of development assistance, and the 'left' argued that aid strengthened the status quo and the vested interests, inhibiting progressive social changes. There were complaints about aid fatigue in the donor countries and the unfulfilled or unmet expectations in the recipient countries. Questions were raised about the limited effectiveness of aid utilization and its inadequate impact on growth, self-help and policy reforms/improvement in the recipient countries. The chapter 'International Economic Assistance in the 1970s: A Critique of Partnership in Development' examines these and other related questions.

What was the impact of aid on savings/investment and growth in the recipient countries? An additional flow of resources was expected to be allocated between investment and consumption, depending upon the time preferences of the recipients. It was fungible; therefore, its direct allocation to investment was consistent with the reallocation of the remainder of the aggregate resources to consumption so that the allocation of the aggregate resources, including aid, conformed to the preferences of the recipient country. Aid, after all, was a small proportion of total investment resources, including foreign exchange resources; its impact was determined by the range of policies and institutions that determined the utilization of aggregate investment resources. Increasingly the role of both macro and sectoral/microeconomic policies was recognized to be very critical in determining the efficient use of resources. After all, development brings about changes in existing patterns of the allocation and the use of resources and income distribution. The process of adjustment to the changes in allocation and distribution brings in uncertainties and risks. It was argued that aid provided a cushion to contain the adverse or destabilizing short-term effects of policy changes and the process of adjustment. This earlier approach appears very similar to the rationale behind structure adjustment assistance of today. The chapter explains why the recipient countries considered 'aid conditionality' regarding policy reforms to be an involvement or interference by the donors in the domestic economy to an extent that was sometimes more than socially or politically sustainable.

Subsequent research showed that 'aid conditionality' did not work unless there was a conviction and politically sustainable commitment by the recipient country to policy reforms. The concept of ownership of the aid programmes by the recipient countries is currently recognized as the *sine qua non* of their effectiveness. The chapter discusses the inefficiencies of tying aid to sources of supply and overemphasis on project aid. On both these issues, a consensus did emerge, even though in recent years there was a backsliding on the tying of aid directly or indirectly. The developing countries were unhappy then, and are unhappy today, about the inadequacy of the volume of aid, noneconomic considerations often determining the intercountry or interproject allocation of aid, and shifting priorities and objectives of donors.

Since the 1970s, there has been proliferation of aid objectives relating not only to such micro issues as participation, women's development and environmental preservation, but also to such broader noneconomic objectives as democracy, human rights and governance, including the elimination of corruption. These objectives are so overwhelmingly sociopolitical in nature that their pursuit or effectiveness in aid dialogue requires a wider degree of international consensus than that exists today on the: a) content and nature of these objectives; and b) on the appropriate policies and measures (country specific) to deal with them. This is a challenge which the 1970's approach to aid (as reflected in the chapter) could not envisage. But once emphasis shifted in 1970s to the crucial role of policies and institutions in generating growth and equity, there was a logical extension to the broader issues of governance. A very significant shift in focus in aid strategy that occurred in 1980s and later related to poverty and human development as a principal objective of aid.

From the perspective of the 1970s, the chapter examines a complementary role between public development assistance and foreign direct investment, the former providing physical infrastructure and human capital and the latter promoting more directly productive projects. It questions, however, whether FDI could/would provide appropriate technology for developing countries, since the most important market for the results of technological research and development lies in developed countries. This is a reminder of the current controversy on biotechnology; it is now widely recognized that appropriate genetic research for crops that are important in developing countries and relevant for the latter's agroecological circumstances, would not attract the private multinational corporations.

Since 1970s when the chapter was written, the climate for FDI in the developing countries has become much more favorable, even though tensions discussed in the chapter have not totally disappeared. Tensions do emerge when the proportion, for example, of the FDI in the total private investment or in what a country considers 'vital' sectors exceeds a threshold; this threshold varies widely among countries, depending on historical experiences and sociopolitical circumstances. The chapter stresses that a vibrant domestic private sector attracts and facilitates successful the FDI partly because it presupposes a domestic policy environment that is favorable to private enterprise; it also encourages joint venture facilitating transfer of technology and expertise. At the same time, new tensions arising out of competition between the FDI and the emerging domestic private enterprise are also real.

The chapter expresses an undercurrent of thinking in those days (not unknown today) that overdependence (somehow defined by each country) on development assistance, including the FDI, constrains the freedom of action, including the freedom to make mistakes, of the recipient countries. Secondly, since there is always a risk of withdrawal or decline in external resource flows, it is imperative that a recipient country quickly develops a degree of self-reliance, so that the withdrawal of external resource flow does not lead to political and social destabilization.

Pakistan was a major recipient of foreign aid from the Western donors during the 1950s and 1960s. The experience with foreign aid in Pakistan illustrates a few of the issues posed above. The chapter 'Foreign Economic Assistance and Economic Development: the Case of Pakistan' seeks to examine such issues as the impact of foreign aid: a) as a supplement to the domestic investment resources; and b) as a source of foreign exchange component of investment. It traces the effect on domestic savings and income growth. In addition, the influence of aid on the sectoral allocation of public investment resources and on policy reforms was analyzed. There was no consistent relationship between savings and foreign aid; there were years when aid flow increased while savings declined and others when they were positively correlated. The savings rate was primarily determined by fiscal policy, on the one hand and, on the other, by domestic financial policy, including its success in mobilizing private savings. There were periods when high growth and investment and the flow of foreign aid were associated with low savings. There were likely adverse effects of food aid on domestic food production in some years; at the same time there were important policy changes introduced by food aid to liberalize food markets and to introduce public employment schemes for the rural poor. Unlike in the recent years, donors in the 1950s and 1960s were not seriously interested either in equity or human development. Very little of the foreign assistance went to social sectors, i.e. health and education; public sector resource allocation pattern often followed foreign aid priorities, because dependence on external resources was very high. The policy of high growth now and equity or poverty alleviation later – a policy that was supported by foreign aid donors – did not work well in the case of Pakistan. To make matters worse, there was a greater emphasis on aid-financial investment in regions that were better endowed with physical infrastructure and a sizable pool of migrant entrepreneurial ability; there was, in fact, discrimination in favor of the more advanced regions since they had the political authority and power. There was an increasing inequality of personal income distribution as resources flowed into large-scale capital-intensive industries as well as to large-scale agriculture dominated by feudal landowners, with a highly unequal land distribution. Income inequity soared and regional disparity deepened, leading to a mounting social tension and regional conflict.

As discussed earlier, the primacy of appropriate policy and institutions in ensuring an efficient allocation and use of resources among sectors or within sectors came to be increasingly emphasized by 1970s and 1980s. The development assistance community started to stress the role of foreign aid in encouraging, facilitating and implementing policy reforms through dialogue with the recipients as well as making aid conditional on reforms. The link between aid and policy reforms through 'aid conditionality' became an objective of the donors and was pursued through a comprehensive, and coordinated efforts by all donors frequently under the auspices of international financial institutions such as the International Monetary Fund (IMF) and the World Bank. The 'aid conditionality' for macroeconomic policy reforms was a prominent feature of assistance from the IMF right from the beginning, as indeed this was its mandate.

The chapter 'Economic Policy Reform and the IMF: Bangladesh Experience in the Early 1970s' depicts such an early experience in Bangladesh – an experience which became very common and familiar for many countries during 1980s and 1990s. It illustrates the kind of issues that came up for negotiation, the arguments advanced on both sides and how eventually they come to be resolved over time. The disagreement with the IMF centered around issues which are now widely discussed and familiar such as timing, speed, extent, and sequencing of policy reforms. What was the relative role of exchange rate policy, on the one hand, and the fiscal (taxes, subsidies, or public expenditure) policy and monetary policy, on the other, in achieving macroeconomic stability? There were disagreements over: a) the extent and timing of devaluation even though there was an agreement on its need; and b) relative primacy of exchange rate adjustment over fiscal and monetary reforms. How they should be sequenced and which should come first? Bangladesh found it politically more feasible and, therefore, was more readily willing: a) to make much needed fiscal and monetary reforms as well as; b) to undertake improvement in infrastructure and institutions ahead of and in preparation for; c) exchange rate adjustment at a later date. It considered that the IMF assistance, accompanied by these earlier policy reforms, would help remove supply rigidities faced by a war-ravaged economy with severely damaged transportation and communications and disruptions in production, marketing, and distribution system. This, the government argued, would have facilitated an effective response to exchange rate adjustment. The IMF disagreed and insisted that devaluation should take place as a priority step.

Subsequently following the donors' consortium meeting, which took place without a preceding devaluation by Bangladesh, access to the credit tranche of the IMF was denied, but some development assistance by other donors was pledged. It was believed at the time that the quantum of commitment of assistance by other donors and the Bank (all of whom accepted the IMF position on devaluation) would have been higher if Bangladesh reached an agreement with IMF. Devaluation took place in early 1995 and not in October 1974 as the IMF wanted. Experience shows that in a given set of circumstances, it is not easy to determine the precise extent of devaluation nor its exact timing. It was possible that, if necessary fiscal and monetary policies were not adopted, the effects of devaluation would have been whittled away. In a particular situation, it was possible to support the fiscal and monetary reforms in advance of and ahead of devaluation. There was a scope for honest differences of opinion in view of margin of uncertainty involved. Furthermore, to what extent and by how much a given degree of nominal devaluation would affect the real exchange rate – the relevant rate that influences the allocation of resources – cannot be precisely determined, depending as it does on a number of other factors. There is a scope for judgment of the policy makers or policy analysts.

References

Lewis, W.A., 'Economic Development with Unlimited Supplies of Labour', *The Manchester School of Economics and Social Studies*, May 1954.

Lewis, W.A., 'Unlimited Labour, Further Notes', *The Manchester School of Economics and Social Studies*, January 1958.

Fei, J.C.H. and Ranis, G., *Development of the Labor Surplus Economy* (New Haven, CT: Yale University, 1964).

PART I

DEVELOPMENT STRATEGY: CONCEPTS IN DEVELOPMENT ECONOMICS

Chapter 1

Development of the Labour Surplus Economy*

The concept of surplus labour has featured prominently in the recent literature on economic development of underdeveloped, overpopulated economies. W.A. Lewis in his two celebrated articles (1954, 1958) attempted a precise formulation of the concept of surplus labour and sought to analyse its implications for the strategy of economic development in the context of a two-sector model of economic growth. Messrs J.C.H. Fei and Gustav Ranis have undertaken in this book[1] an elaborate extension of the two-sector growth model of W.A. Lewis. The major directions in which the authors have sought to extend or elaborate the model and the main results of their efforts in this respect are contained in their earlier article 'Unlimited Supply of Labour and the Concept of Balanced Growth', published in the *Pakistan Development Review*, Vol. 1, No. 3, Winter 1961. The various sections of the article have now been expanded and elaborated into chapters of the book.[2]

The book is divided into eight chapters. The first two chapters are intended to provide a bird's eye view of the main arguments and conclusions. The third chapter analyses the production functions in the two sectors, whereas the fourth and sixth chapters elaborate the conditions of progress in agriculture, and its Impact on industrial wages and profits and hence on capital formation. The fifth chapter discusses the mechanism and process through which the agricultural surplus is transmitted to the industrial sector for investment purposes. The seventh chapter differentiates between stages of economic growth in terms of a gradual disappearance of surplus labour and the eventual emergence of labour scarcity, and analyses their consequences for capital accumulation. The last chapter examines the implications of development of a labour surplus country in the context of an open economy. The main directions in which the authors have sought an extension of Lewis's model appear to be: a) a systematic discussion of the market mechanism which exchanges agricultural goods against industrial goods and determines the intersectoral terms of trade; b) an analysis of the effects of the changes in the terms of trade

* First published in the *The Pakistan Development Review*, Summer, 1965.

on industrial wages, profits, savings and capital accumulation; c) an elaboration of the effects of capital accumulation and technological change, i.e., the effects of a rise in productivity in agriculture and industry on wages, on the supply curve of labour, on profits and investment; and d) the introduction of the foreign trade sector. Furthermore, the authors examine a wide range of policy implications of the existence of surplus labour in respect of choice of techniques of production, intrasectoral allocation of resources within the industrial sector, i.e., between heavy and light industries, and the concept of balanced growth.

The basic features of a two-sector labour surplus economy are all too familiar. There is a small capitalistic industrial sector coexisting with a large subsistence agricultural sector. The sectoral wages are expressed in terms of the products of the respective sectors. This provides a convenient method of handling the analysis of intrasectoral terms of trade as well as the flow of resources. Competitive conditions do not prevail in the agricultural labour market and the criterion of profit maximization does not govern the employment of labour. The wage rate in agriculture is near the subsistence level of the agricultural workers and the prevailing wage rate is kept to this floor by social institutions and convention. This is also the ceiling because surplus labour serves to cushion any pressure for wage increases. The wage rate, though rigid and institutionally determined is, however, above the marginal productivity of labour, which is either zero or very low. In fact, the authors elaborate the distinction between two stages of labour surplus economy which, by the way, is a valuable extension of the concept of surplus labour to be found in Lewis' model. The two stages are distinguished as the stage of labour 'redundancy' and the stage of 'disguised unemployment'. The first stage is characterized by zero marginal productivity of labour so that its withdrawal from agriculture does not reduce agricultural output. Labour is available at the going wage rate, which is higher than the rural wage rate by a given margin.

Once the growth of the industrial sector and the reallocation of labour proceeds beyond the stage of redundancy, the second stage sets in. At this stage labour has a positive marginal product but it is below the institutional wage rate. A reallocation of labour from agriculture to industry reduces agricultural output and reduces the surplus available to the landlord. The terms of trade move against industry. The industrial wage rate goes up to accommodate the higher cost of agricultural product in terms of industrial output. A rise in industrial wages has a dampening effect on industrial profits and accumulation. This can be delayed by a rise in agricultural productivity and a consequent increase in the agricultural surplus, which prevents a deterioration in the terms of trade of the industrial sector. This enables industrial expansion on the basis of constant real wages until the condition of labour surplus disappears. The wage rate, at this point (called the 'commercialization point') is equal to productivity in agriculture and the industrial sector competes with the agricultural sector on the basis of a continually rising supply curve of labour. There is, thus, an asymmetry in the behaviour of the landlord or agricultural employer. So long as the marginal productivity is below the conventional wage rate, the latter sets the floor as well as the ceiling of wage movement. As soon as marginal productivity equals the wage rate, the ceiling is removed and the wage rate is determined, given the total supply of labour, by relative demand in two respective sectors as reflected in their marginal productivities.

Economic development in this model appears essentially as a process of reallocation of labour from an overpopulated subsistence agriculture to a growing industrial sector. Agriculture thus provides labour and manpower for industry as well as the wages fund wherewith to employ them in industry. What are the forces which activate or cause the process of reallocation to set in? The wage differential between agriculture and industry is certainly a factor; the rise in agricultural productivity is another. The increase in the productivity of labour in agriculture helps in two ways; first, it enlarges the amount of surplus originating in agriculture, which is available for investment in industry provided the market mechanism can mobilize and transfer the resources for industrial investment; secondly, the terms of trade move in favour of industrial goods, which increases the real income of industrial workers, widens the gap between agricultural and industrial wages, and encourages movement of labour from agriculture to industry. The rise in real income of industrial workers is temporary and tends to disappear as labour moves to industry and brings down the real wages (in terms of industrial goods) to the previous level. A deterioration in the terms of trade of agriculture reduces the profitability of investment in agriculture but increases in industry. But an increasing allocation of labour to industry after the elimination of labour redundancy, will reduce total agricultural output and will tend to raise the agricultural terms of trade as well as the industrial real wage to the original level obtaining prior to the introduction of improvements in agricultural productivity. Thus, a new level of equilibrium of industrial employment and of agricultural surplus is reached at the old wage rate so long as surplus labour continues to exist.

The major part of investible funds for industry needs to originate in agriculture so long as agriculture constitutes the most important component of national income. When redundant labour as a source of disguised savings dries up an improvement in agricultural productivity is necessary to at least a constant wages fund per head of the allocated labour force and to provide materials for industry without an adverse effect on the terms of trade and without a rise in the industrial wages. This is even more true when unemployment disappears and the wage is determined by and is equal to marginal productivity. A withdrawal of labour from agriculture when the wage is equal to marginal productivity reduces output to the same extent as wages paid to labour. Under these circumstances, the 'terms of trade' effect of the reallocation of labour for industrial expansion is even more serious and a stagnant agriculture brings the process of industrial expansion speedily to a halt. The increase of agricultural productivity appears, thus, to be a precondition of growth in an overpopulated labour surplus economy. To the extent that the industrial sector seeks to grow through a reinvestment of its own profits, the limit to such an expansion is quickly reached through a rise in real wages unless it is offset by a rise in agricultural productivity. The implications are the same as in the classical Ricardian model in which diminishing returns in agriculture eat into the surplus for investment unless counteracted by a rise in productivity. In the classical model, the supply of labour to industry is highly elastic because the expansion of demand for labour in industry is more than matched by the addition to labour supply caused by the increase in population. The classical theory postulated a constant subsistence wage by linking the growth of population to variations in the wage. In the present model, as in the classical model, diminishing returns, consequent on an increasing pressure of population on land, raise the floor (ceiling) of wage rate by raising the cost of subsistence. It is worth pointing out that a case for the reallocation of labour from agriculture to industry as an index of development, and the *sine qua non* of a rise in per capita income can be made without reference to the existence of redundant labour or disguised unemployment. So long as the productivity of labour is higher in industry than in agriculture a reallocation of labour is expected to increase total output. Moreover, an increase in the proportion of the labour force engaged in industry has its basis in Engels' Law, i.e., with an increase in per capita income an increasing proportion of aggregate expenditure is diverted away from foodstuffs to industrial products. A reallocation of labour to industry requires eventually an increase in agricultural productivity in order to compensate for the loss of output consequent on a withdrawal of labour as well as to provide: a) an increasing supply of raw materials for the industrial expansion; b) food for a larger population as well as for an increase in the existing standard of consumption; and c) a market for the output of the industrial sector.

In the light of the above, the requirements for the balanced growth of a two-sector economy are fairly obvious. The changes in productivity in the two sectors should keep in step with each other so that there is neither a shortage of food nor an industrial overproduction. The investible surplus generated in the two sectors should be allocated between them with due regard to the need for absorption in the industrial sector of redundant agricultural labour, which is augmented by improvements in agricultural productivity and increases in population. Furthermore, intersectoral allocation of resources should take into account, among other things, the relative efficiency of investment in the two sectors, interrelations of demand, and intersectoral input-output relationships, including supply of wage goods by agriculture. The level of 'critical minimum effort' in this context[3] is defined as the rate of expansion of industrial investment and employment which is necessary to absorb more than the incremental labour force so that there is a reduction of redundant labour and/or of disguised unemployment in the agricultural sector. These are well known propositions. The authors believe that neither data nor techniques of programming nor the administrative capacity in the developing economies are adequate to permit the governments of these countries to undertake and apply, with any degree of success, the exercise in dynamic programming of intersectoral resource allocation implied in the propositions stated above. Neither do they seem to consider the lack of a successful exercise of this kind a serious limitation in the way of economic development. The more important bottleneck, in their view, is the slowness or the inadequacy in identifying and promoting vigorous entrepreneurial agents, who are able and willing to introduce innovations, productivity changes and to undertake saving and investment. Equally, if not more important, is the organization of the institutional framework, which permits and encourages mass participation in entrepreneurial activities 'along the vast landscape of the less developed economy'. These two approaches do not seem to the present reviewer to be alternatives. They refer to different aspects of a) planning and b) implementation of development programmes and are both important. The emphasis on their relative importance is a matter of judgement.

It is pertinent to examine carefully some of the crucial assumptions of the Fei-Ranis model: redundant labour and disguised unemployment in the agricultural sector, the institutionally-fixed agricultural wage rate, and the constant wage differential between agriculture and industry, leading to a horizontal supply curve of labour in the industrial sector. It may be worthwhile to enquire whether, and to what extent, the model provides any guidance in respect of policy or institutional changes in agriculture for the mobilization of the agricultural surplus for financing industrial expansion. Again, should the labour surplus underdeveloped economies lay a greater emphasis in the immediate future on long gestation projects and heavy capital goods industries, on the one hand, and labour intensive techniques, on the other? Does the introduction of a foreign trade sector, especially foreign aid, alter in any way the conclusions regarding the relative role of agriculture in promoting economic growth? To what extent have trade and trading profits been an important source of capital accumulation in the development of the industrial sector? The trading profits originating in both the domestic and foreign trade sectors, which are substantial in an underdeveloped economy, are an important source of investment funds for industry. Though formally trade is a part of the industrial sector in the Fei-Ranis model, a separate mention seems warranted, especially since trade and services constitute an important source of resources for investment. These are very wide ranging issues. The present review cannot hope to provide any detailed treatment of these issues excepting to raise some questions and to indicate the implications of these questions for the model and policy conclusions derived from it.

The model postulates only redundant labour force and disguised unemployment. The realism of the model would have been enhanced if an explicit account were taken of the fact that in both urban and rural areas there is open unemployment, more so in the industrial than in the agricultural sector. An increasing population in the industrial sector feeds the pool of unemployment and necessitates a transfer of food from rural to urban areas, irrespective of any reallocation of labour from agriculture to industry, with all its attendant consequences for intersectoral price movements. The empirical content of the concept of redundant labour with zero marginal product has been subjected to considerable doubt. The existence of redundant labour in a strict sense may imply that a withdrawal of labour from agriculture without any change in the technique of production, form, or quantity of capital equipment or other inputs and in the organization of production, keeps agricultural output unchanged.[4]

The assumption of unchanged organization and form of equipment can, however, be relaxed. Surplus labour may also be defined as that which can be withdrawn from agricultural work without any diminution of output consequent on necessary changes in the organization and form of equipment but with no addition to capital equipment or other inputs. A consolidation of holdings or a pooling together of holdings for cooperative farming may reduce labour requirements. A change in the form of existing capital equipment to employ less labour but to produce the same output, is in effect a change in the technique of cultivation. Whether it can be effected without extra investment is doubtful. Thus, it is clear that the emergence of surplus labour in this sense is conditional upon organizational and technical changes in agriculture. Moreover the farm workers often adjust themselves to low level equilibrium of more leisurely work, i.e., fewer hours for a lower intensity of work. This could be, however, due to a reduced capacity, which is caused by a low level of subsistence of the farmers, to face the fatigue of agricultural work. This could also be due to the fact that agriculture is not only a means of livelihood but also a way of life. The fact that in most underdeveloped areas the industrial sector does not find any difficulty in hiring unskilled labour often indicates a high rate of population growth which feeds the stream of open unemployment among the agricultural and industrial labour force, as well as a possibly larger participation of urban population in the labour force.

One would have expected from the authors of the volume under review a rationale of their case for disguised unemployment defined as a state in which wage rate is higher than the marginal product of labour. This is crucial to their concept of a perfectly elastic supply curve of labour. Except for a reference to social convention, the authors do not explain why the landlords who in their model (as exemplified by Japan) act as profit maximizing entrepreneurs investing agricultural surplus in industry should act contrary to the principles of maximization of profit in the employment of labour in agriculture. There is an asymmetry in their pattern of behaviour. The concept of disguised unemployment is not a new one, explored exhaustively among others by Leibenstein (1960, ch. 6, pp. 58–76) and Wonnacott (1962). One has to distinguish between alternative systems of organization of agricultural production, i.e., between wage employment under landowners, on the one hand, and, on the other, self- employment, which is usually prevalent under peasant proprietorship or under a system of tenancy where the organization of production is left to tenants, ownership being vested in the hands of absentee landlords earning either a fixed rent or a share of the crop. The system of land management and ownership affects the circumstances under which disguised unemployment may exist.

Under conditions of self-employment, be it peasant proprietorship or a system of cultivating tenants, it is the household which is the unit of employment. It is reasonable to assume that, given the scarcity of land and the lack of alternative opportunities of employment, the production efforts of the house hold are directed towards the maximization of total output. Maximum total output with a given size of the working population in agriculture can be obtained either with a positive or zero marginal product. Under these circumstances the average earnings of the agricultural population would be equal to their average product, which would be higher than their marginal product. Even when the latter is zero, average product, and thus average earnings, of the agricultural households may be positive. Under these conditions, as labour is withdrawn from agriculture, average earnings of the remaining members of the household go up. The supply curve of labour in the industrial sector would be upward sloping as the opportunity cost of additional labour employed in the industrial sector goes up with a rise in the average earnings of the farm labour. The horizontal supply curve of labour for the industrial sector no longer holds true even though the average product and hence average earnings of additional labour continue to be higher than their marginal product. Similar conditions would prevail under a system of land tenure in which the landlord receives rents proportional to the area of land and the tenant cultivators exploit land with the objective of maximizing total output. Even if the rent received by the landlord is proportionate to the value of output, the maximization of output would determine the deployment of family labour, when alternative opportunities of employment do not exist for the farm labour, i.e. the number of members of the household (constituting the individual farm's supply of labour) remains constant. There are no marginal wage costs to offset against the gain in extra output minus proportional rent.

Under conditions of wage employment, the fact that average product as well as the wage rate is higher than the marginal product of the employed farm labour does not necessarily involve a horizontal supply curve of labour unless the additional condition of an institutionally determined wage rate (which may or may not be equal to the average product) is postulated. If it is a subsistence wage, as the authors imply, it is most likely to be less than average product unless the average product is low enough to coincide with a subsistence wage. If social pressure or convention not only requires that the entire available labour force be employed on land but also sets the standard wage rate, then the landlord has no degrees of freedom in deciding the amount of employment or the wage rate. The surplus left to the landlord can vary between zero and a positive magnitude, depending upon the size of the labour force and the product-labour relationship, i.e., the productivity of labour. If the wage rate is equal to the average product of a given labour force then no surplus is left to the landlord.

Harvey Leibenstein (1960) explains the existence of a wage rate at a level higher than the marginal product on two assumptions: 1) a positive correlation between wage rate and productivity; 2) competition among unemployed workers to bring down wages so long as there is unemployment. If productivity is a function of the wage rate, it is possible to increase profits by spending the same wage bill among a smaller labour force, thus raising the average wage rate as well as the efficiency of the workers. But this may not happen, either because the unemployed workers would compete and bring down wages (and thus counteract the increase in efficiency and profits) or because the unemployed workers may continue to share in the consumption of the employed workers, thus preventing an improvement in the per capita consumption and efficiency of the workers. Even if the landlords desist from hiring workers at lower wages for fear of adverse effects on efficiency, the possibility of unemployed workers sharing in the wage goods available to the employed workers can hardly be prevented in view of the family and social relationships in the agricultural sector. Thus, the employers would tend to employ the available labour force at a wage rate, which is higher than marginal product, because an attempt to equate the wage rate to marginal product decreases profits by decreasing efficiency and output. According to this theory the withdrawal of a segment of the labour force from agricultural employment, be it self employment or wage employment, would not result in a loss of output and might even increase it – because it would be expected to lead to a rise in consumption per head, and hence the productivity of those who are left behind on the land. The foregoing analysis is different from the treatment of disguised unemployment in the Fei-Ranis model, since in their system the withdrawal of labour results in a loss of output, though the loss in output is less than the savings in wage cost, which remains constant and institutionally determined so long, as any disguised unemployment exists.

Empirical evidence on redundant labour and disguised unemployment is difficult to come by, and attempts at their quantification have not been very successful owing to various difficulties of measurement. For example, an attempt (Islam, 1965) to measure idle labour time in an operational sense in Pakistan has yielded the following results:

1) the extent of visible or open unemployment varies between 15 per cent and 25 per cent of total man days in a year. This measurement of visible unemployment does not distinguish between voluntary and involuntary unemployment, a distinction which is relevant in the case of rich farmers. There is a wide variation in the extent of unemployment between different seasons. In peak seasons shortages appear – as indicated by the participation of children and women (not usually a part of the labour force) in farming operations – as well as by the reverse movement of labour from urban to rural areas;

2) with a smaller size of farm per active member of the family, i.e., with an increasing pressure of population on land, the employment of man hours per acre, both family as well as hired labour, goes up. Though no direct measurement of the marginal product of labour is available, the smaller farmers are not expected to hire more labour per acre unless it is worth their while in terms of additional output in view of wages, either in kind or in money, which have to be paid. There is a tendency for the yield per acre to rise with an increase in the amount of labour utilized per acre of land, the increase in output per acre being quite significant in the case of smaller farmers;
3) for bigger farmers there is some tendency for there to be an increase in the number of unemployed man days per worker with an increase in the size of holding per active member of the family. This may indicate differential factor proportions, i.e., differences in the relative uses of labour and capital between large and small farmers as well as some amount of voluntary unemployment among rich farmers, since the latter seem to be satisfied with a lower output per acre even though they spend more man days without employment;
4) there is no conventional wage rate. The landless agricultural labourers are employed at varying wages for varying lengths of time. They are the least protected economic group in terms of both income and employment. Social conventions neither protect their wage rate nor guarantee their employment.

A permanent shift of labour from rural areas to urban areas would necessitate, in most underdeveloped countries, measures to meet excess labour requirements during peak seasons by the substitution of capital or other factors for labour. Alternatively, the utilization of idle labour during slack seasons may be undertaken by employing rural labour either in rural industries (which could be contraseasonal in the sense that they can release labour for peak seasons) or in a rural public works programme, i.e., construction of roads, buildings, irrigation canals, embankments and drainage systems, etc. These methods of employment of labour do not imply a reduction in the proportion of the labour force employed in agriculture but only imply a more effective and fuller utilization of total available man days. They do require capital investment from outside the rural economy.

It is most pertinent to observe in this context that Fei and Ranis do not examine the implications of the possibility that the whole of the agricultural surplus consequent on the withdrawal of 'redundant labour' might not be available for use as a wages fund in the industrial sector. The leakages enumerated by Nurkse are well known. An analysis of the implications of such leakages is conspicuous by its absence in the book. The form of land ownership and management which the authors use as the relevant frame of reference for their analysis is wage employment of agricultural labour under a system of landlordism. If the landlords sell (to the industrial sector) the agricultural surplus which emerges as a consequence of the withdrawal of redundant labour, the demand for food on the part of the reallocated labour exactly matches the marketed supply of surplus food and there is no rise in the price of food. There is no net increase in per capita consumption in the economy. Under such a system of wage employment, so long as the excess of output over wages accrues as a surplus to the landlords, the possibility of an increase in per capita consumption on the part of the agricultural labour can be forestalled. Under alternative systems of land ownership and management, the surplus appears wholly or partly (under share cropping) as additional income to farmers., and as an increase in the per capita availability of food. Considering the existing low level of consumption of farmers, a tendency towards an increase in the consumption of food on the part of those left behind on the farms is most likely. Added to this is the fact that the per capita income of that part of the labour force reallocated to industry is higher (expressed in terms of industrial goods), and at least a portion of this extra income will be spent on food. Thus a shift of labour from low income to high income occupations increases total income as industrial output increases and, as a consequence, the demand for food increases while agricultural output remains unchanged (with redundant labour *à la* Fei-Ranis). Under the circumstances, the tendency for agricultural prices, especially food prices, to rise would have all the consequences (although to a lesser extent) on the wages and profits in the industrial sector which have been ably analysed by the authors in connection with the withdrawal of labour under their conditions of disguised unemployment.

It is noteworthy that the authors do not anticipate any spill-over of intersectoral price changes or changes in relative prices into a general price rise. A rise in the cost of living of industrial workers caused by a rise in food prices and a consequent rise in wage costs may be accommodated by an increase in the money supply to enable a rise in industrial prices to keep profits from falling. One may think of various alternative ways of mobilizing redundant labour which may not lead to a rise in per capita consumption of food and may not affect intersectoral price relations.

If agricultural population could be put to work within agriculture itself most of the leakages may not occur. The first alternative is that family farms themselves undertake construction or investment activity through the use of family surplus labourers, who continue to eat the same food in the same kitchen. However, if the surplus labour of one farm family is to undertake construction of a well, a fence, or bounding and levelling of land, the gestation period is likely to be long and the project may be beyond the time horizon of many individual farmers. There are few projects which can be undertaken on the basis of the labour of individual family farms and which can be completed within the time horizon of the individual farmers. Alternatively, if the surplus labour of the farm families can be put together in some sort of work 'brigade' or 'teams' to undertake capital projects on all farms on a cooperative village-wide basis, the gestation lag of each of the project can be drastically shortened. Each will gain from this effort because in exchange for his work on the improvement of other farms, others work for the improvement of his farm. No cash payment and no transfer of food or payment of wages are involved. Some amount of capital has to be supplied either from rural savings or from outside. All these will directly expand employment.

However, if capital projects are to be undertaken by the employment of labour on wage payments away from their own farms, the familiar leakages arising from an increased consumption by those left on land and from the possibility of increased consumption of food by recipients of money income, may occur. This may be prevented if payment could be made in terms of deferred wages, i.e., wages to be paid in cash only when increased output materializes. A significant proportion of increased output will be spent on food, if employment is away from home, and if new homesteads are built up near the projects.

What is happening in Pakistan in recent years in the course of the rural public works programme confirms the above hypothesis. Under this programme labourers are not withdrawn from agricultural operations and only their idle time is being utilized, including unemployed man hours of those who were visibly unemployed. Even then it is necessary to import a wages fund, i.e., foodstuffs from abroad under PL 480 to undertake this programme. The alternative would have been a resort to compulsory labour. Additional food thus has gone to accommodate increased consumption on the part of the farming population as well as of the unemployed rural labour force which is being constantly augmented by an increase in population. The rural workers have continued to live in their old habitat and to share in the consumption of the family. It has not been possible, therefore, to rely entirely on mobilization of the wages fund or the disguised savings implied in the model of redundant and disguised unemployment. An exclusive reliance on disguised savings would have necessitated the employment of labour in the rural works programme without any payment of wages and on the condition that the labourers would continue to receive their previous share of the consumption goods of their families and no more. But the institutional and organizational problem of mobilizing wages funds for the reallocated labour by keeping the consumption of the rest of the rural labour force constant poses an intractable problem in which the market mechanism does not seem to be of much help.

The crucial problem in an underdeveloped economy, where the agricultural sector constitutes the largest component of national income is how, firstly, to generate an agricultural surplus through improvements in agricultural productivity and, secondly, to channel the available surplus to investment in the industrial sector. Fei and Ranis believe that in the improvement of agricultural productivity and in the generation of the surplus, the market mechanism and price incentives can play an effective role. It is indeed difficult to ensure mass participation among the millions of small landlords and cultivators by means of direct intervention by the state in the production process. On the other hand, in the task of channelling investment funds from agriculture to industry the government may play an important role, either by fiscal means or by improving the efficiency of commodity and financial markets to facilitate the intersectoral flow of resources.

Fei and Ranis cite with approval the role of the Japanese landlords, who performed the dual functions of undertaking considerable improvements in agricultural productivity as well as of serving as financial intermediaries to channel agricultural savings to the industrial sector by investing directly in industries or by purchasing industrial shares and debentures. The authors do not throw any light on the best or most appropriate form of organization of agricultural production from the point of view of an improvement in agricultural productivity and of a generation of agricultural surplus. The landlords in various underdeveloped countries have not always behaved in the way the Japanese landlords behaved. The role of the Japanese landlords cannot be divorced from the whole complex of the sociocultural traditions of Japan. This will take us far afield, but one cannot help feeling that there is an oversimplification on the part of the authors. The following features of the Japanese agriculture seem not to have received sufficient emphasis:

1) a land tax-reform accentuated the severe degree of exploitation of tenant farmers who paid rent in kind so that they could not benefit from a rise in the price of their crops; 2) the land tax reform reduced the share of the actual cultivating tenants in the gross product of farm produce to 32 per

cent and increased the share of the state to 34 per cent and that of the landowners to 34 per cent as compared with the previous shares of 27 per cent for feudal lords, 25.2 per cent for landowners and 37.8 per cent for cultivating tenants (Tsuru, 1963, pp. 145–7 and 148–9). The immediate was that a large number of small landowners sold land under the heavy burden of taxation and joined the ranks of tenant farmers. The big landowners who survived the period of severity became richer in the subsequent period. The polarization process in landownership was accelerated (Boserup, 1963);

3) there was considerable sharecropping. The sharecroppers had very tiny holdings, only a fraction of those of sharecroppers in other Asian countries. They could avoid starvation only by cultivating land very intensively (Reynold, 1965). Thus, the intensification of Japanese agriculture followed under the threat of starvation. The intensive application of labour to land resulted in an increase in output per acre. There was no significant increase in the early years in output per man. The increased output resulted in an adverse movement in the terms of trade of agriculture;
4) the dispossessed feudal pensioners, faced with a loss in the value of their commutation bonds, were forced to seek outlets for their enterprise elsewhere. The commutation bonds served as the basis of creation of credit by the banking system;
5) the unfavourable turn in the terms of trade of agriculture coupled with the continued plight of the peasants contributed to the low supply price of labour. The textile industry recruited young girls before marriage in the form of contract labour for which parents were paid a nominal sum (Boserup, 1963). Even in the case of voluntary labour the supply price was not equal to minimum subsistence or a minimum determined by social convention plus a margin to compensate for 'transfer costs' – as Fei and Ranis suggest – but was based on the consideration of supplementing family income and not much more. The survival of traditional family relations in Japan retarded the emergence of the category of wage income as reward for an independent unit of a factor of production. The low wage in the textile industry set the standard for wages in other industries.

It is important to remember that many of these features of Japanese agriculture cannot be reproduced in the contemporary world. One could examine the relevance of the alternative types of land management for present day developing economies. There is the system of the capitalist farmer employing wage labour on large integrated farms, which are probably the most efficient in terms of generating a surplus for investment The system in which landlords act as entrepreneurs in the matter of supplying and supervising the use of resources for investment by tenants and receive a share in output may achieve the same result. Under the system of peasant proprietorship where the owners and cultivators are the same, the peasant households strive to secure the maximum output from the household farm by the fullest application of labour. This results in a wider distribution of income from land, especially when the introduction of peasant proprietorship is associated with a redistribution of holdings and an imposition of a ceiling on landownership. If the units of production are reduced below their optimum size, total agricultural output would be below the maximum. This raises the familiar question of efficiency versus equity. The maximization of output under peasant proprietorship, or with tenant cultivators paying a fixed rent, conceivably could be, consistent with the realization of an adequate amount of investible surplus if price system and the fiscal mechanism could channel a part of the output to investment. It is necessary, however, that peasant or tenant cultivators should have sufficient incentives, i.e., they should be confronted with adequate opportunities for profits from the cultivation of the land. The underdeveloped nations have wavered between various kinds and degrees of land reforms, often faced with conflicting objectives. The trend is towards peasant proprietorship or a redistribution of holdings.

Fei and Ranis have emphasized the role of financing intermediaries which can channel savings from the agricultural sector to investment in the industrial sector. The role of the Japanese landlords as financial intermediaries has been cited in this context, but the authors seem to have paid no attention, to one of the most important consequences of capital formation through direct reinvestment of profits. Owing partly to the inadequate development of financial markets in any underdeveloped countries of today, investment is financed through a direct reinvestment of profits. This gives rise to a concentration of control of wealth and property with its attendant social and economic consequences. When the landlords reinvest their surplus directly in their own industrial enterprises, there is a concentration of big land holdings and large industrial enterprises in the same hands. Again, traders, especially importers, reinvest their high profits in industries, thus combining trading and industrial enterprises in the same hands. Often land holdings, commercial enterprises and industrial enterprises are all combined in the hands of the same group of people. This tendency towards the concentration of ownership and control is aggravated by the role which banking and financing institutions (which are often owned by the same group of commercial and industrial entrepreneurs) play in the financing of investment. Thus, they attain control over the disposition of savings of the rest of the economy to which is added the creation of credit which the banking system undertakes.

The resultant inequality of income and wealth may, however, be conducive to a high marginal rate of saving and hence to a high rate of capital accumulation. An excessive concentration of ownership and control of economic activities has been one of the important features of the Japanese economic growth, which the authors accept as a model of successful development.

The conflict between equity and efficiency can be mitigated if a wider diffusion of wealth and income (which is subject to a greater pressure of consumption demand) can be matched by an appropriate fiscal policy to realize a high marginal rate of savings. These are questions which, in the mid-twentieth century and at the present level of social consciousness prevailing in underdeveloped economies, cannot be easily avoided. To the extent that higher wages or better health services may increase the productivity of the labour force, the conflict between equity and efficiency is reduced. Moreover, if heightened expectations of the masses for an increased share in the fruits of development are frustrated, it may lead to social and political unrest.

The development of the Fei-Ranis industrial sector is postulated on the basis of a horizontal supply curve of labour at a wage rate which is higher than the agricultural wage rate by a constant margin. The authors do not undertake any detailed analysis of the need for an extent of this margin, and they seem to accept the statement of the problem as given by W.A. Lewis. But there are some interesting and plausible explanatory hypotheses. One could argue that the difference in the wage rate partly goes as income to the contractor or jobber, i.e., to the middleman in the industrial labour market, who in spite of apparent surplus labour, does the job of an employment exchange in return for a share in the income of industrial workers. The higher wage rate may also be a partial compensation for the loss of income on the part of those workers of the family (women and children) who contribute to family income in the household subsector of the agricultural sector, but lose their sources of income when they move along with the main breadwinner or head of the family to the industrial sector. With reference to the horizontal supply curve of labour, the industrial wage rate does not necessarily remain constant in the face of increasing accumulation. Autonomous or exogenous factors unrelated to the adverse movements in the terms of trade of the industrial sector may cause a rise in the supply price of industrial labour. The horizontal supply curve may shift upwards through time either because of trade union pressure or because of government intervention in the regulation of industrial wages. The militant strength of trade unionism even in labour surplus countries is not unknown and the governments of developing countries are often obliged to yield to the consideration of social justice and equity. The experience of Puerto Rico in this respect has been delineated in a recent study by Professor L.G. Reynold (Reynold, 1965). The Third Five Year Plan of Pakistan, as another example, postulates that wages in large-scale industry should rise at the same rate as the increase, in productivity which is expected to be 2 to 4 per cent per annum (Pakistan Planning Commission, 1965, p. 157). This would slow down any increase in the ratios of profits to the total 'value added' of the manufacturing sector. One can, however, conceive of possible ways to counteract the consequential adverse effects on saving. The institution of a compulsory provident fund for industrial workers, to which workers contribute the increase in their income, and/or payment of enhanced wages in the form of industrial shares may ensure that an increase in, wages is at, least partly matched by an increase in savings.

If wages were to remain constant at the subsistence level and profits were to go on increasing as a proportion of national income, there would necessarily be a restraint on the growth of consumption demand – which necessitates that industrial investment should have a bias towards capital creating projects, i.e., toward infrastructure and capital goods industries which do not cause an immediate increase in the supply of consumption goods. The more orthodox argument is that relative scarcity of investment funds suggests the adoption of labour-intensive techniques of production which maximize current outputs with given investment resources and which, in the process, lead to a maximization of employment. However, the conflict is usually between the maximization of current output or employment and the maximization of the rate of growth. Techniques which require large capital per unit of output may be more conducive to a higher rate of growth, if the income distribution pattern emerging out of the adoption of capital intensive techniques enables a large saving and investment per unit of additional output. This conflict can be obviated if fiscal measures can be devised to squeeze equivalent savings out of additional income, irrespective of the way in which additional output is distributed between profit and wage recipients. In a labour surplus economy, so long as real wages are constant, the income distribution effects of additional output are conducive to a high marginal rate of saving. The larger increase in output per unit of capital consequent on the adoption of labour-intensive techniques combined with constancy of real wages, may more than offset (up to a point) any adverse distribution effects on the flow of investible surplus.

The authors suggest that industrialization in Japan (at least until 1915) was based on labour-intensive techniques, which accelerated the process of absorption of surplus agricultural labour in the industrial sector. Not all the analysts of the history of the Japanese industrialization agree with this interpretation (Reubens, 1964). The statistical evidence on the Japanese industrial sector in the Fei-Ranis model relates to the whole of the nonagricultural sector, consisting of both secondary and tertiary industries which were an amalgam of industrialized and pre-industrialized sectors. They include traditional small-scale enterprises in the service and trade sectors as well as the government services, all of which are highly labour-intensive activities. An analysis of manufacturing industries alone reveals that in spite of the existence of an abundance of labour, technological innovation in Japan consisted in the introduction of capital-using techniques. This is what is corroborated by the data relating to the period 1905–15 as presented by Watanabe (1965). The adoption of labour-saving techniques was based on borrowed technology from advanced countries. The Japanese, according to this interpretation, had a high ability to understand and to digest foreign, advanced techniques and to operate and maintain imported equipment without any serious trouble; but they had a low ability to develop new techniques in the early stages of industrialization. The ability to meet the requirements, of skill for advanced technology was based on a sufficient supply of a reasonably well educated labour force, which subsequently underwent intensive technical training. The relatively high capital requirements of capital intensive technology were met by a relatively high rate of saving. The socioeconomic character of the agriculture and service industries was such that they absorbed a considerable proportion of the labour surplus. While borrowed technology was mostly confined to larger industries or firms, attempts were made to ensure a reasonable increase in the levels of productivity in the residual sectors small firms, proprietors and farmers. The smaller firms specialized in products and processes which were particularly suitable for them. The old capital assets, which were considered to be inefficient for the higher level of modem technologies, were transferred from the larger firms which owned them to smaller firms. The smaller firms often used second-hand capital assets which, associated with an elastic supply of labour enabled them to earn almost the same rate of return on capital as the larger modern firms.

On the question of choice of techniques, it is worth emphasizing that the great dependence of developing economies on imported equipment embodying capital intensive techniques of advanced countries is a reality and increasingly so, in the face of tied-aid financing of most of the imports of capital equipment.

This eliminates whatever differences and, therefore, choice there may still remain between the techniques of individual advanced countries. Moreover, only a limited range of techniques are alleged to be available, i.e., the prevalence of fixed coefficients of production is often postulated. This would suggest the necessity of invention of labour intensive techniques which would in turn require a considerable investment on research in the development of techniques.

The authors' advocacy of a relative emphasis on the capital goods industry and projects with a long gestation lag in the early years of development of a labour surplus economy is based on the assumption of low and a constant real wage exercising a restraint on the growth of consumption demand. In many underdeveloped countries of today the emphasis on heavy industries has been justified not so much because of a lack of demand for consumers' goods as because of the relative ease of restraining consumption and mobilizing a higher marginal rate of savings from additional output when it accrues in the form of nonconsumable investment goods. If in this situation the propensity to consume tends to outstrip the supply of consumers' goods, a restraint on the rise in wages or appropriate fiscal measures are necessary. Otherwise increased savings are realized through a rise in profits caused by a rise in the prices of consumer goods under the pressure of excess demand. The relative emphasis on heavy industries is also justified on the ground that low income and price elasticity of primary exports, the chief source of foreign exchange for developing economies, severely limits the possibility of raising investment via import of capital goods. The probability of loss in real income from adverse terms of trade caused by attempts to push inelastic exports in world markets may justify the domestic production of capital goods at a high cost. Moreover, while admittedly capital goods industries do not always require capital intensive techniques, it is equally true that requirements of relative skills significantly differ between different industries, and capital goods industries tend to require a higher level of skill. The relative abundance of an undifferentiated mass of unskilled labour favours only those industries which require relatively low levels of skill. The formation of skill requires capital investment and returns on such investment are expected to be high. A labour surplus economy has an additional incentive to invest in skill creation because of the pressing need to provide employment through the development of a wider range of industries, which require the development of skill. Moreover, in a capital and natural-resource poor and labour abundant country (such as Pakistan, for example), one may suggest a relative emphasis on industries which have a high ratio of 'value added' to total output, i.e., industries which depend relatively more upon human labour and skill than on materials which in the context of relative poverty of resources need be imported.

In an open economy the transfer of resources from the agricultural sector to the industrial sector may take the form of an export surplus originating in the agricultural sector and financing the import of capital goods and raw materials for feeding the expansion of the industrial sector. Thus even in an open economy improvement in agriculture holds the key to industrial expansion by providing a large export surplus as well as raw material inputs to the domestic industries. Most of the underdeveloped, agricultural export economies have used import and export controls to obtain an agricultural surplus for industrial expansion. Increased incomes in agriculture in an export-oriented economy appear as increased cash crops exports which can be siphoned off by the mechanism of foreign trade more easily than would be the case of increased incomes were to accrue in the form of increased food production which may be consumed by the farmers.

The introduction of the additional variable of foreign aid in an open economy has important implications for the pattern of development which do not appear to have been adequately explored by the authors. If foreign aid supplies are investible surplus in the form of a wages fund, as well as of capital goods, and raw materials for industrial expansion, the role of agriculture is limited to the supply of labour for industry. Imports of food under foreign aid compensate for any fall in agricultural output consequent on the withdrawal of labour from agriculture, leaving the intersectoral terms of trade and industrial wages unchanged. If the manpower for industrial expansion is supplied by a net increase in population, there is no net withdrawal of labour from agriculture and no fall in agricultural output. If the expanded industrial output can find markets abroad, it can overcome the limitation of a narrow domestic market arising from constancy of real wages and from an absence of any increase in agricultural productivity.

If industrial expansion is large enough to draw on disguised unemployment an increasing withdrawal implies a fall in agricultural output at an increasing rate which, unless matched by an increasing flow of imports, requires an improvement in agricultural productivity. Moreover, an expansion of the industrial sector based on the export market seems to ignore the difficulties which exports of light consumer goods face in the world market in view of their, low income and price elasticities. World trade in light manufactures grows at a slower rate than that in heavy producers and intermediate goods, the very fields in which underdeveloped countries are in a position of disadvantage (at least initially) to compete in the world market. Therefore, the expansion of a home market based on an increasing agricultural productivity seems to be a more reliable and surer base for industrial expansion. Moreover, as the experience of countries with considerable industrial exports shows, industries based on a large domestic market are the most successful ones in export markets as well. A large domestic market not only enables the realization of economies of scale and reduces cost, but also provides the necessary incentive and confidence, mainly springing from the stability and assurance of a domestic market as against the vicissitudes of a world market, for the exploitation of technological improvements. The enterprises which successfully meet the competitive struggle in the home market are more likely to succeed in the world market. The world market for manufactures is as imperfect as the domestic market and requires high-pressure salesmanship and development of consumer preference. Those who have done it well in the domestic sphere are likely to do it well in world trade.

While the authors recognize the role of foreign aid in supplementing domestic resources they assign an additional, at least equally important, role to foreign aid in: a) freeing the foreign exchange market from direct controls; and b) in the creation and strengthening of domestic financial intermediaries via the use of counterpart funds generated from the flow of foreign aid. Foreign aid has two distinct functions which are not often adequately stressed in the literature on aid and development. Foreign aid not only supplements domestic resources but also augments the supply of scarce foreign exchange. Even though in an *ex post* sense the savings-investment gap is equal to the foreign exchange gap (i.e., the gap between foreign receipts and payments) the *ex ante* gap between savings and investment need not be equal to the *ex ante* foreign exchange gap. An excess of estimated requirements of foreign exchange over anticipated resources may be a more serious limitation to growth than the insufficiency of anticipated domestic savings. Owing to the immobility of resources between the foreign trade sector and the rest of the economy, and to the imperfections of the price mechanism (and inelasticities of supply and demand), resources released by savings accruing in the domestic sector may not be transformed into an expansion of exports or a curtailment of imports. In a typical underdeveloped economy where the capital goods sector is strictly limited and the resources base is narrow, most investment projects have a foreign exchange component. They cannot materialize, even if domestic savings are available, unless a part of savings accrues in the form of foreign exchange to supply the essential imports for investment. In an *ex post* sense the equality between the two gaps may have to be reached via a frustration of potential savings which fail to be invested; and the gaps are then (*ex post*) both equal to the amount of foreign economic assistance which is available. An increase in foreign economic assistance under these circumstances may raise the level of total investment by enabling domestic savings to be invested with the help of imports of critical inputs.

Fei and Ranis lay great stress on the opportunity which foreign aid offers for greater use of the price mechanism in the foreign exchange market. But a free foreign exchange market is desired not so much for the superior allocative efficiency of the price mechanism as for the possibility it offers for mass participation in entrepreneurial decisions via free access to the foreign exchange market. It provides for an improvement in entrepreneurial resources as experience accumulates with the opportunity for participation. However, the authors do not explain whether or to what extent intergovernmental 'tied' loans (which are the most predominant form of international economic assistance) permit the fulfilment of these functions. It should be pointed out that this particular role of foreign aid can be fulfilled only if, foreign aid takes the form of 'free' foreign exchange making a net addition to the foreign exchange reserves of the economy. It should be 'free' in the sense that foreign exchange so supplied is not tied to the purchase of specific commodities, to specific projects, or to purchases from a particular country. With any of these forms of tied aid, government intervention in the administration of foreign exchange becomes necessary. The tied aid, in the sense of compulsory purchases from an aid giving country, is best administered by means of country quotas. It is, however, possible to resort to the free market in a limited sense, insofar as the distribution of aid among individual importers is concerned, by auctioning to intending importers the specific aid licences for specific countries or for specific projects. The use of foreign aid for freeing the foreign exchange market from direct controls presupposes an agreement on the part of the donor countries to guarantee a certain minimum amount of foreign aid to add to the pool of foreign exchange reserves, taking into account the foreign exchange requirements of a country's investment programme and prospects of its own foreign exchange earnings. The increased flow of commodity aid, as the recent experience of Pakistan shows, does facilitate a liberalization of imports from direct controls. In addition to the flexibility which untied commodity aid imparts to the implementation of development programme, its liberalizing impact on the foreign exchange market is an important argument for increasing the ratio of commodity aid in the total flow of aid to developing countries. The counterpart funds can play a useful role in strengthening domestic financial institutions and have often been so used. This is a subject which, though mentioned by the authors, is a complex one, and which needs a much more thorough analysis than that contained in the book, nor can it be explored in this review.

To sum up, the authors have made a contribution in working out a wide range of possible implications on the basis of a given set of assumptions. But in some instances one wonders whether they have based their conclusions on sufficiently realistic assumptions or on adequate empirical evidence. For the basic core of their analysis, the distinction between the two stages of labour surplus is not strictly necessary. Moreover, the distinction is not operational. The authors do not provide any explanation as to why the equilibrium wage rate in the agricultural sector may be higher than the marginal product. What is relevant in most of the underdeveloped, overpopulated economies in Asia is self-employment and not wage employment, which is the only frame of reference of Fei-Ranis model. Again, the assumption of a constant wage rate in urban areas is not always justified, and the market prices of factors may diverge from their scarcity prices. The implications of these departures from their assumptions are not examined. Fortunately, neither the assumption of constancy of the wage rate, that of redundant labour, nor that of disguised unemployment is essential for the analysis of the interrelations between development in the agricultural and industrial sectors in terms of productivity changes, technological innovations, and movements in intersectoral terms of trade.

This is an important book, dealing with important issues. The authors have certainly put in a sharper focus many of the assumptions which are stated, but not rigorously demonstrated, in the Lewis model. They show considerable skill in reducing the complicated general equilibrium equations to geometry and relatively simple, familiar diagrams. The book certainly contains, in many respects, the neatest and the most skilful pedagogical device so far employed ,in the exposition of the two-sector allocation model. Their exposition of the interrelated effects of technological improvements in agriculture on output, wages, profits, employment, intersectoral terms of trade and industrial investment is most admirable in its clarity. Though the Fei-Ranis analysis of foreign trade and aid is not peculiarly relevant to the labour surplus economy, it does contain valuable observations which apply to all underdeveloped economies. Despite the many reservations expressed here, the book, especially chapters 3, 5 and 6, must be considered essential reading for students of development theory.

Notes

1 J.C.H. Fei and G. Ranis, *Development of the Labour Surplus Economy* (New Haven, CT: The Economic Growth Centre, Yale University, 1964).
2 It is curious that not even a passing reference is made to this article anywhere in the book.
3 The concept of 'critical minimum effort' originates with Harvey Leibenstein (1957). There are important differences, between the concept as developed by Fei and Ranis and Leibenstein's neo-Malthusian conception.
4 Professor T. Schultz (1964) refers to the case of heavy losses of rural manpower in India in 1918–19, without any loss in any of the other factors of production such as animals, etc. which caused a considerable decline in agricultural output as well. While agricultural labour was reduced by 8 per cent, area sown under crops fell by 3.8 per cent. The provinces with the highest death rates also had the largest percentage declines in acreage sown. On the basis of a labour coefficient of about 0.4, which is held to be a plausible value for coefficient of labour in agricultural production in India, the predicted fall in agricultural production of 3.2 per cent compares favourably with the observed reduction of 3.8 per cent, which is well within the standard errors. Moreover, examples of labour shortage in agriculture consequent on the movement of labour from agriculture to industry have been frequently cited. One may, of course, in such cases, suggest that the extent of withdrawal or movement of labour was more than what is warranted by the quantum of surplus labour.

References

Boserup, M., 'Agrarian Structure and Take Off', in W.W. Rostow (ed.), *The Economics of Take Off into Sustained Growth* (London: Macmillan, 1963), pp. 201–24.
Islam, Nurul, 'Concept and Measurement of Unemployment and Underemployment in Developing Economies', *International Labour Review*, March 1965.
Leibenstein, Harvey, *Economic Backwardness and Economic Growth* (New York: Wiley, 1957).
Lewis, W.A., 'Economic Development with Unlimited Supplies of Labour', *The Manchester School of Economics and Social Studies*, May 1954.
Lewis, W.A., 'Unlimited Labour, Further Notes', *The Manchester School of Economics and Social Studies*, January 1958.
Reubens, E.P., 'Central-Labour Ratios in Theory and History: Comment, *The American Economic Review*, December 1964, pp. 1052–62.
Reynold, L.G., 'Wages and Employment in the Labour-Surplus Economy', *The American Economic Review*, March 1965, pp. 19–39.
Rostow, W.W. (ed.), *The Economics of Take Off into Sustained Growth* (London: Macmillan, 1963.
Schultz, T.W., *Transforming Traditional Agriculture* (New Haven, CT and London: Yale University Press, 1964).
Tsuru, S., 'The Take Off in Japan (1868–1900)', in W.W. Rostow (ed.), *The Economics of Take Off into Sustained Growth* (London: Macmillan, 1963).

Pakistan Planning Commission, *Third Five Year Plan* (Karachi: Manager of Publications, May 1965).

Watanabe, T., 'Economic Aspects of Dualism in the Industrial Development of Japan', *Economic Development and Cultural Change*, Vol. XIII, No. 3, April 1965, pp. 293–308.

Wonnacott, Paul, 'Disguised and Overt Unemployment in Underdeveloped Economies', *Quarterly Journal of Economics*, Vol. LXXVI, May 1962, pp. 279–97.

Chapter 2

Concepts and Measurement of Unemployment and Underemployment in Developing Economies*

This chapter attempts to analyse the concepts of underemployment and unemployment in the context of an underdeveloped, overpopulated country. It focuses attention on the agricultural sector, explains in, detail a number of attempts made to measure unemployment and underemployment in the rural areas of Pakistan, especially among the agricultural population, and examines the results obtained and problems encountered.

Concepts of Rural Unemployment and Underemployment

The problem of unemployment and underemployment can be approached from two different standpoints: i.e. from that of the individual members of a community or from that of the community as a whole.

The former approach is related to the felt needs of the individual members of a community and is expressed in their search for work opportunities in order to increase their income. When work opportunities are not available, there is enforced idleness or unemployment either for a whole or a part of a year or some part of a month in the course of a working year. This kind of unemployment may be called visible unemployment.

From the point of view of a community as a whole, the concept of underemployment or of unemployment is geared to the requirements of the mobilization of manpower in the context of economic development. In this context one is concerned with the net marginal contribution to social product of the existing labour force and hence the appropriate concept is that of disguised unemployment or, more accurately, underemployment.

Disguised unemployment in agriculture has been said to exist if a section of the labour force can be withdrawn from work, if there has been a change in

---

* First published in *The International Labour Review*, March 1964.

organization and the form of equipment (involving little or no addition to capital equipment), without any diminution of output. This may be true when the marginal product of labour is negligible. But then the exact nature and extent of the contemplated reorganization of agricultural production needs to be known before evaluating the extent of surplus labour. For example, a consolidation of scattered holdings which involve a waste of labour time in going back and forth between plots or a pooling together of holdings for cooperative farming may reduce labour requirements. Similarly, if some workers leave the land and those remaining on the farm work longer hours or with a greater intensity, the effective supply of productive effort may remain unchanged. A change in the form of existing equipment to employ less labour but to produce the same output is in effect a change in the technique of cultivation and whether it can be effected without an extra investment is doubtful. Once an additional investment is allowed, the range of alternatives widens and the possible magnitude of surplus labour becomes conditional upon an exact specification of changes in these variables.

Disguised unemployment may even be consistent with a positive marginal product of labour on the following two assumptions: (a) there is a positive relationship between wages and efficiency of work; and (b) a number of institutional factors result in a level of wage rates which is higher than the marginal product of labour in agriculture (Leibenstein, 1960; Wonnacott, 1962). It is believed that with the very low level of per capita consumption prevalent in underdeveloped economies, a rise in wages enables a rise in per capita consumption of agricultural workers and increases their vigour and consequently their productivity. Thus the supply of 'work units' or productive effort per man may be presumed to increase with an increase in wages per man. Under conditions of wage employment when employers or landlords strive to maximize profits, it is necessary to introduce additional assumptions to demonstrate why wages may be higher than marginal productivity. For one thing, social and political considerations may prevail upon the landlord to employ a fixed proportion of total labour force so that the only way he can vary the supply of productive effort on his land is by a variation in the supply of work units per man in response to a change in the wage rate per man. With a very heavy pressure of population on land, it is quite conceivable that wage cost per work unit may be higher than the marginal product of a work unit and the corresponding wage rate per man will be higher than the marginal product of labour (Wonnacott, 1962). This is because a reduction of wage rate may lower the supply of work units, and hence of output, more than it saves on wage cost.

When a landlord is free to vary the amount of employment, he may maximize his profits by varying the employment of labour until the marginal product of labour per work unit is equal to the wage cost per work unit.[1] He can increase his profits by discharging workers and spreading the same wage bill over a smaller labour force; per capita consumption and hence efficiency of workers goes up, with the result that an increase in output owing to an increase in efficiency may more than offset a decrease in output consequent on the employment of a smaller labour force. However, the employers may be deterred from following such a course either because: (a) unemployed workers may compete in a free labour market to bring down wage rates, and hence the marginal productivity of labour, to such, an extent that the gain from a lower wage rate is more than offset by a fall in efficiency; or because (b) unemployed workers, even if they do not bring down wages by competition in the labour market, may start sharing in the wages and the available wage goods of the employed labour force and thus prevent a rise in the per capita consumption of the latter.

The situation postulated in (a) may not materialize if the employers desist from hiring labour at lower wages. But the alternative possibility is more real, as employed workers are usually responsible for supporting the unemployed members of their households so that the per capita consumption of employed workers may remain the same even at a higher wage rate. Therefore, it may be a second-best alternative to employ a larger labour force at a wage higher than its marginal product, because of adverse repercussions of a lower wage on efficiency. This may not happen, however, if the surplus labour is not only removed from employment but also prevented from sharing in the consumption of the employed.

When analysing either visible or disguised unemployment it is important to identify the labour force in an agricultural economy and to decide whether and to what extent women and children, for example, should be included in it. Moreover, the extent of participation by persons of working age in agricultural work may vary at different seasons so that the supply of effective man-days fluctuates over the year. Similarly, the number of hours constituting a man-day must also be specified. But if the measurement of underemployment is to serve as a basis for policy decision it will not be enough merely to define such labour norms; it may also be necessary to enforce them, especially if they differ from existing practice.

Any attempt to measure surplus labour should take into account the overall pattern of utilization of total man-days supplied by agricultural households which, in Pakistan for example, are available not only for work on their own farms and for work as hired labourers but also for various nonagricultural activities such as cottage industry, handicrafts, trading and other ancillary activities. Even when the members of a household remain idle for a larger part of a day or month, they may not feel that they are unemployed and thus may not actively seek additional employment. Even if they are available for additional employment, they may like to work only within their own households or only within precincts of their own village. These factors enhance the difficulty of identifying rural unemployment.

Early attempts at measurement of unemployment and underemployment in Pakistan were mainly concerned with deriving some aggregate and global estimates of surplus labour on the land. One such attempt used the concept of a 'standard holding' which is assumed to provide full employment for a cultivator the latter being defined as 2,500 man-hours of work (cf. Ministry of Economic Affairs, 1953, p. 77). In a way this procedure aimed at measuring both visible unemployment and disguised unemployment, i.e. underemployment. It was estimated that a farmer would require four acres in East and six acres in West Pakistan, besides his cattle, to keep himself fully employed.[2] It was further assumed that one acre of double-cropped land is equivalent to two acres. On the basis of the above assumptions the amount of surplus agricultural labour in Pakistan was estimated to be about 20 per cent of the total agricultural labour force in 1949–50.

The concept of the standard holding based on an estimate of labour requirements per acre for different crops, however, misses two vital aspects: firstly, the seasonal variations in agricultural employment and, secondly, the relationship of labour requirement to output. It lumps together the phenomenon of labour shortage during the peak season and that of unemployment during slack seasons. An estimate of surplus labour so derived is not operational unless a surplus can also be demonstrated during the peak seasons or special measures are adopted to meet labour requirements of these seasons. Moreover, given the fertility of a piece of land and the nature of crops, labour requirements per acre are assumed to be fixed and invariant. It is assumed that an acre of land under rice, for example, can use only 500 man-hours of work and no more. It is not clear whether this is the maximum number of man-hours of work per acre so that the employment of extra labour will produce no additional output. If not, a farmer can increase his output by employing more than 500 man-hours per acre and by varying the ratio between labour and non-labour inputs. Whether the additional output consequent on employment of additional labour is adequate or not depends upon whether there are alternative opportunities of employment yielding a higher marginal product of labour elsewhere in the economy and whether the farmers are willing to avail themselves of the opportunities. Besides, estimation of surplus labour merely with reference to purely farming operations rather than to the totality of activities may yield an exaggerated figure, in view of nonagricultural subsidiary occupations which often provide farmers with many man-hours of gainful employment.

In 1956 an expert team from the ILO conducted a manpower survey which produced some information relevant to the present discussion (ILO, 1956). It covered rural households but did not distinguish between hours of work in agricultural and nonagricultural occupations. Moreover, the data were collected only for a week preceding the time of interview and thus could not convey a faithful picture of the situation throughout the year, specially regarding the seasonal variations in hours of work. According to the survey self-employed or independent workers and unpaid family workers constituted about 60 per cent of the total rural labour force. A person was considered to be employed if reported to be working for four out of seven days before the interview. The percentage of the rural labour force looking for work was very small and varied between 2.2 and 3.1 in different regions, thus indicating that independent workers and unpaid family workers do not consider themselves unemployed even if they are working only a few hours a week. Their working hours were 18 per cent less in East and 9 per cent less in West Pakistan than those of wage and salary earners.

In 1956 a sample survey was carried out in four different regions of East Pakistan to investigate a few limited aspects of rural unemployment (Dacca University Socio-Economic Survey Board, 1956, pp. 102–3). The farmers were asked to report man-days actually utilized by active members of their families in farm and non farm activities. No attempt was made to determine, either from actual observation or from interviews, the total number of available man-days, including those which either were lost owing to sickness and other reasons or were available for additional work. Since the number of hours of work constituting a man-day was not stipulated, a full man-day might have been reported where much less was actually spent in work. In addition information was sought for the period of the last year and, since farmers kept no record of their days of work, considerable errors in reporting could not be avoided and no information could be obtained on seasonal variations in employment. It was assumed that 250 man-days represented the total annual potential labour supply per active male. Visible unemployment in the case of farm families was then estimated as the difference between potential man-days and actual utilization of man-days expressed as a percentage of potential man-days (ibid., p. 103), giving the following results for the four regions: Naryanganj, 41.3; Rangpur, 11.5; Rajbari, 26.1; Feni, 45.2. It is interesting to observe that the percentage of total active members who reported as being completely unemployed throughout the whole year and as seeking a job in the labour market was very small – 0.27 per cent in Naryanganj, 0.48 in Feni, 0.6 in Rajbari and 0.09 in Rangpur.

Subsequently, more intensive surveys were carried out in two small selected areas of East Pakistan, one in the northern region consisting of 12 villages and the other in the southeastern region consisting of only one village. In both cases predominantly agricultural families were investigated and investigations were conducted continuously throughout the year.[3] The findings of the two surveys are discussed below.

Survey No. 1[4]

The selected families were interviewed every two weeks and data were collected on the utilization of total labour time in terms of hours of work in various occupations. The survey indicates that an active adult male suffers from visible unemployment almost two months in a year, i.e., about 62.7 working days are lost through lack of work. The average number of working days per active member in a year is 270 and average working hours per active member per day are about eight. Thus, the days lost through lack of work constitute about 19 per cent of the total man-days available in a year, excluding days lost because of sickness, rains and social functions. Data on the distribution of unemployed man-days over the different seasons are not available. The seasonal pattern of employment can be seen from the extent to which actual hours of work in individual two weekly, periods differ from the average hours of work for all such periods throughout a year. In some seasons the former falls as much as 48 per cent below the latter insofar as purely agricultural work is concerned and in some other actual hours of work rise as much as 54 per cent above the average. Employment in nonagricultural activities tends to offset partially the seasonal variations in agricultural employment.

Generally speaking, with an increase in the size of holdings per household the proportion of total man-hours spent on self-employment in agricultural operations goes up and the proportion of total man-hours available for work outside the household goes down. This is shown in Table 2.1 (Raishahi University Socio-Economic Research Board, 1963).

However, in the case of holdings which are between one and two acres in size the inverse relationship between the size of farms and the proportion of

Table 2.1 Percentage distribution of total hours of work of households, by size of holding per household

Size of holding (acres)	Agricultural activities (self-employment)	Employment outside household	Marketing household and other gainful activities
0.01–1.00	38	28	34
1.01–2.00	30	47	23
2.01–3.00	40	31	29
3.01–4.00	51	12	37
4.01–5m	51	11	38
5.01 and above	53	9	38

labour time spent on work outside the household does not seem to obtain. The household cultivating less than one acre spend a larger proportion of labour time on the farm than those cultivating between one and two acres. The same relationship holds on a per capita basis.

As regards the number of unemployed man-days per active member per year, there seems to be a clear-cut distinction between those who cultivate less than 1.5 acres per head and those whose per capita holdings are larger. Among the first group of farmers, the number of unemployed days per head decreases with an increase in the holding per head and among the second group the reverse is true. This difference in the incidence of unemployed man-days between the rich and the poorer farmers can partly be explained by the fact that the survey did not distinguish between voluntary and involuntary unemployment. The rich farmers may be idle because they may not want to work more. The survey has not explored the content and nature of 'days without work' from the point of view of their availability for additional employment as well as the terms on which they would be available.

From the results of the survey it is difficult to measure the extent of underemployment or disguised unemployment. One may suggest that, other things being equal, in a situation of disguised unemployment a household puts in more man-days or man-hours for doing a job than it would otherwise do, since it has surplus labour in the family. However, total hours of work per active member in both agricultural and nonagricultural operations do not seem to be consistently related to the size of holding per active member.[5] Total hours of work per active member when the size of holding is 0.41 acres per active member are about the same as the total hours of work in the case where the size of holding per active member is five acres and more. In the case of purely agricultural operations there appears to be a positive relationship between the size of holding per active member and hours of work per active member, except as between the first two groups (i.e., 0.01–1.00 acres and 1.01–2.00 acres). With a smaller size of holding per head, i.e., with a larger working population per acre, less time seems be spent per head on the farms and thus more workers can be accommodated per acre. This may indicate a certain amount of work spreading to provide employment for a larger active population so that if employment is provided for some of the members outside the farm to a larger extent than at present, then the remaining labour force may work more hours on the farms. However, the fact that the smallest farmers work as a whole for fewer aggregate hours per head on all kinds of activities may raise doubt as to whether fewer hours per head on the farm are entirely a device of work spreading and not partly a matter of lower physical vigour and capacity for work. No direct evidence on this latter aspect is, however, available.

Utilization of labour hours per acre goes up with a decrease in the size of holding per active person, i.e., with an increase in the pressure of population on land. Even though on smaller farms each active member works fewer hours, this is more than offset by the dependence of a larger number of persons on an acre of land. Whether a direct relationship between pressure of working population on land and utilization of labour per acre is evidence of disguised unemployment or underemployment depends on other things remaining equal as between different farms. Labour requirements per acre of land even for a given amount of output depend on a number of factors such as the crop pattern, quality of soil, the extent of fragmentation of land and the nature of rotation of crops, which would vary not only as between the 12 villages from which the 45 households in question were selected but also as between different farms in the same village. However, to the extent that these characteristics are randomly distributed between the different farms or holdings of different size, one may trace the increase in the hours of work per acre on smaller farms to a greater pressure of working population on land. But the larger number of man-hours per acre cannot be considered redundant unless output per acre is invariant with respect to the utilization of man-hours per acre. No data are available on output per acre for farms with a more intensive utilization of land. However, it may be noted that the utilization of hired labour per acre goes up in the same way and almost to the same extent as the utilization of family labour goes up with a decrease in the size of holdings. From this it may be inferred that smaller farmers would not have hired more labour unless the additional output more than offset the wages, kind or in cash, paid to hired labourers.

Survey II[6]

In this survey data on the utilization of total labour supply in agricultural and nonagricultural activities were collected by means of weekly interviews of households, supplemented by continuous observation of the activities of the households throughout the whole year on the part of the interviewers who were stationed in the village. A man-day was defined, for the purposes of investigation, as consisting of eight hours of work, so that four hours of work were recorded as 0.5 man-days and 10 hours as 1.25 man-days. The supply of man-days varies from one week to the next because of the movement of the working population in and out of the village and because of changing hours of work per active member.[7]

For the year as a whole rural households were without any work for 25 per cent of the total supply of man-days and this may be taken as an index of visible unemployment. But the percentage of unemployed man-days varies widely at different seasons of the year. Table 2.2 indicates the pattern of utilization of total man-days for each of the 13 four-week periods of the year. The percentage of man-days of unemployment rises up to about 47 in the fourth and fifth periods and drops down to as low as 6 in the eleventh and 1 in the twelfth periods (columns 11 and 12). During the ninth and tenth periods it varies between 17 and 13 per cent. It is to be noted that during the last five periods farmers often work overtime in the sense of working more than eight hours a day.[8]

In order to meet excessive labour requirements during peak seasons the farmers reduce the use of labour in nonfarm occupations. Moreover, during the busy seasons both women and children are engaged to a varying extent in agricultural activities. The total man-days recorded in the table include both self-employment and wage employment (hired labour). The utilization of hired labour reaches its peak in the last four periods, but its distribution between agricultural and nonagricultural activities varies with the different seasons. The shortage of labour in the peak seasons is also indicated by a rise in wages in these periods which go up by about 50 per cent (Habibullah, op. cit., Appendices B–11 and C–3).

These estimates of visible unemployment do not indicate whether all employed man-days are available for work and if so, to what extent they are available for work outside the village and on what terms.[9] Moreover, if there is unemployment for 25 per cent of total man-days in a particular season, then 25 per cent of the individual workers can be withdrawn for work elsewhere only if: (a) each member of the remaining working population works eight

Table 2.2 Seasonal pattern of employ
June 1962

Utilization of labour supply	1	2	3	4
Total agricultural work (self- and wage employment)	116.0 (37.0)	1326.0 (42.0)	1911.0 (29.4)	59 (19
Garden work (self- and wage employment)	133.0	26.0	65.0	15
Total farm work	1249.0 (41.0)	1352.0 (43.6)	976.0 (31.5)	74 (24
Total non-farm work	736.5 (24.0)	567.0 (18.3)	694.0 (22.4)	75 (24
Total gainful employment	1985.5 (65.0)	1919.0 (61.9)	1670.0 (53.9)	150 (48
Not available for work	270.0 (8.9)	342.0 (11.0)	218.5 (7.0)	19 (6
No work found	767.5 (26.1)	833.0 (27.1)	1210.5 (39.1)	141 (45
Overtime work (more than 8 hours a day)	–	4.5	–	
Total labour supply	*3023*	*3094*	*3099*	*31*

Note: The figures within brackets are percentages
therefore they do not indicate the actual utiliz

Source: Habibullah, 1962, Appendix C–1.

hours a day (which is the norm for a full man-day accepted for the study); and (b) workers left behind are available without restriction for all the economic activities in the village, including all kinds of farm and nonfarm work on the pre-existing terms and conditions. In other words, individual components of total available man-days must be completely substitutable for each other in order to enable a certain portion of the labour supply to be withdrawn for employment elsewhere.

In view of the excess requirements of labour during the peak seasons, i.e., in the eleventh and twelfth periods, an attempt to withdraw a uniform percentage of labour force for the whole year, let us say even 10 per cent would involve a reorganization of work on the part of the remaining labour force. As a matter of fact, if the excess labour requirements of the ninth, tenth, eleventh and twelfth periods (i.e. when labour requirements are in excess of 83 per cent of available man-days) can be taken care of, presumably as much as 17 per cent of the labour force can be withdrawn from the village. There are in fact several different ways of increasing the supply of man-days in peak seasons consequent on the withdrawal of a segment of working labour force: (a) a readjustment of labour time between agricultural and non-agricultural work; (b) overtime work; and (c) a larger participation of women and, children in the labour force. No data were available on the magnitude of these readjustments, however, certain broad implications may be pointed out. There is already a tendency to a decline in man-days spent in other kinds of work during peak seasons in agriculture. Whether more can be achieved in this direction would depend upon whether the nonagricultural activities can be postponed during the peak seasons in agriculture without a reduction in income derived from nonagricultural work over the year as a whole.[10]

Secondly, the supply of man-days from a smaller working population can be increased by overtime work. In the busiest period, i.e. the twelfth, the number of available labourers is about 120 persons who undertake overtime work on an average of one-and-a-half hours a day per head. If 10 per cent of the force is withdrawn, roughly 12 farmers will become unavailable for employment in the peak period. The remainder of the labour force (108 persons) will have to supply an additional 96 hours of work each day which, together with the existing overtime, works out at two to about three hours average overtime a day. This amount overtime may also be supplied by the third source mentioned i.e., an additional participation of women and children in the labour force. The data on the employment of women and children in different seasons show that the supply of labour on their part rises during the peak seasons to as much as 155 man-days a week.

Insofar as disguised unemployment is concerned, no attempt at measurement has been made, but certain indications along the same lines as the previous survey may be pointed out. With an increase in the pressure of population on the land, i.e., a decline in the size of holding per active member and therefore an increase in supply of family labour per acre, there is a tendency towards an increase in the amount of family labour used per acre of land and in the number of unemployed man-days per worker as seen in Table 2.3 (columns 2, 3 and 7). This is particularly noticeable in the case of smaller farms.

The differences in the utilization of labour per acre may partly be traced, other things being equal, to a need to provide additional employment (work spreading) on small farms with a larger labour force per acre. But then the utilization of hired labour (column 5) also generally increases with an increase in the pressure of population on the land. And there is a tendency for yield per acre to rise with an increase in the amount of labour utilized per acre of land, this rise being quite significant in the case of smaller farmers, as is seen in Table 2.4.

It is, of course, not possible to identify the marginal product of labour on smaller farms. This would necessitate the estimation of a complete production function.

The growth in the number of unemployed man-days per worker with an increase in the size of holding per worker (Table 2.3, column 7) may also indicate the differential factor proportions respect of labour and non-labour inputs as between big and small farms. Moreover, the evidence of some amount of voluntary unemployment among rich farmers may not be entirely ruled out Since they seem to be satisfied with a lower output per acre even though they have more unemployed man-days per head.

Conclusion

The two surveys indicate that the extent of visible unemployment in the rural areas of East Pakistan seems to vary between 15 per cent and 25 per cent of total man-days in the course of a year. However, there is a wide variation in the extent of unemployment between the different seasons resulting in a shortage of labour in peak periods, when children and women, not usually a part of the labour force in these areas, are also employed. There are instances when labour moves from urban to rural areas in peak seasons. A permanent removal from rural areas of a segment of agricultural labour force necessitates special measures to meet the excess requirements of labour during peak

Table 2.3 **Size of holding per wor[illegible] man-days**

Holding per worker (acres)		Famil[illegible] labour su[illegible] per acr[illegible]
Class intervals (1)	Average (2)	(3)
(0.01–0.50)	0.22	3,186.3[illegible]
(0.51–1.00)	0.79	419.5[illegible]
(1.01–1.50)	1.19	322.9[illegible]
(1.51–2.00)	1.66	257.7[illegible]
(2.01–2.50)	2.03	179.0[illegible]
(2.51–3.00)	2.67	130.7[illegible]
(3.01–4.00)	3.49	106.5[illegible]

Source: Habibullah, 1962, Appendix C–2.

Table 2.4 Utilization of labour and output per acre

Class interval of man-days per acre	Total labour used per acre (man-days)	Output per acre (Rs)
0–50	27.54	324.04
51–100	74.95	355.17
101–150	121.64	408.16
151–200	169.64	400.20
201–250	224.11	425.81
251–300	283.00	381.80
301–350	322.25	775.25
351 and above	1,414.25	1,018.00

seasons. This may involve a reorganization of agricultural as well as of non - agricultural operations or longer working hours for the remainder of the labour force or a larger participation of women and children in the labour force. However, the provision of employment opportunities in the rural areas by means of public works programmes or development of cottage industries helps to reduce unemployment in the slack seasons, when the percentage of total man-days employed drops as low as 45 or so.

The measurement of visible unemployment is best attempted by means of a direct observation of farming operations continuously throughout the year, supplemented by interviews of the farmers. Both surveys have identified only the number of days without work, and have not distinguished between voluntary and involuntary unemployment – a distinction which is of some relevance in the case of rich farmers. Neither have they clarified whether days without employment are available for work outside the village either temporarily or permanently and, if so, on what terms and conditions.

The identification and measurement of disguised unemployment which is related to the social marginal product of labour is more difficult. The smaller farmers seem to employ more man-hours per acre than the bigger farmers and usually get a larger product per acre. The rather fragmentary evidence available at present indicates that the average income of a self-employed farmer working on his land is often less than the wage earned by the ordinary labourer (Habibullah, op. cit., Appendix C–2). This may be due partly to a lack of free mobility of labour between these occupations, since self-employed labour may consider agricultural employment as a way of life, and partly to the possibility of an inadequate valuation put on household and garden work in the measurement of income of a self-employed person.

Notes

1 Under conditions of self-employment the productive efforts of a household hold may be assumed to be directed towards the maximization of total output, given the supply of land and equipment, so that it goes on utilizing its labour force so long as the total output can be increased. Maximum output with a given size of the working population of a household may be obtained with either a positive or even a zero marginal product of labour the pressure of population on land is very heavy. The average product and average earnings of a household may still be positive and higher than the marginal product of labour, even though it would be operating on declining portion of the average productivity curve. The above situation can arise even if self employment and wage employment coexist in agricultural economy when, for whatever reasons, there is no free mobility of labour between wage employment and self employment.
2 It was assumed that one single acre of rice crop would yield employment of 500 man-hours in East Pakistan whereas in West Pakistan a single crop per acre would require 244 man-hours for wheat, 209 for maize, 667 for cotton and 472 for rice. In addition a farmer would need about 500 man-hours in West Pakistan and about 300 in East Pakistan to look after his 2.5 and 1.7 cattle respectively. It is not quite clear how the differential labour requirements, for individual crops were weighted in order to arrive at an estimate of standard holdings for East and West Pakistan separately. Though the assumption of 'existing manual technique' is implied in these estimates. No mention is made as to whether any reorganization of agricultural operation considered necessary, and if so, what is its nature. The standards of labour requirements are not based upon any actual investigation of the techniques of agricultural operations in Pakistan.
3 In the first case, 45 randomly selected families from the 12 villages having only 88 active members were covered. In the second case, one whole village consisting of 129 families and 164 active members was surveyed. Active members in both cases excluded children, invalids and women.
4 Raishahi University Socio-Economic Research Board, 1963.
5 The average working days per active member in a year generally varies inversely with the average hours of work per day per active member.
6 Habibullah, 1962.
7 Almost one-third of the active adults move in and out of a village temporarily for short periods in search of jobs as millhands, domestic servants, office menials, bricklayers, etc.
8 Some farmers may prefer to undertake overtime work instead of hiring labour which may be under-utilized elsewhere. Busy hours in one farm may not coincide with slack hours in another. If the total overtime work deducted from the aggregate unemployed man-days for the village as a whole, then net unemployed man-days are considerably less in peak periods and in the twelfth month there is a net shortage of labour. The shortage of labour in peak seasons can be further illustrated by estimating unemployed man-days net of Sundays. In that case there would appear to be a considerable labour shortage in the tenth, eleventh and twelfth periods and unemployment in other periods is also considerably reduced.
9 However, there are indications, in view of a movement of labour out of the village in slack seasons in search of jobs, that farmers in this particular village are in general willing to move out if they are offered employment elsewhere.
10 The occupational distribution of total man-days devoted to nonfarmwork in the survey area is as follows: 67.36 per cent of man-days were to trading in various agricultural products

such as rice, jute, fish, fruits and vegetables and 27.51 per cent were spent on retail trade in groceries. The existing pattern of distribution of labour time between kinds of activities in each of the seasons may be presumed to have been based partly on a comparison of the relative profitabilities of the alternative deployments of labour and partly on institutional factors. A change or an adjustment in this distribution may affect total income if there is no change in the technique or organization of agricultural operations.

References

Dacca University Socio-Economic Survey Board, *Rural Credit and Unemployment in East Pakistan* (Dacca, 1956).

Habibullah, M., *The Pattern* of *Agricultural Unemployment: a Case Study of an East Pakistan Village* (Dacca: Dacca University Bureau of Economic Research, 1962).

ILO, *Report to the Government of Pakistan on Manpower Survey* (Geneva, 1956), pp. 23–42 (roneoed).

Leibenstein, Harvey, *Economic Backwardness and Economic Growth* (1960).

Ministry of Economic Affairs, Pakistan, *Report of the Economic Appraisal Committee* (Karachi, 1953).

Raishahi University Socio-Economic Research Board, *The Pattern of a Peasant Economy – Puthia, A Case Study* (1963).

Wonnacott, Paul, 'Disguised and Overt Unemployment in Underdeveloped Economies', *Quarterly Journal of Economics*, Vol. LXXVI, May 1962, pp. 279–97.

Chapter 3

External Economies and the Doctrine of Balanced Growth*

This article is an attempt to follow up and examine some of the points in the two recent illuminating articles (Fleming, 1955 and 1956) on the doctrine of balanced growth and explore further some of its implications.

The obstacles to increased investment in undeveloped economies operate from the side of demand as well as from the side of supply. The proponents of the 'balanced growth' doctrine seek to overcome or obviate the inhibiting effect of limited internal market on the inducement to invest by emphasizing the complementarity of a wide range of industries, which, if simultaneously established, would provide market for each other mainly via the expenditure of income generated in one industry on the products of another. Such analysis, however, has neglected the competitive relationships between different industries the course of industrial expansion, especially if supplies of labour and capital are limited. The articles mentioned above are an excellent attempt to put the nature of industrial expansion in its truer perspective by providing a penetrating analysis of the nature of the limitations on the supply side, their implications in terms of competitive relationships between different industries in the process of growth, and their consequences for the doctrine of balanced growth.

It is true that an inelastic supply of an essential factor like skilled labour, entrepreneurship or capital would restrict the simultaneous expansion of a wide range of industries. But the magnitude of the constraint set by a fixed factor is a function of the degree of variability of the technical coefficients of production. In case of a fixed factor available in limited supply and fixed coefficients of production, expansion of one industry would draw away resources from another, involving a diminution of the latter's output. With variable factor proportions, the increased use of the scarce factor by the expanding industry may be compensated in other sectors by the substitution by other factors so that expansion of output on many lines is possible subject to this limitation that substitution beyond a certain point may lead to a rise in costs.

---

* First published in *The Indian Economic Journal*, 1957.

The introduction of more efficient large scale methods of production in the place of older methods in underdeveloped economies would involve a saving or an economizing of inputs necessary for a given amount of output. However, the saving may not occur necessarily in terms of all the inputs; the new methods may employ more of one and less of another than before for a given amount of output; or it may involve less of both. But insofar as factor proportions are variable and substitution is easy, even an economy in the use of only one factor or input may, within limits, be as effective as the economy in the use of more or all inputs. At the same time, however, as Mr Fleming rightly emphasizes, the introduction of more efficient modern methods in an industry would not necessarily involve a net release of resources from that industry. It will depend upon the scale of output of the industry produced by the new methods. If the efficient scale of output of the new plants is smaller or equal to the previous output of the whole industry, then there would be a net release. If it is larger than the previous output, then there is, on the one hand, the economizing effect of more efficient methods of production and, on the other hand, there is the increased employment of inputs for larger output. The net result can go either way. However, insofar as the efficient scale of output produced by the new methods is equal or smaller than the previous output of the industry, the demand-factor is no more a limiting factor on the introduction of modern large-scale methods. Instead of being produced in a large number of scattered, small workshops run on older methods, the whole output would now be produced in a few large scale, more efficient plants. Whether the introduction of large scale methods of production, in order to be efficient, would involve a scale of output larger than the previous output of the entire industry would depend upon the circumstances of each industry. The logic of the balanced growth doctrine would tend to imply that the efficient scale of output to be produced by the modern methods is larger than can be sold in the pre-existing market and this necessitates the expansion or creation of market via the establishment of other industries so that the income generated in the latter provides market for the former and vice versa.

The interrelationships between different industries arising from the side of demand as well as of supply are many and various. With their emphasis on the complementary effects of the simultaneous establishment of a wide range of consumers' goods industries, the proponents of the balanced growth have inadequately integrated, at least in the formal body of their theory, the possibilities of a wider range and types of interdependence between different industries. Investment in the industry X not only increases the profits of those industries the products of which are bought by those employed in X but also, by reducing the price of X, can increase the profit of industries using X. It can also increase the profitability of those industries the products of which are used as inputs in X industry, or of those industries whose products are used in conjunction with X as joint inputs in some other industry, or of the industry whose product is a substitute for a factor used in X industry. If, owing to such external interdependence, profits of one industry are increased consequent on increased investment in another and vice versa, dynamic equilibrium in the context of industrialization of backward economies would be reached when investment in both the industries is carried to the extent that a further increase in one does not create additional profit opportunities in the other.

As an important illustration of external interdependence, Mr Fleming emphasizes the interdependence between consumers' goods and intermediate goods industries. The vertical propagation of external economies from the consumers' goods industries to the intermediary industries, or vice versa, counteracts to a certain extent the competitive effects arising from the inelastic supply of primary factors of when simultaneous expansion of various industries is undertaken. Simultaneous expansion of both would release primary factors from the consumers' goods industries in the production of which the primary factors would be substituted by the output of the intermediate industries in which the primary factors would produce a more than proportionate increase in intermediate products. This would not only leave the factors employed in other consumers' goods undisturbed but also make an increase supply of intermediate products available all around. However, the intermediate industries and consumers' goods industries are competitive in terms of demand for primary inputs or factors. Therefore, it is not obvious how there can be a net increase of primary factors available for consumers' goods industries unless, of course, efficiency in the intermediate industries is increased, so much as to require a smaller net amount of primary factors for a larger total output of intermediate products.

The doctrine of balanced growth in underdeveloped economies has frequently been conceived in the context of disguised unemployment, mostly in the agricultural sector, which is expected to counteract the inhibiting effect on balanced growth exercised by the inelasticity in the supply of labour. Mr Fleming's doubts as to the possibility of utilizing disguised unemployment so as to provide labour supply for a wide range of expanding consumers' goods industries seem to be based upon the following reasoning: (a) expansion of say, an industry A would, by increasing real income, increase the demand for food and the necessity of increased food production would tend to counteract the release of labour from agriculture; (b) industries whose products are competitive in consumption with A would decline under the impact of competition of the cheap substitute, i.e. A; therefore, expansion of A involving an increased employment of labour would withdraw labour from these substitute-product industries rather than agriculture; (c) the labour unions in the urban centres would force up the wage-rate as the demand for industrial labour increases.

As for (b), if the advocates of balanced growth ever happen to suggest that all consumers' goods industries are mutually self-supporting – providing markets for each other – it is a useful reminder to point out that there may be some pre-existing industries whose products would be substituted by the newly established industries. Even if the nature and composition of the industries whose simultaneous expansion is planned are so chosen as to render them mutually complementary, there may be industries outside this range which may be adversely affected and which consequently would release factors. But insofar as they do so, this would help (and not hinder) the expansion of the range of mutually supporting industries which are conceived in the process of balanced growth. As for (c), the strength of labour unions in undeveloped economics in raising wages in response to increased demand is a factual question and the general observation seems to be that lack of strong and highly organized labour unions is a feature of such economics and this lack is itself a counterpart of economic underdevelopment. The necessity of increased food production would probably release less labour from agriculture than would otherwise be the case. However, this hinges on the question whether more efficient methods of production would be introduced in agriculture and insofar as they are introduced, agriculture would be able to release labour.[1]

However, the source of underemployed labour lies not only in the agricultural sector but also in the casual workers in the different urban occupations, the petty traders who are in excessively large numbers primarily because of lack of alternative employment opportunities, the redundant servants and retainers in the domestic household sector or even in the commercial enterprises. This last sector is overinflated partly because the line between dependants and employees is often blurred and partly because charity of relatives is the only form of unemployment insurance available. Historically the supply of labour has been augmented in the course of industrialization by the industrial employment of wives and daughters of the household sector. Over all these is superimposed the fact of large population growth in underdeveloped economies. In the initial years of industrial development improved health measures tend to reduce the very high death rates whereas the effect on birth rate remains uncertain so that there is an increase in the rate of population growth. The primary cause of underemployment in the non-industrial sector is the lack of demand for labour emanating from the industrial sector owing to the absence of complementary factors to enable industrial expansion. It is, of course, granted that expanding industries cannot draw upon the disguised unemployment at the same rate of wages as that prevalent in agriculture but has to pay a higher rate. The wage differential has to be sufficient to overcome the resistance of the workers to the unfamiliar surroundings and the discipline of a factory life and also to compensate for the security afforded by the traditional pattern of life and family relationships. The existence of a wage differential to accomplish such a transfer has been a perennial feature of the process of industrialization.

Alongside the pull of higher wages and expanding employment opportunities in the industrial sector, the push from the rural sector is accelerated by a change in the organization of agriculture, including systems of land tenure, the size of holding, etc., which renders the potentially surplus labour actually and operationally redundant. This has been the historical process of labour supply for the expanding industries in countries like England, Japan, etc. It is to be noted, however, that the supply of labour which comes from the rural areas is essentially the supply of unskilled labour. Insofar as capital is available, the training of skilled personnel can be accelerated. The tempo and scale of industrialization may have to be attuned to the necessities of evolving labour force.

It is recognized on all hands that the prerequisite of a process of balanced growth is a sufficient volume of capital mobilized internally or from external sources. As Mr Fleming points out, if the simultaneous expansion of many industries runs up against an inelastic supply of capital and raises its price as a consequence, balanced development would be retarded. It has to be added, however, that the doctrine of balanced growth seeks to emphasize that the profitable application of whatever capital that can be internally or externally mobilized is enhanced by investing them in mutually reinforcing industries. One of the important reasons why private foreign capital is not attracted to underdeveloped economics is precisely the limitedness of internal market. Though a single foreign investor contemplating investment in one industry is discouraged by the limitedness of market for it, simultaneous establishment of many mutually supporting industries would provide the inducement to invest. While the unregulated price and market mechanism may fail to achieve such an end, external borrowings by the government or semi-public investment institutions may be utilized to promote a balanced structure. Also, the government, by a combination of tax and subsidy policies and other forms, of inducements can attempt to channel and dovetail both foreign and domestic capital into a balanced and well-integrated industrial pattern so that each separate act of investment, in the light of the accompanying investment similarly induced and undertaken, would find itself profitable. However, the author argues that as the servicing of foreign capital necessitates encouragement of export industries, this would detract from the usefulness of foreign capital in facilitating balanced growth of the home market industries. In the first place, it can be said that the distinction between export industries and home market industries is not always rigid. There are export industries which also supply internal market and in these cases, external and internal markets combined provide greater facilities for the realization of economies of scale. In the second place, the home market and export industries can together form parts of a mutually reinforcing process of balanced growth. The income generated the export industries can provide market for the home industries whereas some part of the output of the export industries can be used as inputs in the home industries or vice versa. Once the multiplicity of the types of interrelation ships between different industries is recognized, as illustrated earlier, the possibility of integrating the export and home industries in a mutually supporting pattern is considerably enhanced. In the third place, servicing of foreign capital can be accomplished by the replacement of imports, i.e., by the development of home market industries. The concentration of foreign capital in the export sector historically has been to a large extent due to the lack of internal market. The substitution for imports does not have to be via physically similar goods and substitution can be both price substitution and technical substitution. In the fourth place, depending upon the conditions of demand and technical conditions of production, the opportunities for increased economies of scale in the export sector may expand. It has also to be noted that whatever may be the source of capital, i.e., even if the whole of the capital is from domestic sources, so long as dependence upon foreign capital equipment or technical skill or essential raw materials from outside is necessary, foreign exchange has to be earned either by expansion of exports or replacement of imports. It is to be noted in this connection that the take-off into the process of cumulative expansion has been historically marked and initiated by an expanding foreign trade sector, i.e., expansion in the export industry or in the replacement of imports. In many cases the development of export industries has supplied the initial push and led to the development of other industries. However, in many underdeveloped areas the growth factors originating in the export sector, initiated by the foreign capital and enterprise, have not spread to the other sectors of the economy.[2]

The basic argument for the doctrine of balanced growth has two aspects which are closely interrelated and interdependent. Besides the need to overcome the limitedness of market by the establishment of mutually reinforcing industries, there is the need for an integrated programme because such a programme by the nature of the case would be large enough to cause a significant rise in national income and in the rate of saving and investment as a consequence. Such a rise in the rate of capital formation consequent on a rise in the national income would push the economy out of the stagnation and enable a take-off into a self-sustained process of growth. In the course of the development of the economy along the path of cumulative expansion as well as in the earlier stages, there is a continuing need to watch and to keep proper balance between the different industries and sectors of the economy. That income generated in an individual enterprise or industry may not automatically and immediately take up the corresponding output is obvious, above everything else, from the mere fact that there is a leakage from this income in the form of savings, taxes and imports. This leakage is there even if the expenditure out of this income is not directed towards other commodities. This particular enterprise or industry can, of course, capture some part of the demand enjoyed by the other producers by offering a new good or a more convenient or attractive service or a lower price. However, the immediate depressive effect of the leakage may be compensated, after a time lag which may be substantial depending upon circumstances, by an increase of exports, investment and government expenditure. Even without this tendency towards deficiency of marginal demand, the maintenance of intersectoral and interindustrial balance in the process of development is of prime importance. An expansion of manufacturing production without the simultaneous expansion of agriculture, unless it is intended for export markets, would be checked because prices of agricultural products would rise consequent on increased demand for them and this would lead to a rise in imports creating in turn balance of payment difficulties. There would also be the problem of finding markets for manufactures in the absence of a rise in income in agriculture. Alternatively, increased manufacturing production could replace imports and consequently release foreign exchange for the import of food and raw materials. Even in this case, however, if the output of the technically and economically efficient units of industry engaged in replacing imports is larger than the existing level of imports of that particular type of industry, the inadequacy of internal demand would start acting as a limiting factor and would need to be supplemented by the expansion of other sectors. Similarly, agriculture can expand if manufactures or exports expand or if import replacements are possible.[3]

A programme of industrial investment has to check the intersectoral or interindustrial balance by comparing the projections of planned output against the estimates of final demand, for example, by means of family budget studies. It has also to check the projected inputs against the estimates of planned outputs of the supplying industries and vice versa, for example, by means of input-output studies. In an unregulated private enterprise economy one enterprise is not aware of the production plans of other enterprises which may create markets for it or supply inputs to it. These plans do not get reflected in the price mechanism until the plans are actually executed and therefore cannot influence the investment decisions of the former. It is true that one enterprise may anticipate that its own expansion may stimulate some other industry which in turn would justify its own expansion.

However, this is too uncertain a guess on the part of one individual entrepreneur. To a certain extent individual industries do make some attempts, especially in advanced countries to analyse and to predict general growth of the economy, growth of demand and the pace and rate of development of related industries. By the nature of the case an individual industry cannot get an overall comprehensive view in the same way as a central organization can. The proposition that a set of uncoordinated investment decisions in the face of uncertainty may result in a smaller total investment than that undertaken under one integrated, well-informed programme of investment does not imply that the state has to undertake directly all investments. It does imply that private enterprises should work within the framework of a comprehensive programme, i.e., in the full knowledge of interrelations of different industries, their respective rates of growth and the rates of final demand. If necessary, the state has to assist the attainment of such a balance by regulation or by the offer of inducements and incentives.[4]

However, even if a balance is struck between various sectors and industries in a planned programme, the rates of growth of different industries may have to be adjusted as actual final demand and its components change as also the various interrelations, including technological input-output relations, change as between industries in the process of development. It is also worth reminding ourselves that the projection of these interrelations and changes therein becomes complicated partly because they depend upon the effectiveness of measures undertaken to execute the programme and the measure of such effectiveness cannot be foreseen. Besides the changes in final demand and technique, there are also the possibilities of substitution – price and technical substitution – as between various inputs on the one hand and various outputs on the other so that any future excess or shortage which may develop in one sector may partly or fully be adjusted through substitution. If it were possible to predict and foresee for a sufficient number of years these various changes and possibilities of substitution, a fine balance could be worked out and the various industries could be allowed or induced to develop according to such a predetermined pattern. Under these circumstances it may be said that investment is the most efficient and optimum if it takes place and grows in the future along such predetermined lines. All we can do in the actual circumstances of many underdeveloped countries, however, is to collect the requisite information and statistical data on the relevant aspects of the economy and fill up the gap in our knowledge as far and as fast as possible, to try to reach some sort of balance and consistency in the plan or programme in the light of existing knowledge and continue to adjust and readjust the original programme as our information accumulates and as the underlying factors change.

Notes

1 Rural unemployment may be viewed from different standpoints. In many underdeveloped countries such unemployment may, however, be of predominantly seasonal character. Insofar as this is true, this does not provide net addition to the permanent industrial labour force in a programme of balanced development. The real surplus manpower available in the rural areas would be the excess of labour supply over labour requirements in the peak seasons. The magnitude of this excess would depend upon the technique, equipment and organization prevailing in agriculture. The extent to which any net release of labour force can be effected from agriculture is a function of the speed and magnitude of changes in these latter factors.

2 It is true that export industries possess certain advantages in the context of a developing economy. The market for them being abroad, they neither cause competitive struggle at home for the disposal of their output nor do they depend upon appropriate growth of demand in other sectors. They may provide stimulus for other industries by providing market through, generation of income or through creation of services like transportation and engineering services etc., which others may utilize. They may also stimulate changes in methods and technique in the home market industries.

On the other hand, in many underdeveloped areas of the world concomitant in other sectors of the economy seems to have been retarded by: (a) the lack or inadequate development and diffusion of habits of saving, investment and enterprise in the domestic sector; (b) transfer abroad of income generated in the trade sector; (c) limited impact of the foreign trade sector because of foreign trade sector because of inadequate size vis-à-vis the total economy.

3 'When we speak of manufacturing and agriculture growing "together", or "at appropriate rates", or "in balance", in a closed economy, we refer to the rates which are determined by the community's marginal propensity to consume agricultural products as compared with manufactures. The open economy is more complicated, since the growth of manufactures for home consumption can be balanced by the growth of manufactures for export, instead of by the growth of agricultural production (or vice versa, substituting "agriculture" for

"manufacture", so in the real world we have to keep balance between imports, exports manufacture and agriculture, and not just between any two of them' – Lewis, 1955. p. 278.

4 If the state undertakes all investments, there is no guarantee that all inconsistencies would be abolished and that a perfect balance would be obtained. After all it consists of various departments and sections pulling their weight in different directions. Out of their conflicting viewpoints a neatly balanced investment programme may not always emerge.

References

Fleming, M., 'External Economies and the Doctrine of Balanced Growth', *Economic Journal*, June 1955.

Fleming, M., 'External Economics and the Doctrine of Balanced Growth – A Rejoinder to Prof. Nurkse,' *Economic Journal*, September, 1956.

Lewis, W.A., 'Economic Development with Unlimited Supplies of Labour', *The Manchester School of Economics and Social Studies*, May 1954.

PART II

DEVELOPMENT STRATEGY: PLANNING MODELS AND IMPLEMENTATION

Chapter 4

The Relevance of Development Models to Planning*

Planning models have been increasing in number and complexity in the developing countries in recent years. Strictly speaking, a model is a hypothesis involving a relationship between variables which intends to explain and predict events, past and future, or prescribe policy. For policy purposes, a model distinguishes between target variables which constitute the objectives of economic policy and controlled variables which are instruments affected by policies and which, through the variations occurring in their levels affect the levels or magnitudes of the target variables. A broader classification of variables is between exogenous and endogenous variables. The model specifies relationships, technological, institutional and behaviourial, between variables. In its reduced form, a model can always be expressed in terms of number of exogenous variables, some of which are predetermined for the purpose in hand, whereas others are policy or instrument variables; the values of the endogenous variables are then solved or determined by the interactions postulated in the equation system of the model.

Models can be sector-oriented or can encompass the entire economy. A restrictive model can be designed to explain or forecast the developments or prescribe policy in a particular sector, such as agriculture. It can be directed towards illuminating narrowly defined problems such as the optimum cropping pattern, given the prices of alternative crops and the prices of inputs and given a set of technical-production relations between inputs and outputs in agriculture. The first attempt in a planning exercise starts from an assessment of prospects in the individual sectors of an economy. Information on individual projects is often more readily available, especially in the early stages of a country's progress in economic planning. However, the sectoral approach has the drawback that it often results in assorted projects which are not part of an

---

* First published as 'The Relevance of Development Models to Economic Planning in Developing Countries', *Economic Bulletin for Asia and the Far East*, June–September 1970, pp. 56–61, 64–5. Reprinted in *Leading Issues in Economic Development*, 3rd edn, ed. G.M. Meier (New York: Oxford University Press, 1976) (paper for the UN/ECAFE Seminar on Economic Planning, USSR, 1970).

integrated total. If planning ends with project analysis only, it is not possible to evaluate the projects on a consistent basis or to check the total inputs against the overall availability of resources. Models can also encompass the entire economy; here one can mention the highly aggregative Harrod-Domar type of model, where the operations of an entire economy are conceived in terms of aggregate saving, investment and income relationships, which can further be extended to include the effects of external economic relationships in terms of interactions between income and imports or exports. The two gap model which seeks to investigate the gap between savings and investment, on the one hand, and that between exports and imports (i.e., the demand for and supply of foreign exchange) on the other, is an obvious example. While the original Harrod-Domar model treats capital as the only relevant scarce factor of production, the elaboration of the two gap model introduces an additional scarce factor, foreign exchange, which is crucial in many developing countries.

There is a variety of planning models which, while global in character in so far as they encompass the operations of the entire economy, are highly disaggregative in the sense that they postulate quantifiable relationships between the various sectors of the economy. A simple taxonomy of the planning models of this type is: 1) input-output models; 2) simultaneous equation models of a behaviourial nature; and 3) models which combine features of input-output models and of behaviourial models. The planning exercise involves the need to discover a consistent set of economic aggregates over the plan period, which are economically, feasible and politically acceptable; it is necessary to connect these macroeconomic aggregates with the sectors and projects which constitute the operational content of the development effort. As a general rule, input-output models provide the sectoral details and consistency checks among sectoral targets in terms of output and investment targets in the various sectors. An open input-output model seeks to provide a consistency check between the overall import requirements and export possibilities arising out of or associated with various sectoral outputs. The input-output models provide estimates of sectoral outputs and import levels which are consistent with each other and with the estimates of final demand. They aid in the allocation of investment to achieve the projected sectoral outputs and provide a more accurate test of the adequacy of the available investment resources. The requirements for skilled labourers, if a set of skill coefficients for individual sectors is available, and for imports can be estimated; and the possibilities of import substitution can be analysed. The indirect requirements for scarce inputs such as skill, capital and imports can also be estimated. Regional input-output models help in the regional allocation of investment. The main difficulty in the use of input-output models in developing countries concerns the treatment of structural changes such as substitution of domestic production for imports, changes in technology and other shifts in the structure of intermediate demand.

An input-output model usually does not handle questions of price formulation, except indirectly by postulating an equality between costs and prices in each sector; it does not deal with questions of fiscal and monetary policies, which are among the principal instruments of economic policy making. The standard input-output model deals with questions of fiscal policy and monetary policies as a subsequent or subsidiary exercise, once it has established the intersectoral flows. There is a class of macroeconomic aggregative model which does not directly, incorporate input-output relations among the various sectors of the economy. Such models deal with behaviourial relationships such as consumption and investment functions and institutional relationships such as those relating to tax yields, money supply, price formulation for broad classes of commodities such as agricultural and industrial commodities. They are usually expressed at high levels of aggregation. The basic elements of a traditional input-output model are the exogenously determined final demand, consisting of consumption and investment demand and the technical coefficients which quantify the intersectoral flows. The behavioural models concentrate on explaining behavioural relationships which are estimated from time series or cross-sectional data. They incorporate both linear and nonlinear relationships among variables whereas the input-output models are linear models and postulate a relationship of proportionality between inputs and outputs. These two types of model can be and are often combined, though they run into the problems of compatibility when the behaviourial relationships are nonlinear.

Over the years, the input-output model has been extended in various directions, especially in the direction of explaining the determinants of final demand by means of behaviourial and technological equations purporting to explain consumption and investment behaviour. The input-output model concentrates on consistency among sectoral outputs and inputs as well as between the former and the components of demand. In some formulations of the model, a capital output matrix is added so that investment by sector is related to the outputs of investment goods. The input-output models do not provide solutions for the best combinations of sectoral outputs or optimal combinations of techniques of production in each of the sectors. The important problems in economic planning are the allocation of scarce resources among competing sectors and, within each sector, among alternative techniques of production. Furthermore, in each sector, there is a choice between providing the required output either by domestic production or by imports and it is necessary to find the optimum pattern of import substitution. The linear programming models provide the necessary extension of the consistency models of the input-output variety to optimization criteria. Given the objective of economic policy, which could be maximization of income or employment or any other quantifiable economic objective, the linear programming models provide optimal solutions for combinations among sectors or techniques, including optimal combinations of domestic production and imports in each sector. Optimization is subject to the restraints inherent in technological production possibilities and limited resources, which can be distinguished in terms of foreign exchange, domestic savings and skills or any other scarce resource as indicated by the policy makers.

The use of models to illuminate discussions on economic policy making and economic planning has been facilitated in recent years by an increasing sophistication on the art of policy makers in handling economic tools. In the initial years, the discussion between model builders and policy makers in the planning agencies often threatened to become sterile since there was a difficulty of communication between them. As the number of model builders proliferated, the number of planners who were given training in quantitative economics increased correspondingly. As the planners and economists began to be taken seriously by the governments and faced the incredible complexity of the economics they were asked to understand, they began to look for ways to judge the possible consequences of alternative policies or actions. It was increasingly realized that models do not provide solutions but only assist in finding them. The models were as good or as bad as the assumptions on which they were based; real life complexities could hardly be all encompassed in models. They included the critical variables and key assumptions and provided the range of choice or alternatives within which the forces of political bargaining and judgement of the policy makers could function. The effectiveness of models depends in this context upon whether they are overloaded or pushed beyond their capacity.

Furthermore, model building and model using became increasingly a part of a dialogue between the model builders and those engaged in planning and policy making. Modifications are wrought as new knowledge accumulates; new decisions or analysis designed to understand the economic process or economic behaviour or effects of policy on human economic actions are derived from quantitative work outside the model. In the planning process, model builders have to be in constant touch with conventional economists, technicians and administrators. They have to check their assumptions and postulate relationships among variables by a constant dialogue with the specialists in the field.

The relevance of models in development planning may best be discussed in connection with the use of planning models in the actual planning process and with the discussions of development problems in the South Asian countries. The experiment with econometric models in the context of actual experience in development planning has thrown up issues and problems which have illuminated both the usefulness and the limitations of models in development planning. The following discussion of the subject is inevitably influenced by the experience in Pakistan.

In the use of models in development planning in a concrete context, several steps can be distinguished. These steps are not necessarily planned in any strict time sequence; there is always a feedback between one step or stage. However, a comprehensive linear programming model, which adequately disaggregates the economic sectors and which in addition contains dynamic elements, seeks to compress several steps into one grand exercise. In the absence of an all embracing model used right from the beginning of the planning process, several steps can be distinguished as follows: At the first stage of plan formulation, the model can assist in producing the macroeconomic framework. At this stage, interest centres mainly on broad magnitudes and their interrelationships in order to determine the tentative size of the plan or its broad orders of magnitude and to explore the relationships among growth targets, savings and the balance of payments.

At the second stage, the macroeconomic framework can be further elaborated to illuminate a few critical issues such as the interrelations between agricultural overall growth, the implications of alternative rates of savings or exports, the implications of import propensities for the balance of payments and the effects of alternative levels of foreign assistance. At the third stage, by, the elaboration of an input-output model in a macroeconomic framework, the implications of sectoral balance (the relationships among sectoral targets), and savings and the balance of payments can be explored. Even a relatively aggregative input-output model can help in clearing up intersectoral relations, the balance of payments implications of alternative growth patterns and final demand structures.

In Pakistan, for example, the first and second five-year plans were not based on any detailed input-output model. However, a macroeconomic framework of the Harrod-Domar type was used in discussions relating to the broad size of the plan as well as saving and investment implications. No model was ever formally presented, but various calculations regarding the size of the planned investment requirements were conceived in terms of such a framework which the planners had, so to speak, at the back of their minds. The elements of such a Harrod-Domar type of model can be found not in the form of a set of equations but in the form of analysis of investment requirements and role of savings, etc. as indicated by a careful reading of the plan documents. One can say that something like the following model was implicit in the discussions relating to the plan formulation.

$$I_t = K_1 (y_t - y_{t-1}) \quad (1)$$
$$D_t = d_{t-1} + K_2 I_t \quad (2)$$
$$GDP_t = y_t + D_t \quad (3)$$
$$GI_t = I_t + D_t \quad (4)$$
$$\bar{S}_t = GI_t - A_t \quad (5)$$
$$S_t = K_3 (y_t - y_{t-1}) + S_{t-1} \quad (6)$$
$$C_t = GDP_t - \bar{S}_t - G_t \quad (7)$$
$$M_t = K_4 \, GDP_t \quad (8)$$
$$\bar{E}_t = M_t - A_t \quad (9)$$

Symbols

t indicates the end, and t-1 the beginning of the plan period.

I = net investment
K_1 = net capital/output ratio
y = national income
D = depreciation and D_{t-1} is depreciation on the old stock of capital
$\bar{E}$ = required exports
A = foreign aid
K_2 = the share of investment required to account for the depreciation on the increase of the capital stock during the plan period
GDP = gross domestic product
GI = gross investment
$\bar{S}$ = required savings
S = savings
M = total imports
C = consumption
G = defence expenditure
K_3 = marginal savings rate
K_4 = import coefficient

The logic of the model is as follows. The planner sets a target in national income (y_t); on the basis of the capital/output ratio, this yields required investment (I_t) which, together with depreciation requirements, yields required gross investment and the resultant gross domestic product. Given foreign aid, required savings to match required investment is derived. Given initial savings, additional expected savings are thus determined. If required savings are not equal to expected savings, the marginal savings rate (K_3) needs to be adjusted; K_3 is a policy variable. Consumption is a residual. Given import requirements, which with varying K_4 (that is, the degree of import substitution) and foreign aid, required exports would be derived. The balance between required exports and imports can be adjusted by either import substitution or export promotion.

Such an aggregative framework is not of any relevance in the matter of choice between alternative sectors, projects or techniques. Nor does it encompass the consideration of the alternative objectives of economic planning such as growth of income, expansion of employment, more equitable distribution of income, or a more equitable balance in regional development.

The overall analytical framework discussed above provided the planners with a guide to the broad orders of magnitude of the investment programmes and the relationships with savings, growth of income and balance of payments. These orders of magnitude provided the first tentative step in the actual formulation of the plan, which consisted of sector programmes and of individual projects in each sector.

At the next stage, lists of projects by sector were drawn up, on the basis of technical feasibility studies and market demand. These were related to all public sector projects and to the big projects in the private sector. The small projects in the private sector escaped scrutiny. The total claims on domestic resources and foreign exchange exceeded the availabilities of these two scarce resources, as estimated from or as assumed in the aggregative framework. Priorities were allocated to various sectors and to projects in each sector in order to arrive at a feasible programme. The criteria were the contribution of the project to national income and the balance of payments. In the case of conflict between the two criteria, the latter received greater weight. Further adjustments were required to make the outputs of different sectors consistent with one another and to avoid bottlenecks in the production of power, cement and other critical materials. This called for an analysis of demand and supply of these key commodities and for the establishment of physical balance sheets.

The investment and foreign exchange requirements emerging from the analysis of sector programmes and individual projects are not necessarily consistent with the aggregative projections of savings, investment and balance of payments. Any such consistency has to be matter of trial and error in a process of feedback between the two sets of exercises. The assumptions made at the start of the aggregative model have to be re-examined. The various methods of assessing priorities require the estimation of future value to the economy – as determined by aggregate demand and total availability of labour and foreign exchange – and the prospective demand for individual commodities. These magnitudes can best be derived from an intersectoral general equilibrium model, but approximation can be attempted on the basis of partial analysis.

Thus in Pakistan in the matter of sectoral allocation of investment and choice of projects in the broad sectors, it was sought to apply some forms of cost-benefit analysis. In an ideal situation, the planning model is expected to generate the shadow prices for the relevant scarce factors of production, which should then provide a guide for the solution of projects, the aggregation of which will eventually produce the sectoral allocation of investment. The derivation of scarcity prices for scarce factors by considering the supply and demand for them, without direct relation to a detailed intersectoral model, is at least a partial equilibrium approach. After all, in a general equilibrium framework, the factor prices and commodity prices are to be simultaneously determined, given the supply of scarce resources, and given the structure of demand and of consumption or investment expenditures which are the functions of the overall development programme, including its sectoral composition. Some crude estimates were made of the scarcity, price of foreign exchange on the basis of a judgement regarding the possible overvaluation of the exchange rate. This kind of experiment with a simple, aggregative planning model seems to have been common to many developing countries. In India, it was the extension of the Harrod-Domar model in terms of the consumption and investment goods sectors, with different capital output ratios, which played a crucial role in the choice of investment decisions during the second five year plan. The model was first formulated in an Indian context by P.C. Mahalanobis.

It was during the formulation of the third five-year plan in Pakistan that an attempt was made to formulate a formal planning model based on: (a) an input-output model consisting of the current flows; (b) a highly aggregated seven-sector capital-output matrix; and (c) a set of expenditure functions. The seven sectors as distinguished in the model represented an aggregation of a much larger input-output model for Pakistan's economy. However, since the objective of the model was to illuminate the implication of choices in terms of some key exogenous variable such as the rate of growth of GNP, the agricultural growth rate, the savings rate, and nonagricultural exports, the more detailed input-output matrix was not used in the formal model. The sectors were: agriculture (including forestry and fisheries); manufactured consumer goods; intermediate goods; investment goods; construction, transportation and communication and other services. In the final demand sector, a distribution was made in terms of consumption, fixed investment, inventory investment, exports and import substitution. The model consists of a set of input-output relations, consumption functions, investment functions (relating the flow of investment goods by sectors of origin and destination), inventory adjustment functions, indirect tax functions and import functions. The reduced form equations of the model express agricultural exports, imports, fixed investment, balance of payments surplus and deficit, import substitution, consumption, inventory investment and seven sectoral outputs as functions of GNP, agricultural output, savings, and four different nonagricultural exports.

In the reduced form of the system of equations, the sectoral outputs, investment (fixed and inventory), agricultural exports, imports, import substitution, consumption and the current account of the balance of payments are endogenously determined; the exogenous variables are income, agricultural output, savings and nonagricultural exports. The model, therefore, permits an examination of the implications of the changes in the exogenous variables for the rest of the economic system.

The model helps planners to examine the implications of alternative values of exogenous or predetermined variables. Since the model generates a unique solution for any set of values of parameters and exogenous variables, the changes brought about by alternative parameters or exogenous variables can be deduced. Given the rate of growth of GNP, the implications of different rates of growth of agriculture on the rest of the economy, that is, on the GNP, different industrial sectors as well as on the savings investment gap or the foreign exchange gap can be readily estimated. A slower rate of growth of agriculture reduces the agricultural export surplus, increases the balance of payments deficit, and necessitates a larger import substitution in the nonagricultural sector, which increases investment requirements, in view of the higher capital requirements. The model makes it possible to arrive at feasible targets of overall GNP growth and of agricultural growth and nonagricultural exports which are consistent with the maximum or feasible degree of import substitution or foreign assistance. In other words, the model can answer the following, question. Given feasible levels of deficit in balance of payments, of exports, and given the ability to achieve import substitution, what growth rate of GNP can Pakistan achieve and what marginal savings rate does this growth require?

The answer – or more correctly, the best guess at the answer – to any such single question can probably be fairly quickly provided by a planning commission economist without such an elaborate and time-consuming model; but for the answers to dozens of such questions, the model is an extremely efficient device. Beyond that, moreover, is the fact that time and skilled manpower in a planning commission are so scarce that, without such a model, most of the interesting alternatives would never be explored; if they were, the research would be done at different times by different people, who would almost inevitably apply different assumptions, procedures, and degrees of optimism and therefore produce incomparable results.

There are two sorts of policy variables (or policy parameters) in the model, namely direct and indirect. The direct are those which are fixed by government policy, for example, the income tax rate or the size of government expenditures; the indirect are those which can only be influenced by government policies, such as the marginal savings ratio or the level of GNP in some future year. In the model discussed above the policy variables (or parameters) are entirely indirect with the result that the precise implications of alternative policies cannot be calculated without further work with information from outside the model. Nevertheless, a great deal of insight into direct policy problems can be gained from the model. Indeed it is often analytically convenient to break up a policy analysis into two stages, and the model may provide the answers at one stage of the work: for example, when faced with the question of the influence of a change in a direct policy variable on gross savings (which probably requires that its effect on corporate, household and government savings be discovered) and the effect of a changed volume of savings on the economy has to be examined. The model will not help at the first stage, though that stage may require a fairly straightforward supplementary analysis, but it will be of much assistance at the more complex second stage of the analysis.

The policy changes about which the model offers insight are those which operate through either the exogenous variables or the parameters of the technological behavioural relations. It is clear that many aspects of government policies operate through their impact on the exogenous variables of the model (GNP, savings, agricultural output and exports). For example, the level of savings (or alternatively the marginal savings ratio) may be influenced by excise, income and corporate tax policies, income distribution and social welfare policies, interest rates and other monetary policies, regional growth policies and the like. Similarly, many of the parameters of the behavioural relations are influenced by various government policies. For example, one of the relations of the model gives the consumption of each sector's product as a linear function of the aggregate consumption; but this propensity to consume from a particular sector can be altered by government price policies or excise taxes, the availability of imported substitutes (or complements), the redistribution of income, and so forth. There are very few parameters in the model which cannot be reached through government policy. Manipulation of the model will give information as to the size and direction of the impact of any changes in a parameter – and may thereby help to decide about policies which affect that parameter.

Furthermore a model of this kind permits a wide variety of sensitivity tests, tracing out the effects of changes in exogenous variables on the endogenous variables. The sensitivity tests free the model builder from the constraint of using a single value for any variable and permits him to try out a range. But it is important to remember that the range within which various values of exogenous variables are tried is largely a matter of judgement on the part of the model user. He cannot be entirely freed from the need for exercising judgement. In determining the plausible range of values with which to experiment, history and analogy are standard guides to a policy. But both history and analogy are less reliable guides in a developing country than in a developed country. As R. Vernon puts it, 'typically, the history of development in the underdeveloped areas is brief; typically, it is grossly and imperfectly recorded. Even if recorded with great accuracy, however, one would expect to find highly unstable and swiftly changing patterns at even modest levels of aggregation' (Vernon, 1966, p. 64). Therefore, judgement plays a vital role. Moreover, a wider range of alternative values has to be tried out in the particular situation.

What combination of ranges of different exogenous variables can be tested together? The variables which formally appear exogenous and independent in a model may not be and are not in many cases independent. A sensitivity test for one variable at a time may not be appropriate. Sensitivity tests on the basis of alternative sets of values of exogenous variables generate a wider range of choice than the sensitivity tests of one variable at a time. It is desirable that the results of sensitivity tests should not cover a very wide range, because in that case the policy maker's scope for judgement is not appropriately reduced. The purpose of a policy model is to expand the role of quantitative analysis and to reduce as far as possible the role of intuitive judgement.

Conclusion

In most of the developing countries, consistency models consisting of input-output tables have been most widely used. Their use has been supplemented by material balances for key commodities. Formal linear programming models have been less extensively used because of the data requirements. Moreover, quantification of the multiple objectives of economic policy has been far from satisfactory. The most important problem in many developing countries has been the determination of the weights to be assigned to growth of income on the one hand, and to creation of employment and attainment of a more equitable distribution of income on the other. Even when supplementary objectives such as employment creation and income distribution can be treated as constraints to a linear programming model, while the maximization of income is treated as the objective function, the problem of weighting is not to be avoided. The levels at which the constraints regarding employment and the distribution of income have to be put depend basically upon social value judgements and hence on the relative weights which are to be accorded to conflicting objectives. Most of the optimizing models have concentrated on scarce inputs of capital and foreign exchange and have assured labour as a free input in view of the disguised and open unemployment in many countries in the ECAFE region. The most important problem in these countries is, however, the provision of employment. This has implications for the other sectors in terms of supply of wage goods, even if the objective is to put labour to work in construction projects in the rural areas. In the formal models, one could incorporate the rural public works programme as an activity and consider as its major input the supply of wage goods and a limited amount of capital goods.

Again, many models are basically one-period or terminal-year models, in which the increase in income is considered for the plan period as a whole. The models are usually conceived in terms of maximization of terminal year income or consumption or terminal stock of capital or increase in income or capital stock. The distribution of year by year growth of income to be achieved during the plan period necessitates a dynamic programming model which, in terms both of data requirements and of mathematical properties or ease of computational manipulation, is not entirely satisfactory. In the actual planning process, the annual phasing of the investment programme is of crucial significance; it has implications for the generation of savings and hence for the requirements for foreign aid. The rate at which investment can be accelerated over the plan period is closely related to the absorptive capacity of an economy, conceived broadly in terms of technical and managerial ability to generate, prepare and execute development projects. This is partly an institutional problem and partly a problem of a package of appropriate policies, including the training of manpower in the various aspects of development activity. Any dynamic model has to incorporate some assumption regarding the rate at which investment can be accelerated from one year to the next during the plan period. The annual rate of acceleration which is feasible requires analysis of institutions and policies outside the formal structure of a model. In addition to postulating a feasible rate of acceleration of invest merit and income, such a model has also to incorporate the time lag between investment and income. The generation of income in any plan period is partly the consequence of investment which has taken place in the last planning period and partly of what is invested during the plan period itself. Again, a part of the investment in the present plan period will come to fruition and yield income in the next plan period. While the capital output relations may be satisfactory for a relatively general discussion of long-term perspectives, detailed knowledge of the projects under implementation, their completion times, the requirements for their future operations and capacities which might become critical in the immediate future will play an important role in a five-year plan period.

It is remarkable that, in most of the developing countries, the use of models for short period planning has been very limited. The concept of annual planning except in the sense of an annual government budget, elaborating government tax and expenditure policy, has only very recently been introduced in the planning process in a few developing countries. Even though an annual plan is the instrument through which five-year plans can be implemented, little effort has gone into the elaboration of models for planning for a year. The technical and behavioural relations which may hold for a five year period can seldom be used for annual plans. An annual plan has to be in much greater detail in terms of projects and programmes as well as in terms of fiscal, monetary and commercial policy which need be integrated with the investment and income targets. Import and export projections have to be made every year to be realistic and worthwhile, and so do the savings and investment projections. So far, input-output models have been fruitfully used in forecasting annual import requirements. The stability of import coefficients is much greater in the short run than in the medium or long run.

The optimizing models which consider only the choice between domestic production and imports for each sector implicitly assume that the currently or historically prevailing technical coefficients of production are the optimum ones, given the prices of factors and outputs. In view of the known imperfections in the factor and commodity markets, the prevailing technology is far from optimum. Unless comprehensive linear programming models – which specify alternative techniques in each sector and which encompass the problem of choice between sectors and techniques simultaneously as an integral part of the solution – are used, it will be necessary to undertake supplementary analysis for the choice of the appropriate techniques in each sector in the light of the assumed opportunity cost of the scarce resources. In the absence of a formal model incorporating the choice of techniques, there is need for a process of trial and error, since the selection of techniques for each sector on the basis of the assumed scarcity prices, and the subsequent derivation of the output and investment figures, may not be consistent with the given quantity of scarce factors, which will necessitate a revision of the original prices. In a few steps of iteration a reasonably consistent solution may be found. The important supplementary analysis at the sector or project level is therefore the analysis of alternative techniques and a sensitivity analysis of how the optimal choice of techniques varies in response to change in scarcity prices.

Reference

Vernon, R., 'Comprehensive Model Building in the Planning Process: the Case of the Less Developed Economies', *Economic Journal*, March 1966.

Chapter 5

Development Planning and Plan Implementation: Lessons of Experience*

Preface

Readers will, perhaps, be forgiving if this prefatory note largely assumes the form of a biographical sketch. Amiya Kumar Das Gupta was born on 16 July 1903 in what is now Bangladesh. A childhood spent in a rural setting amid lush tropical surroundings, a rudimentary education along lines laid down, absent-mindedly or otherwise, by the British for the subject people, his early upbringing was no different from that of others who grew up in that particular milieu during the first decades of the century. Das Gupta, however, came from a sect which, by tradition, combined the pursuit of the medicinal science with a passion for classical grammar. Deriving pleasure from logical propositions thus came easily to him. And once he joined the University of Dacca as an undergraduate in the Department of Economics in 1922, a new world was suddenly revealed. It was a quiet university, for Dacca was a quiet colonial town. A recently built secretariat, intended for the stillborn provincial government of East Bengal and Assam; had been handed over to the university in the suburb of Ramna across the railway track. The architecture was half Gothic, half Moorish, the indolence of *casuarinas* and *gulmohurs* filled the air. There was a simulation of the British pattern of university education, with halls of residence and all that. In today's cliché, it was an elitist institution; such elitism somehow fitted snugly into the context. Das Gupta was able to extract the most from the university. A remarkable collection of dons had gathered at Dacca: for example, the Hartogs, the Langleys, the Wrenns. You had personages like S.K. De, A.K. Chanda, Mohitltal Majumdar, A.F.A. Rahman in literature; R.C. Mazumdar, Susovan Sarkar, K.R. Kanungo in history; S.N. Bose, J.C. Ghosh, K.S. Krishnan in mathematics and the physical sciences. The university's Department of Economics (which at the time, was not separated from Politics) had its fair share of interesting men. S.G.

* First published in *Economic Theory and Planning*, ed. A. Mitra (Oxford: Oxford University Press, 1974).

Panandikar and J.C. Sinha were competent teachers. K.B. Saha knew his Alfred Marshall, and succeeded in transmitting to his students some of his quiet, but fierce, loyalty, to Book V of the *Principles of Economics.* There were others, who were at least aware that the science of economics went beyond totting up the number of tubewells in a district or estimating the yield from land revenue in a given year.

Development Planning and Plan Implementation: Lessons of Experience

Planning for social and economic development has been widely accepted in the developing world since the 1950s. Increasingly a large number of developing countries have formulated five-year or longer term plans for economic development. In the last two decades or so, considerable progress has been made in devising planning methodologies, techniques of formulating plans as well as appropriate planning machinery. There has been a reappraisal in recent times of the objectives of development planning – a trend away from overemphasis on income objective towards a greater awareness of the problems of employment and income distribution. However, in the actual formulation of the plans, the reflection of the new awareness has been slow. Similarly, while the methodology and techniques of plan formulation have attained a high degree of sophistication, their application in practice in the poor countries has been handicapped by a variety of factors. Furthermore, considerable divergence between the plans and their actual implementation has frequently been noticed. This paper presents, in a preliminary fashion, a few thoughts on these problems in the light of experience of a development planner.

One of the basic problems posed to a development planner in a poor country in the task of plan formulation is that there is invariably an acute dearth of reliable factual information which is relevant for the purposes of planning. Improvement in planning techniques and methodology has progressed far ahead of the availability and quality of data. Unfortunately, statistical requirements of a development plan are both enormous and very expensive; they require a large number of trained people and a long time to collect, analyse and make them usable for planning purposes. The concepts and methods of measurement used in collecting data in many developing countries are inappropriate for policy formulation and quite frequently a vast amount of data are collected in a routine fashion as a legacy of administrative requirements inherited from the past before the advent of economic planning as a tool of policy formulation.

Very often there are insufficient data on the socioeconomic characteristics, at both micro and macro levels, an inadequate national accounts system and scanty information on the magnitude and pattern of employment or income distribution, specially in the non-monetized sector which is significant. The interrelations between the monetized and non-monetized sectors are necessarily ignored; the contribution of the latter to output and employment is omitted in the plan projections. In the formulation of investment projects, a planner is handicapped by the lack of knowledge about alternative production possibilities or of techniques; he is ill-equipped to judge the impact of information on the technological and behaviourial relationships in the various sectors of the economy.

The planner does not have before him all the alternatives or the full range of investment projects; nor does he have the benefits of a systematic analysis of the effects of alternative policy packages. The latter takes time in view of the need to ask right questions as well as to collect and analyse the relevant data. Since he is under constant pressure to recommend within a short time-frame investment projects and economic policies, he is often forced to make snap judgements.

There is a great pressure for taking immediate decisions. In a situation where there is a paramount need for quick action or decisions, to advise caution and to slow down action is likely to be considered retrogressive. Inadequate information about alternative investment projects or the effects of alternative policies has consequences for planning in two directions. Firstly, the time horizon of the planner has to be kept within narrow limits so that commitments involving large expenditures for many years ahead are avoided; secondly, projects and programmes should have sufficient flexibility so that revisions are possible and do not become costly. To illustrate, in Bangladesh a development planner faced with a paucity of data and full range of feasibility studies necessary for the formulation of projects had to initiate investment projects simultaneously with studies and collection of data. Many past studies about investment projects were brought up-to-date on the basis of whatever recent data were available and in the light of detailed discussions with persons and experts actually involved in the preparation and execution of projects. But then two steps were taken to minimize the risks of planning without sufficient knowledge. Firstly, whenever the area of uncertainty was large, the investment package was kept at the minimum so that one could start from a small programme and build up to a larger one over time as one felt one's way through the programme, so to speak. In the transportation sector, for example, a large-scale inter-model study was under way, but investment decisions were kept down to a hard core of projects which would broadly remain unaffected when full-scale study would be available and where adjustments could be made without significant costs. Projects involving large expenditures and involving considerable inter-relationship with projects in different modes of transport and with projects in the other sectors were postponed. The general shortage of investable resources and the pressure of competing demands from other sectors to finance more fully worked out investment projects succeeded in keeping down the pressure for investment in this sector. Similarly, in the power sector, a need was felt for a major energy study, spelling out the long-term requirements of power and alternative ways of generating it; major decisions on new power projects were kept pending.

Secondly, it was decided to build into the investment programmes and policies a large component of continuous research and evaluation of the effects of programmes and policies, as well as sufficient flexibility so that adjustments could be made as facts were accumulated and analysed. The most important illustration of this in Bangladesh was the investment programme in the sector of health and family planning. There was considerable uncertainty as to the appropriate mix of technology in terms of choice between various terminal and non-terminal methods of family planning for an effective family planning programme; there were unresolved issues regarding the proper extent and nature of integration which should be effected between the health and family planning programmes. Past experience in Bangladesh was very scanty; existing programmes were not successful. Experience elsewhere in the developing world was very limited and its relevance to the socioeconomic background of Bangladesh was not established. Neither was it easy to decide on the appropriate institutional framework for launching an effective motivational programme for population planning, including the use of educational system, agricultural and rural development programmes, social welfare agencies and information media. Therefore a large component of experimentation was built into the programme. Learning by doing was accepted as the keynote of progress in this area; hence not only research and evaluation was made an integral part of the programme, but also it was started in a modest way, including a number of pilot projects to demonstrate the effectiveness of various kinds of institutional arrangements as well as the use of different techniques of family planning.

In spite of considerable advance in the technique of plan formulation in recent years, one aspect of intersectoral consistency in planning exercise remains very poor. This relates to the integration and consistency between investment in social sectors, such as education, health, manpower and population planning, etc., on the one hand, and investment in the directly productive sectors on the other. The demand for investment in these sectors continues to be exogenously determined on the basis of what is socially desirable as well as what is feasible in terms of actual programmes; they are very loosely integrated with the targets in the physical sectors. The impact of investment in these 'exogenous' sectors is subsequently related to the demand for inputs from the rest of the economy. This was, for example, the case with the Bangladesh Five Year Plan. The two-way relationship has yet to be built at the operational level into the plan formulation. The consistency models, such as input and output models, are primarily built around the directly productive physical sectors. The requirements of manpower by types and categories which are needed in the investment projects in the various sectors are often dealt with in a very general way. The appropriate degree of specialization in manpower training is as yet an unsolved issue, at both theoretical and operational levels. In most cases, there is an inadequate inventory of existing skills and professions in the country. Moreover, the executing agencies are unable to forecast their requirements of manpower until the projects and programmes are formulated in detail. Every plan at the time of its formulation has a large component of projects and programmes which are not fully worked out and the input requirements, including manpower requirements, are not known in any great detail. The input-output exercises are usually formulated in terms of broad sectors of an economy varying between 30 to 40, depending on the availability of data. The input coefficients in terms of broad categories of skills are seldom worked out. Even if they are available, they are highly unreliable. The substitutability between sub-components of skills in a particular sector is often considerable, depending upon the degree of in-service or on-the-job training which is visualized as well as upon the nature of training which is imparted initially.

Manpower planning adds to the complexity of the planning exercise in yet another way. The training of skills encompasses more than one plan period since the gestation-lag in manpower training is very long, covering several stages of education from primary through secondary schooling to institutions of higher learning and specialized training. The longer time-horizon in manpower planning, combined with the fact that long-term projections of the rate and pattern of economic growth are highly uncertain, necessitates that a considerable flexibility gets built into any reasonable programme for manpower training. In view of the fact that the most critical bottleneck in the implementation of a plan in developing countries is the shortage of trained manpower, this shortcoming in the planning procedures is a very serious one. The historical experience of the private enterprise economies is not particularly appropriate, and the experience of the centrally planned economies is inadequately understood, especially with reference to its relevance in the mixed economies prevailing in the majority of the developing countries.

While on the subject of planning methodologies, it should be mentioned that of late questions have been raised regarding the appropriateness of the traditional methodology of planning in view of the emphasis now being placed on income distribution and employment. In determining the size and sectoral composition of a plan, the starting-point is often the target rate of growth of income and final demand. The employment implications are derived from the choice of the sectoral composition of output, on the one hand, and on the choice of techniques in each sector, on the other. In this view, maximum employment consistent with the composition of final demand is secured, if appropriate scarcity prices of labour, foreign exchange and capital are used at the time of the selection of techniques as well as of the specific investment projects in each sector. However, the level of investment in any sector can be raised above the level which is suggested by cost-benefit analysis based on an appropriate shadow pricing of labour, if it is decided to accord predominant consideration to employment. The cost in terms of output forgone is clearly indicated by this approach.

However, what is important in this context but is often neglected is the fact that composition of final demand is not neutral in terms of its effects on employment. The pattern of final demand which is geared to the requirements of the lowest 30–40 per cent of the population in a poor country often tends to consist of the kind of goods which are more easily amenable to labour-intensive techniques. This involves a prior decision as to the pattern of income distribution and as to the policy packages and institutional changes necessary for attaining the desired distribution if income or appropriate composition of final demand, which is consistent with the employment-oriented production programme.

There are those who argue that planning should not start with income target but should start with employment target and work out its implications in terms of total national income and its composition rather than the other way round. Given the employment target in the aggregate and its distribution between sectors, output is derived by the use of optimum labour-output ratios. The sectoral distribution of the aggregate target of employment depends on the choice of technology and composition of demand for output, which in turn is determined by the level and distribution of overall income. No matter where one starts, income and employment are inter-related and have to be consistent in terms of composition of output and sectoral distribution of employment. An iterative process will be necessary to make them consistent irrespective of the starting point. In a programming exercise, once income and employment targets with clearly specified, objective functions and constraints are postulated, consistency between the two is assured in course of the solution of the programming model. To start with employment as a target and to derive output as a dependent variable rather than vice versa, even though both the exercises finally end up with the same result, may help to highlight the enormity of the employment problem and its implications in terms of growth of output, scale of investment and institutional requirements for the mobilization of labour.

In most planning exercises, no attempt is made to integrate income, employment and investment in the monetized sector with those in the non monetized sector. Admittedly, very little information relating to this sector is currently available. What is more important, however, is that whatever information is available, for example, in the numerous agricultural or village surveys which have been undertaken in some of the developing countries, has not received adequate attention in either economic analysis or planning. A villager often undertakes investment in the repair and construction of his own dwellings with his own labour and materials which he either produces in his homestead or on his farm or collects from the fellow-villagers by way of barter or common sharing arrangements. He builds stores for his produce, repairs embankments and makes small investments for the maintenance of his farm. Also, a village trader builds his godown for storing his merchandise, which is often located on his homestead. The resources for such investments in kind are generated in the non-monetized sector; they do not pass through the financial institutions or monetary system and, even it they do, they are only marginal flows. The fact that there are such investments in kind and also economic activities associated with the non-monetized sector partly explains how one reconciles the very low level of estimated or visible monetary income and investment in the poor countries with the fact that the poorest sections of the population in these countries at all manage to survive or carry on their living in the way they do. The analysis of the behaviour of the non-monetized sector is important in determining trade and monetary policies, specially in countries with a significant non-monetized sector. The rampant hyperinflation in some developing countries bypasses this sector; that is why they have not faced the kind of social upheaval, dislocations and distress which a similar degree of inflation would have produced in a fully monetized economy prevailing in the developed world. In the Bangladesh Five Year Plan, the investment in the non-monetized sector was explicitly included in the estimated development outlay. It was about 11 per cent of the total development outlay. More than 80 per cent of this outlay in the non-monetized sector is expected in rural housing and construction, in agricultural projects and in village trading activities.

Planning is not merely an exercise in economics but also in sociology and politics. Any plan worth its name has to be a sociopolitical document. A plan is not merely a programme of allocation of resources; it must recommend appropriate policies which need be pursued and institutions which must be built up so that the planned investment programme yields expected results. Policies and programmes incorporated in a plan affect different sections or groups of people differently. In a democratic political system the conflicting interests of the divergent groups have to be reconciled. Moreover, new interest groups emerge and they with the support of the groups already participating in the system must satisfy all or a good part of these demands. Admittedly, even in a limited form of political democracy, a policy of compromise is necessary, involving permanent negotiations, fleeting agreements, controversial arguments and, in general, a search for the consensus. But once the politicians publicly accept and genuinely believe in the goal of economic development under a system of planning, the negotiations and political compromises are confronted with constraints; they are to be conducted within the limits set by the national objective. The limits to the fulfilment of political objectives are set by a consideration of their economic implications, that is, their effects on growth and efficient allocation of resources.

The interplay of politics and economics does not permit of an easy solution. The role of the planners is particularly difficult in this context. It is for them to analyse the implications for growth and equity of the non-economic, that is the political and social, objectives. The feasibility constraints, political and social, on the economist's maximizing or optimizing models have to be clearly spelled out. The task is rendered more arduous by the fact that the non-economic feasibility constraints or objectives are often multiple. The political leadership is required to choose between them because the realization of any of the non-economic objectives involves economic cost and requires resources. Since resources are limited and not all non-economic goals can be realized, the politicians have to exercise a choice in the light of their evaluation of the relative costs of the different non-economic goals. Moreover, given a specific non-economic goal, there are different ways of achieving this goal; some methods are more costly and less efficient than others. There is thus the problem of choice between the alternative ways of achieving a non-economic goal. Independence from foreign domination in nation-building activities or integration of different ethnic or social groups or more equitable regional distribution of income may be highly desirable or acceptable social and political goals. At the same time, there are efficient as well as inefficient ways of achieving each of these objectives. Therefore, the planners as social scientists cannot always blindly or passively accept the social and political feasibility constraints. It is necessary for the planners to examine and analyse the circumstances and factors governing the non-economic constraints and to suggest policies which may enable the political leadership to eliminate or ignore some of these non-economic constraints. The task of planners in suggesting appropriate policies or institutional changes is complicated by the fact that politicians seldom clearly articulate all their non-economic objectives and distinguish between short-run and long-run political goals. Nor are they always determined to follow the full implications of these objectives.

The impact of politics on economic planning is nowhere more obvious than in the determination of the size of the investment programme. A large investment programme and a high target rate of growth of income in poor societies is often a political imperative. A plan ideally should serve as a rallying point around which popular enthusiasm is to be generated and national efforts are to be integrated and mobilized. A modest plan may not succeed in enthusing the people and fire the imagination of the masses. No plan can or should be based on a mere extrapolation of the past trends or the past levels of performance which are in any case very low. A plan to be acceptable can seldom project a decline in per capita income or a rate of growth of gross national product or employment which is lower than what was achieved in the past. It is not politically feasible for the government of a developing country, however autocratic it is and whatever its constraints are, to make a plan which does not provide for an improvement in the standards of living. On the other hand, a very large plan runs the risk that if it cannot be realized for one reason or the other, it may lead to frustration, breed cynicism and generate a lack of confidence in planning efforts as such, which are criticized as being unrealistic.

To implement a large investment programme in a poor economy involves considerable sacrifice on the masses in order to restrain present consumption in the expectation of future gains. There is a great pressure for increasing present consumption in view of prevailing low standards. A large investment programme not only imposes severe constraints on the level of present consumption but also thrusts a serious responsibility on the planners and administrative agencies to choose highly productive projects and implement them efficiently. Low productivity projects, inefficiently executed, impose present sacrifices without adequate compensation in terms of future gain in the consumption. There is yet another constraint on the size of a plan. Planned development implies a change in existing productive structures, inter-group relationships, methods of work and organization. The higher the size of the investment programme and the higher the rate of growth, the greater is the speed of change involved in the existing institutions and productive structure. In poor societies with meagre resources and limited room for manoeuvre, the scope for discontinuous changes or disruptions in the existing system is limited since they often involve heavy costs of dislocation, and even sufferings, if policies do not yield the expected results. Therefore, the speed of change has to be kept within manageable limits.

In the light of the foregoing circumstances, the size of a development plan needs to be determined by a careful evaluation of at least three sets of considerations, amongst others. Firstly, the planners must evaluate the administrative ability of the government – including the ability to implement programmes in the light of past experience as well as the expected improvement or growth in its performance. A plan must suggest policies and institutions which will lead to improvement in standards of performance. Secondly, the scope for mobilization of domestic financial resources for development must be assessed. Thirdly, it must provide for a proper balance between domestic resources and foreign assistance or net capital inflow. The costs and benefits of different levels of net capital inflow, including its composition, have to be properly appraised. The relative magnitude of dependence on foreign assistance is not merely a function of narrow economic criteria, but also involves important political considerations including considerations of both domestic politics as well as foreign policy implications in so far as the geographical composition of available foreign assistance is concerned.

The inter-relationship between domestic resources and foreign capital inflow is not merely that of one supplementing the other; the latter may under certain circumstances negatively affect the former. Moreover, there are political limits in many countries to the extent of dependence on foreign capital inflow which a plan postulates. Given the urgent economic need and political imperative of aiming at a positive rate of growth in per capita income, a shortfall in domestic resources mobilization results in a greater dependence on foreign assistance. In many countries, however, a high proportion, let us say 40–50 per cent of the total investment programme to be financed through foreign assistance, is considered politically unacceptable. If one reasonably expects a certain absolute amount of foreign assistance during the plan period, any lowering of the investment target, because of shortage of domestic resources, will increase the ratio of foreign financing of the plan. In many countries, this is a politically unacceptable proposition for fear of foreign domination over the economy. This implies that in order to utilize available foreign assistance, the investment target needs to be kept high enough to keep the proportion financed by foreign assistance within politically tolerable limits; if raising additional domestic resources to match available foreign aid is not feasible, both have to be reduced, thus reducing the rate of growth below acceptable limits. Faced with this dilemma, most countries would keep the plan size and foreign assistance unchanged, letting domestic resources fill up the gap, without at the same time pursuing vigorous policies to mobilize domestic resources. The *ex post* investment size eventually becomes smaller with a higher percentage of foreign assistance. Experience suggests that an *ex post* higher rate of dependence on foreign assistance is politically feasible, but a planned high rate of dependence is not. The First Five Year Plan of Bangladesh is an example of how these multiple, often conflicting, considerations had to be weighed in an attempt to determine the appropriate size of the programme. The levels of per capita investment in Bangladesh in the last years of the pre-liberation period, the subsequent rise in investment costs and the imperative needs for investment in those critical sectors which were severely neglected in the past, the urgent need to supply the minimum needs of a few essential items, the politically tolerable limit of foreign assistance, even if it is available, all these factors were brought into focus. The prevailing very low levels of poverty necessarily put a floor to the size of the programme. That there is a need for strenuous efforts for: (a) mobilization of domestic resources, including unconventional methods of mobilization of labour, in an overpopulated country; and b) for significant improvements in the administrative and implementation machinery were clearly emphasized in the plan. The social and political consequences of a smaller-sized plan, in the opinion of planners, could not be viewed with equanimity. There was no way but for the plan to produce a basis for sociopolitical mobilization of the country: (a) to bring the economy back to, the level of performance reached in the last normal year 1969–70; (b) to provide a rate of growth in per capita income, which was positive but low, that is, about 2.5 per cent per annum; and (c) to provide a minimum increase in employment expansion to take care of the addition to the labour force during the plan period. Considering the meagre resources, absolute poverty and limited scope for flexibility and manoeuvrability, costs of experimentation and innovations were sought to be kept down in view of considerable institutional changes in the economy which were already undertaken in the past two years partly as a matter of positive decision and partly as a consequence of the liberation war and its aftermath.

Frequently a plan, even after being fully approved by the political authorities, is deviated from in course of its implementation due to political considerations. The fact that a plan once approved to be meaningful implies a self-imposed discipline on the political authorities is inadequately appreciated; the fact that *ad hoc* deviations in one sector distort priorities and affect the targets in other sectors is often not realized. The political pressure for changing intersectoral allocations must be, as far as possible, exercised before a plan is formulated. The need for the observance of the discipline of the plan in course of its implementation in terms of sectoral priorities is crucial to its success in achieving the results postulated in the plan. Here also, the task is not an easy one in the framework of democratic politics and the constant need for compromise. The political authorities should have a global view of their political considerations to be reflected in the plan at the time of its formulation. In practice, however, *ad hoc* pressures are generated and are intermittently brought to bear all throughout the plan period and are reflected discontinuously every year at the time of the annual budget or the formulation of the annual plan. This severely distorts consistency. The pressure from interest groups and constituencies on the political ministers is inevitable. In a democracy each minister feels that he has obligations to various interest groups on the support of which he relies for his political existence. What is required is a systematization of these political pressures.

While much has been written on this subject over the past years, the appropriate machinery for consultation with and the flow of information to and from the public and private sectors has yet to be devised. This brings us to the issue of the optimum degree of decentralization in the formulation and implementation of the plan, both in terms of levels of governments, that is, local versus national governments, and between different ministries. Should not the local governments be associated with the formulation of projects relating to their area or region? Should they be merely executive agents for the implementation of the programmes or projects decided centrally by the national governments? Logic would suggest that local governments are best conversant with the local conditions and, therefore, their association in the formulation of projects relating to their area would only improve the content and quality of the exercise of plan formulation. Moreover, they might be required to suggest and undertake modifications in projects and programmes in course of their implementation as new facts come to light and unforeseen circumstances intervene. In a realistic situation, where full information relating to all the circumstances of a region or project is not available centrally, delegation of authority to the local level and continuous feedback between the local and national governments are imperative. Ideally, the two-way process can in part be implemented by local governments formulating their targets and consequential programmes and projects, and then sending these to the national government for ensuring coordination and consistency. The national government, in the light of programmes of other local governments, national agencies and the private sector as well as the overall availability of resources and priorities, can send them back for adjustment down to the local level. This exercise involves a flow of information and suggestions back and forth between the two layers of governments until a consistent plan for the country as a whole emerges.

There are several problems in the realization of such a plan of action. Firstly, the local governments are often inadequately endowed with professional and administrative manpower. Secondly, there are political problems of relationship in a democratic political system between national and political leaders, members of parliament and leaders of the local government. There is competition and political rivalry between the political leaders at two levels. The problem is aggravated if the national government is controlled by one political party whereas local governments are controlled by different parties. Thirdly, there is the problem of relationship between the local officers who belong to the national government and those who are the functionaries of the local government. Fourthly, there is often a reluctance on the part of the officials of the national government to share decision-making functions with the officers at the lower tiers of government. The problem of decentralization is particularly serious in a big country with a large population and a large size.

A satisfactory system of local government, with appropriate delegation of authority as well as resources, is yet to be evolved in the developing countries. It is not necessary for the development of a particular region to have the same number of schools, or roads or industries; what is necessary is to have an optimum combination of projects which contribute to the overall development of a region. Otherwise the consideration of economic and cost conditions in the location of individual investment projects is thrown to the winds and considerable inefficiency creeps in the use of limited resources. It is unrealistic to assume that such pressures and consequent distortions in the economic or least-cost location can be completely eliminated. The optimum policy is to keep the distortions to the minimum.

In Bangladesh, local development institutions are sought to be built in an integrated framework at the level of the *thana*, which is a conglomeration of villages comprising a population ranging from 150,000 to 200,000. The rural development activities, including mainly small scale irrigation, supply of modern inputs such as fertilizer, pesticides, improved seeds, agricultural credit and extension services, are to be coordinated through cooperatives which will undertake group action in using improved agricultural technology. It is also expected that local governments would play a major role in the planning and execution of rural roads, drainage and irrigation, and rural housing. The detailed plans or schemes in these sectors are to be worked out by local governments while the resources will be channelled from the national government which will also provide technical and specialized services in the aid of local governments for carrying out these functions. Till now the initiative in this programme has emanated from the national government which is trying to induct the local governments in development functions. What remains as yet unresolved is the extent to which the allocation of resources from the centre would be linked up with the mobilization of resources by the local governments either in kind, that is, land and labour, or in cash. The problems faced in this task include appropriate combination of guidance and control from the national government with the village leadership which is expected to emerge through the democratic method in the village cooperatives which will depend upon the emergence of an appropriate relationship between local government institutions and cooperatives as well as between the officials of the central government and the locally elected government. Apart from departures from the plan priorities under the *ad hoc* pressures of the interested groups, there is the problem of an adequate planning and administrative machinery to ensure the proper implementation of a plan. There is also the problem of the appropriate role of the Planning Commission – or whichever machinery is responsible for planning – in the entire governmental machinery. It should be more than a mere advisory body to make its recommendations once in five or 10 years; it needs to be continuously involved in policy-making as well as evaluating and monitoring the performance, of the economy and the fulfilment of the plan targets. As has been discussed earlier, a plan is not merely an expenditure programme but also a set of policies and institutional arrangements for implementing the expenditure programme. The latter is in a sense more important than the former. It is necessary to associate right from the beginning the executive agencies or ministries in the task of plan formulation because the implementing ministries must fully appreciate the implications of the plan in all its aspects. Moreover, it is absolutely crucial that the Planning Commission and the Finance Ministry jointly participate in the exercise for resource mobilization and spell out the whole gamut of policies implied in the assumptions regarding domestic resources. Secondly, it is essential that the planning machinery actively participates in the actual formulation of short-term or annual policies in all sectors of the economy. The short-term policy formulation ranging from exchange rate changes to price and subsidy policies are integrated to the fulfilment of the plan. In most instances, the planning machinery is very inadequately consulted in the detailed policy formulation which is undertaken annually or even at shorter intervals. The task of the coordination of economic policies and programmes is inter-related and unless the planning machinery, with its overall view of the entire economy and its inter-relationship, is involved in this task on a continuous basis, departures from plan priorities are inevitable. Thirdly, it is highly desirable that: (a) the private sector in a mixed economy: and (b) the general public are associated in the process of formulation of the plan and associated policies.

A five year or long-term plan needs to be implemented through the mechanism of the annual plan, which can readjust the time phasing of the projects and programmes in the light of availability of resources, both domestic and foreign, and which can also formulate concrete short-term policies in the light of the longer-run perspective. Annual evaluation of the programmes and policies is also an integral part of the annual exercise. Most planning bodies are not equipped with a proper evaluation machinery; this is also true of the administrative ministries. Research and evaluation are two integral components of annual planning and these are more often than not ignored. Annual planning assumes critical importance in most developing countries because of several reasons. Foremost amongst them is the predominance of the agricultural sector in the national economy and the role of unpredictable weather in determining the performance of the agricultural sector. The fluctuations in agricultural output have important implications for both the balance of payments and the domestic resource mobilization. Secondly, the dependence on foreign trade is often significant and changes in the external sector, especially import prices and exports, may upset plan performance. With a large foreign exchange component in the investment programme, changes in import prices change the costs of the investment programme as well as foreign exchange resources which are available for financing the investment programme. A five year plan of necessity is formulated in terms of constant prices, mostly prices prevailing in the base year. However, changes in domestic as well as international prices take place as between annual plans and sometimes within the same year. Most developing countries suffer from various degrees of inflation arising from both internal and external factors. A rise in the cost estimates of the investment programmes requires a reassessment of the requirements of resources, including reordering of priorities and rephasing of investment programmes over time. An appropriate balance has to be kept between short-term needs and requirements of long-term growth, especially the continuance of ongoing projects or projects which are under way. Yet another element of uncertainty is that which surrounds the flow of foreign assistance, depending on a large number of imponderables and on shifts in international political and economic relations. The uncertainty is particularly disturbing for a country with heavy dependence on foreign aid and is aggravated by the fact that on account of legislative requirements at home, the donor countries seldom make commitment for more than one year.

A basic response to the uncertainty should be to formulate a hard core of projects in the plan, which can be protected in the event of a shortfall in aid and which can be financed largely from domestic resources. It is not an easy exercise, especially because projects in various sectors are interrelated; moreover, it is tantamount to making two plans – inter-relationships between each of which are minimal so that one could be added on to the other. Each such set of programmes and projects would have to be a consistent whole; if projects in the first set are closely inter-related with the second set, the yield from the first set would depend on the investment in the second. In practice, all the projects which are inter-related are included in the plan and, as resources fall short, they are spread widely and thinly over the entire range of projects with the result that the completion of all the projects is delayed. Ideally, all the projects which are not closely related to the hard core should not be started until resources are in fact committed. But pressure for investment in a poor country is so high that all projects get started initially, and as resources fall short a uniform cut is applied throughout or in *ad hoc* fashion as a result of inter-ministerial or inter-agency negotiation and bargaining. Bangladesh, for example, in her short period of planned economic development, has faced the impact of uncertainties in external factors on her planning exercise. She depends largely on imports for the supply of essential raw materials in both the agricultural and industrial sectors; these include fertilizer, pesticide, raw cotton and cotton yarn, crude oil and petroleum products, iron and steel products and cement. In addition, she requires a large quantity of imports of foodgrains and edible oil. Rising prices and uncertain supplies of imports in the last two years have been compounded by the uncertainty in the flow of aid, especially commodity aid, on the one hand, and by a lack of any compensating rise in the volume and price of exports, on the other. Consequently, she has been compelled to interrupt and revise the cost estimates and the size of development programmes more than once in one year. Since the predominant share of tax revenues is derived from custom duties and sales tax on imports, a shortfall in imports reduces tax revenues and increases the pressure for deficit financing or reductions in local expenditures on projects. The counterpart funds which are generated through the local sale of commodity aid also fall and this further restricts the availability of domestic resources in the public sector.

The reduction in expenditures on development projects often takes place across-the-board. This is largely because, in the case of projects under way or ongoing, expenditures cannot be reduced below the level which is necessary for the payment of wages and salaries to all categories of labour force employed in them. However, in a situation in which expenditures on development projects consist largely of payment of wages and salaries, the impact of such expenditures is purely inflationary in the short run; the lag between investment and output is further lengthened. Worse still, in Bangladesh, owing to the physical and geophysical characteristics of the economy, all construction work comes to a standstill during the monsoon; it is concentrated in a short period of five months or so during the cooler season. Interruption in the flow of inputs during the construction season implies a delay of one year in the time phasing of the development projects. The seasonal variations in the supply of necessary inputs in the world market, especially agriculture-based inputs, rarely coincide with the strictly limited construction season in Bangladesh. This has two serious implications for planning specially at the micro level. First, an advance programming for obtaining supplies from abroad assumes critical importance so that they are available in time during the winter months. Second, larger inventories than are normally necessary need be held in order to match the availability of supplies in the world market with the requirements of development projects at home. This is especially so in a situation of scarcity in the world market.

The experience of the past two decades with planning illustrate that unresolved problems, both in terms of theoretical analysis and programming procedures, as well as in terms of close correspondence between implementation and planning, are many. How to formulate a development programme successfully under conditions of uncertainty is an issue which have not been satisfactorily resolved in operational terms in spite of the considerable advances in theoretical analysis in these directions which have taken place. The range and quality of information which such inter-temporal dynamic programming exercises under conditions of uncertainty require are often far beyond the range of practical realization. Also, the amount and quality of highly specialized manpower required for such exercises are hardly available. Questions have often been asked about the desirability of investing scarce manpower in designing such complicated tools of analysis, since the scope for their practical application is severely limited in view of inadequate information as well as the fact that these sophisticated programming techniques require drastic simplifications of reality in order to make them manageable. It has often been argued that in many countries scarce manpower limits a systematic analysis of a wide range of simple policy tools, which are often seriously deficient, such as fiscal and monetary policy. One also wonders whether scarce manpower should, in addition, not be devoted to the exploration of such broad policy issues as well as to specific problems as the analysis of the major investment projects, the consistency between major sectors and the balance of supply and demand of critical input. In view of the fact that the most sophisticated plans, fulfilling both efficiency and consistency criteria, will mean little if they are not successfully implemented, considerably more attention need be given to monitoring and expediting significant projects and programmes, as well as to the continuous *ex post* evaluation of these projects and programmes and the examination of the optimum degree of decentralization to the executing agencies in decision-making and implementation.

PART III

DEVELOPMENT STRATEGY: LESSONS OF PAST EXPERIENCE

Chapter 6

National Import Substitution and Inward Looking Strategies: Policies of Less Developed Countries*

Import substitution has been and continues to be a major development strategy in the poor countries. This is in part a reaction to export pessimism which has been widespread among the developing countries in the last two decades. The slow growth of agricultural exports from the poor, developing countries, some of which are dependent on one or two major agricultural crops, and the limited progress in the diversification of the agricultural sector in these countries, have strengthened this trend. The growth of demand for major agricultural products has been slowed down by technological changes involving synthetic substitutes and increasing economy in the case of raw materials. The protectionist policies in the developed countries, combined sometimes with inappropriate export policies, including exchange rate, internal pricing and production policies, have no doubt contributed their share.

In such an environment, both national and international, import substitution strategy appeared an appropriate policy as a step towards diversifying the structure of a developing economy. Domestic production which relates to the replacement of imports is quite obviously based on easily identifiable investment opportunities. By shutting out or reducing imports, the domestic market is assured to the domestic producers.

The import substitution strategy usually receives its impetus during periods of balance of payments crisis brought about either by internal or external policy or by a combination of both. An inflationary situation at home or a policy of export restriction for ensuring domestic consumption of exportables as well as collapse of the export markets, as happened after the Korean boom, are examples of internal and external factors which contribute to the balance of payments difficulties.

Import substitution can also be associated with the pursuit of a policy of 'autarky'. Usually, however, hardly any country pursues a policy of autarky

---

* First published in P. Streeten (ed.), *Trade Strategies for Development* (London: Macmillan, 1973).

in respect of all commodities. There are strategic items or essential items of mass consumption like food grains in respect of which dependence on foreign sources of supply is frowned upon on the basis for non-economic considerations.

Import substitution in the sense of replacement of imports by domestic production has various connotations. It is undertaken in various degrees under different circumstances. What is important and interesting in the analysis of national import substitution strategy is an examination of the activities, techniques and policies which are pursued in order to implement the strategy.

How does one measure import substitution?

If import substitution is merely to imply that a few items which were previously imported are now produced at home, this is an inevitable phenomenon associated with the diversification of an economy in the process of economic growth. Import substitution is, however, sought to be measured more precisely. One straightforward measurement is a decline in the share of imports in total supply (including imports and domestic production). Other measurements are variations of the same theme. Taking this straightforward measurement as a starting point, it is worth noting that a decline in the ratio of imports to total supply is consistent with a constant, a decreasing or an increasing volume of imports. This is true in the case of one particular item of import or a large number of individual items of import, as well as in the case of aggregate volume of imports for the country as a whole. Import substitution is also measured in terms of the ratio of imports to GNP. Here again, a decline in this ratio is quite consistent with an increase in the aggregate absolute volume of imports.

The pattern of import substitution is critically related to the pattern of growth and changing structure of an economy. Since developing countries are overwhelmingly agricultural and their imports consist mainly of manufactured goods, any significant step towards diversification consists primarily of industrialization based on import substitution. The evidence of the historical path of industrialization indicates that in the developing countries import substitution initially takes place in the consumer goods industries. This is partly because in the early stages of development most of the imports consist of consumer goods. Also, for many consumer goods industries, technology is relatively simple, capital requirements are not large and existing domestic demand is adequate relative to the economic size of an industry. A young, industrializing country with low per capita income, small savings and limited familiarity with advanced technology expects to find it easier to develop comparative advantage in the field of such consumer goods industries.

The next step in the import substitution process relates to the domestic production of consumer durables. This is facilitated by a highly unequal distribution of income which prevails in many a developing country. Such inequalities often get accentuated during the early stages of growth, owing to the initial pattern of capital ownership and government policies. In private enterprise economies, a limited number of entrepreneurs, in view partly of scarcity of managerial and entrepreneurial skills, tend to own and control a disproportionate share of investment and productive activities. The government also frequently plays a significant role in providing financial and other incentives for the development of private entrepreneurs through direct controls over the distribution of scarce inputs such as capital and foreign exchange. The mechanism of control, by its very nature, favours a few rather than many.

However, import substitution policy, directed primarily towards consumer goods industries, quickly reaches its limit as set by the narrow domestic market. It then gradually extends in successive steps to intermediate and capital goods. By then the newly developed domestic consumer goods industries tend to provide a market for intermediate goods and capital goods. The pattern of import substitution moves in this direction because these are the sectors which provide obvious opportunities for profitable investment. To be sure, the size of the domestic market is frequently inadequate for the production of the intermediate and capital goods on an efficient scale. However, there is no alternative if industrialization is to proceed primarily for the domestic market and in substitution for imports.

As the experience of many developing countries demonstrates, the limits to the expansion of markets are set by two particular features of the development path followed by many developing countries. Firstly, since import-substitution strategy has been closely identified with industrialization, agricultural development in the poor countries has not received adequate attention. Import substitution in the industrial sector proceeded apace in many overpopulated and poor countries, while at the same time imports of agricultural products, especially of foodgrains, increased rapidly. Agricultural stagnation hampers the growth of primary exports and in some cases has turned a net exporter into a net importer of agricultural products, including food grains. Since the large majority of people are employed in agriculture, which also contributes the major proportion of national income, stagnation in this sector limits the growth of demand for mass consumption goods. In view of the inequality of incomes between the agricultural and non-agricultural sectors, agricultural growth contributes not only to the growth of aggregate demand but also to its more equitable distribution. Growth in agricultural income, especially where such growth is associated with a more equitable distribution of the benefits of agricultural progress, enlarges the market for mass consumption goods. This in turn provides the basis for an expanding consumer goods industry which can also reap substantial benefits from economies of scale. Moreover, many consumer goods industries are based on the supply of domestically produced agricultural raw materials, which depends on agricultural growth. Therefore the success of and limits to the import substitution strategy in the industrial sector are closely linked with agricultural growth.

The second factor which limits the progress of import substitution in the industrial sector is the slow growth of exports of manufactured goods. This in turn is related to: (a) the choice of specific activities which are selected for import substitution; and (b) the techniques and instruments which are used to implement the import substitution strategy. If import substitution is indiscriminate and without any regard for a country's comparative advantage, the exports of the industrial sector are unlikely to find export markets, not to speak of the barriers put up by the advanced countries in the way of expansion of manufactured exports from the developing countries. Moreover, the instruments which are used to regulate imports also discriminate against exports. The incentives provided to investment in import substituting activities often exceed those afforded to the export promoting activities. The nature and extent of discrimination against exports, which follows from the particular pattern of exchange rates, tariffs and quota restrictions adopted by the developing countries, have been the subject of considerable discussion in the literature. Briefly, the argument that a strategy of industrialization, which is nondiscriminatory between import substitution and export expansion, ensures a more efficient allocation of resources has not been properly appreciated. The participation by domestic industries in the export markets exposes them to forces of competition; it also provides access to and familiarity with the advanced methods of marketing and organization. Furthermore, access to world markets provides scope for specialization consistent with a country's endowment of resources. Above all, the domestic industries are no longer constricted by the limited size of the domestic market; instead they can reap the benefits of economies of scale.

This brings us to the examination of the instruments of policy used to implement a strategy of import substitution. Tariffs and import licensing are both used to implement this strategy. For the purpose of import control, tariffs are redundant if, under a regime of quantitative import restrictions, domestic prices of imported goods exceed the world price by more than the tariffs.

However, in addition to being a major source of government revenue, tariffs also serve as a cushion to the import substituting industries against sudden exposure to foreign competition in the event that fluctuations in the balance of payments cause corresponding changes in the intensity of such import restrictions. The structure of tariffs and that of import licensing are usually such that they impose higher restrictions on the imports of consumer goods than on those of raw materials and capital goods. As between intermediate imports and capital goods, the degree of restriction is usually lower on the latter than on the former.

The rationale of this approach appears to be an attempt to ensure that import substituting consumer goods industries are able to secure intermediate inputs and capital goods at a rate cheaper than that which a uniform degree of restrictions on all imports would imply. This reduces the cost of investment as well as the requirements of working capital: on the one hand, therefore, in financial resources required for investment; on the other hand, profitability of domestic production is enhanced through the imposition of a higher degree of restrictions on the competing output than on imported inputs used in the industry. This policy, then, results in increasing the effective rate of protection for the import substituting consumer goods sector, correspondingly reducing the degree of effective protection for the domestic intermediate and capital goods sectors. Thus the maintenance of a structure of import restrictions, which discriminates heavily in favour of consumer goods, militates against the establishment of intermediate and capital goods industries, even when these sectors may develop long run comparative advantage. However, as import substitution extends to the intermediate and capital goods sectors, the intensity of import restrictions is increased on them and the relative degree of effective protection to them is thus increased.

Over time, the structure of effective protection, which results from the differential structure of high nominal protection, results in a range of effective rates of protection, often rising up to 500–1,000 per cent. Moreover, instances of value added in domestic activity, when measured in world prices, being negative are also not infrequent. Domestic production in such cases as these is rendered possible or found profitable only because of the very high rates of effective protection. Under a regime of import restrictions, there develops a structure of domestic prices for inputs and outputs and a range of techniques which, if evaluated at prevailing world prices, would have been found unprofitable. The cost of import substituting activities, evaluated in terms of scarce domestic resources, thus turns out to be very high and, in the limiting case, extends to infinity.

The question may legitimately be asked why a regime of import restrictions with its attendant shortcomings is so popular in the developing countries. This is mainly because the reaction of the policy makers to a situation of shortage is very strongly in the direction of price control and rationing. If the supply of foreign exchange falls short of demand, the predilection is towards controlling the price and rationing of foreign exchange. This suggests a confusion in thinking about the nature of shortage. It is somehow believed as a matter of habit that shortage is a short-term phenomenon which corrects itself, once the 'circumstances' improve. There is inadequate understanding that circumstances, in the sense of an improvement in the supply of foreign exchange, do not improve by themselves; it is not independent of policies which are pursued during the period of shortage. That the price of foreign exchange should be raised in order to adjust demand to the level of supply is not considered advisable, because supply and demand of exports from developing countries is somehow considered inelastic. The 'inelasticity' assumption and the assumption of the non workability of market mechanisms are somehow believed to be the primary characteristics of developing countries. While there is a considerable degree of truth in the assumption of inelastic demand for many raw materials and foodstuffs provided by developing countries taken as a whole, this is not necessarily true for individual countries and their export performance in the world market. The experience of the Green Revolution has provided examples of supply functions of agricultural output which are responsive to improvement in income and profits of the agricultural sector, provided a proper price support programme is adopted to cushion the fluctuations in agricultural prices. The developing countries themselves are beginning to question the assumption of universal inelasticity of supply and demand for agricultural products, although old habits of thinking die hard.

The 'elasticity' pessimism militates against adjustment of exchange rates for dealing with foreign exchange shortage through a straightforward devaluation. But then why is the effective rate of exchange for imports not allowed to rise for the purpose of allocation of foreign exchange, which could be done, for example, by auctioning of import licences? One version of the answer is that this will result in the foreign exchange being channelled to the highest bidders, who will be the richest groups in any society. In other words, it will be inequitably distributed. So long as income distribution is unequal, the argument runs, to let the price of foreign exchange reach the level warranted by scarcity and have it distributed to the highest bidder is to let the richest get the largest allocation of foreign exchange. This is sought to be mitigated by distributing import licences among the widest section of the population, with the stipulation that no one gets more than the maximum. The distribution of import licences necessarily has to be limited to the traders, i.e. those who are familiar with foreign trade or, at least, with domestic trade, with the hope that the latter will learn the techniques of foreign trade soon enough. However, given the aggregate shortage of foreign exchange and the limitation of a maximum value of licences which could be issued to one licensee, without sacrificing substantially economies of scale in the handling of imports, the potential number of import licences far exceeds the potential number of import licensees. Therefore there has to be some arbitrary method of rationing import licences, even if they are distributed in small amounts. In this exercise it is usually the small man, away from the seat of government, the small importers in the distant, small cities and the small or cottage industries dispersed all over the country who are left out. Special organizational and institutional arrangements are necessary in order for them to benefit from the import and financial facilities, be it under a system of import licensing or under a more realistic exchange rate and freer exchange market. The disadvantages which a small enterprise in trade or industry suffers are multifarious, including inadequate credit facilities and limited access to physical infrastructure, transportation and communication facilities, etc., to which a wide distribution of import licences provides no solution. An the absence of concomitant facilities, the small trader is more likely to sell the import license at a premium rather than utilize it himself.

It is also said that if import licences are not specified in terms of commodities, only luxury goods required by the rich or raw materials which are used by industries making high profits would be imported, and the requirements of the poor or of the essential industries will go unsatisfied. The remedy of restricting luxury imports by direct financial measures for redistribution of income and by means of heavy consumption taxes is seldom resorted to. Instead, it is found easier to control distribution of income and consumption by controlling imports according to broad criteria of social priority. Import control authorities decide not only the eligible importers but also the amount of licence for each importer and the commodity composition of his licence. The basic objection against restoring the price mechanism in the allocation of imports is the realization that the existing income distribution is inequitable, but there is inability or unwillingness on the part of the political authorities to deal with the adverse effects of unequal income distribution by direct taxes or domestic income policy. It is felt that there is a time lag in the adjustment of demand to higher price because of pent-up demand, so that any relaxation of controls will lead to speculative purchases, even at high prices. Stocks will be built up following the liberalization of controls, the foreign exchange shortage will be aggravated and the import restrictions would have to be tightened again. In a situation of acute shortage of foreign exchange and considerable overvaluation implied in the prevailing rate of exchange, a sudden rise in the price of foreign exchange, up to the level which equilibrates present demand and supply, is neither desirable nor efficient. Supply takes time to respond to price changes and hence the present scarcity value of foreign exchange does not or may not reflect the long run equilibrium rate. It is therefore widely held that the movement towards an equilibrium rate has to be a gradual one, with a progressive dismantling of controls.

As already mentioned, earlier import restrictions are not only often high but also discriminatory. It is not that import restrictions discriminate only between broad categories of imports; they also discriminate between individual commodities in each category. Frequently, imports are banned or high tariffs are imposed as soon as there is adequate domestic production capacity, irrespective of costs. Not infrequently, tariffs are imposed to equalize landed costs of imports with domestic costs, plus an adequate margin of profits for the domestic industry, independently of any yardstick by which the degree of protection granted to an industry is to be evaluated.

Duties are also imposed for revenue purposes, without any regard for the pattern and degree of protection which they generate. Faced with a deficit in the balance of payments, luxury imports are either banned or very heavy duties are imposed on them, affording, in the process, a very high rate of effective protection and thus a considerable degree of incentive for the indigenous production of such goods. The intensity of import restrictions is often determined with a view to domestic availability; in other words, the larger the domestic availability, the greater is the degree of restrictions. In fact there is almost an automaticity in higher rates of protection being awarded to the industry which provides a larger proportion of total supply (imports plus production) from domestic sources.

The relative degree of restrictions on various categories of consumer goods is also based on 'essentiality' criteria; imports of commodities which are somehow considered essential and whose domestic production is inadequate are usually more liberally licensed. These 'essentiality' criteria necessarily reflect, except in the case of mass consumption goods, the preference pattern and value judgement of the ruling group and are sociologically and politically determined phenomena. The essential industries are given preferences over the nonessential industries for the purpose of raw material licensing. The raw materials for 'nonessential' industries thus tend to receive a higher degree of protection under this system than those for essential industries.

The brief outline of the structure of import restrictions given above is primarily based on the experience of South Asia. The differences from other parts of the developing world may not be so considerable as to cast doubt on the relevance of this kind of analysis. The intensity of import restrictions determines the relative profitability of output expansion and investment in the different economic activities. The profitability of industries is a function of the size of the market as well as of the magnitude of differentials between the c.i.f. and domestic prices of imports. However, it also depends on the availability of imported raw materials which determine the rate of output and therefore affect the level of costs. The relative profitability of domestic industries is also closely related to internal policies, most important among which are: (a) financial and credit policies; and (b) domestic investment licensing policies and controls of various kinds. These domestic policies exercise considerable discrimination between alternative economic activities. Under these circumstances, the observed levels of effective protection, as indicators of relative profitability, are not necessarily determined by restrictions on imports. The domestic policies and controls also determine the level of domestic output and hence influence the relative activity levels of different industries and the levels of their prices.

One can reasonably postulate that industrialization, in the context of an overwhelmingly primary producing country, requires in general a certain degree of protection. This is necessary in order to accommodate external economies which follow industrialization and which market prices do not adequately reflect. Industrialization contributes to the formation of skills and the development of attitudes and habits of work and discipline which are relevant to urban, 'machine paced' societies. It generates willingness to take risks and to experiment with new ways of doing things, not always associated with traditional and agrarian societies. These changes are brought about over time, and these external economies of industrialization are seldom reflected in market costs and prices. In order for a country to reap these benefits of industrialization, it is necessary for it to discount, by a certain factor, its high current costs in excess of international prices. That this requires protection for domestic industry against foreign imports is obvious and it is a necessary cost of industrialization in so far as the society sacrifices cheaper imports in favour of higher priced domestic output. It is therefore for the society to decide the margin of excess costs over international price which it must impose upon itself in its attempt to derive the benefits, both economic and otherwise, of modernization and transformation of the traditional economy which industrialization brings in its wake. However, it is not only the magnitude of the cost but also the time period for which this must be borne which should be decided by the society.

The rationale for tariff protection, discussed above, establishes the case for a generalized, uniform rate of protection for securing the general benefits of industrialization. It does not provide rationale for a discriminatory, differential protection. Over and above this general range of protection, providing – let us say – a 20 or 30 per cent uniform rate of protection, the basic rationale for discrimination as between individual import substituting industries has to be sought elsewhere. The differentiation between industries can be justified on the basis of an 'infant industry' argument for protection or on the basis of an 'optimum tariff' argument used to exploit inelasticity of the foreign supply demand situation. This would ideally require analysis of the time profile of progressive reduction in costs through learning by doing, and the determination of the appropriate rate of interest at which savings in costs in the future years are to be discounted. Ideally, the unit costs in the future must fall below the level of world price and the discounted value of future saving in costs should equal, if not exceed, the excess of current costs over the international level. This is a difficult exercise from the quantitative point of view, involving as it does not only a projection of future domestic costs but also, strictly speaking, the future movements in the world prices of competing products.

That in a developing economy there is a justification for providing protection to the domestic industry at a uniform rate over its entire range, and that, in addition, there are specific cases which deserve a higher rate of protection on the basis of an 'infant industry' argument, does not imply that import substitution must be associated with discrimination against exports. An industry that is protected in the home market must not get a lower return if it attempts to sell in the export market. But this is usually the consequence of an import-substitution policy by means of tariffs or quantitative restrictions which does not provide equivalent incentives for export expansion. Import restrictions without corresponding and equivalent export incentives, cause the investment of domestic resources to yield a higher return in production for the home market than for the export market. The high costs of import substituting industries, in so far as they sell inputs to the export industries, also raise the costs of the export industries and reduce their competitiveness in world market. Furthermore, the high profits, nurtured under import restrictions, in the import substituting industries enable them to pay higher wages which spread to the rest of the economy, including the export sector, and thus raise their costs. The adverse consequences of import restrictions on export expansion can only be counterbalanced by corresponding incentives for export expansion. There may, indeed, be exceptional instances where discrimination in favour of exports against import substitution may be justified. This relates to circumstances when private evaluation of: (a) uncertainties of selling in the export market, which is subject to interruptions due to factors beyond the control of the exporting country; and (b) the additional private costs of finding profitable export markets, exceed social costs.

Limits to and shortcomings in the import substitution strategy, pursued via conventional techniques of import restrictions, have been referred to in the foregoing analysis. A few other important limitations of this approach are summarized below. The widespread prevalence of excess capacity in the manufacturing sector is one of the important features of the developing countries. Apart from overvalued exchange rates, this phenomenon is caused by the market assumptions of the import substitution strategy. The degree of capacity utilization, as permitted by the size of the domestic market, often falls short of the installed capacity in many industries because of technical indivisibilities of investment in such industries. If investment is to take place only in areas where the domestic market is adequate, while exports are discouraged, the total investment in industrialization would be very limited indeed. The need for expanding industrialization necessarily implies that instead of concentration in a few selected sectors, industrialization is spread over the entire industrial sector. Since industrialization in a few selected sectors cannot be carried far enough, for reasons explained above, the alternative to calling a halt to the process of industrialization is to establish enterprises wherever there is room for one or two plants. This pattern of industrialization prevents specialization by product or processes and therefore stands in the way of realization of economies of scale or benefits of specialization.

As already stated, the prevalence of widespread excess capacity in the manufacturing sector is partly due to the underpricing of imported capital goods because of low effective rates of exchange for the import of capital goods. The system of licensing of industrial raw materials which is related to installed capacity implies that the manufacturing concerns, in order to obtain additional licences for raw materials, must install additional capacity. It thus pays to increase output by creating additional capacity while the existing capacity is initialized. Again, it is not just the import licensing system which favours a more liberal import of capital goods in relation to raw materials; the additional foreign exchange through external aid in most of the poor countries is supplied relatively more in the form of 'project aid' (i.e., assistance in the form of capital equipment and plant rather than intermediate inputs) than in the form of commodity assistance. Aid giving countries and agencies demonstrate a distinct preference for project aid rather than for programme loans, since development or capital accumulation is mistakenly associated with the import of capital goods and consequent increase in investment rather than with a fuller utilization of capital and consequent increase in output and income, domestic savings and employment. A more liberal licensing of capital goods is combined with a lower rate of tariffs to cheapen capital goods vis-à-vis imported raw materials. Quantitative restrictions on imports, which are coupled with the persistence of overvalued exchange rates make imported inputs cheaper than domestic inputs. Thus import incentive techniques and projects are encouraged. This, combined with liberal licensing of capital goods to the users of capital goods, i.e., industrial enterprises, also leads to a preference for capital intensive techniques and projects.

In most poor countries suffering from heavy pressure of population, the choice of capital intensive techniques and projects militates against the growth of employment. The restrictions imposed by advanced countries on labour-intensive exports from the developing countries aggravate the difficulties in the development of labour-intensive industries. The factor price distortions, i.e. provision of capital and imports to those with privileged access to imported capital and raw materials at prices below their scarcity levels, are aggravated by the domestic monetary and fiscal policies in the developing countries. The rate of interest is kept artificially low by the cheap money policy and capital is rationed by financial institutions, under governmental controls, to those who receive industrial licences to expand existing capacity or install new capital. Since licensed imports and rationed capital funds are available to a limited few, such a policy encourages, as a by-product, the concentration of income and wealth. Moreover, the limited size of the domestic market contributes in yet another way to the concentration of income and wealth. There are only a few firms in each industry, which limits the degree of competition. Thus the domestic enterprises are, on the one hand, immune from foreign competition owing to import restrictions and, on the other hand, face a very limited competition within the domestic economy. Because of the way in which import licensing for raw materials is related to already installed capacity, the more efficient firms in an industry cannot expand at the expense of the less efficient ones. Similarly, an industry which has adapted to the factor endowment of the economy cannot expand its output by having a more liberal access to the imported raw materials on the basis of its own appreciation of the cost and demand conditions.

Under a regime of import restrictions, the major source of capital accumulation is derived from the high profits which accrue to those who receive foreign exchange at a rate lower than its scarcity price in the domestic economy. They are mainly traders or importers, or industrial units, which may directly obtain import licences for raw materials and equipment. These profits accrue to a small section of the economy, where export sector and the agricultural sector are correspondingly deprived of high profits. There is thus a shift in the distribution of income to the import substituting sector, away from the rest of the economy The concentration of high income and profits in a few hands also leads to a concentration of investments in a few hands. Reinvestment of profits provides the major source of accumulation capital and of the financing of development activities. Growth through reinvestment of profits necessarily implies that the existing owners of capital or existing productive enterprises either expand along the lines they are already familiar with, or they themselves branch out into new enterprises, either horizontally or vertically. In this scheme of things, the financial intermediaries, which traditionally serve the function of mobilizing savings and channelling funds for investment, are thwarted. The habits of saving and investment are not encouraged among a wider range of the population. The low interest rate policy, aided by capital rationing through direct licensing, adds to the difficulties of mobilizing savings from the larger segment of the population.

The strategy of mobilizing savings and channelling investment through a limited number of high profit earners, brought into being and nourished under a regime of import restrictions, is not conducive to growth. It inhibits the growth of entrepreneurship and the diversification of the economy. If the limited number of profit earners are the only source of entrepreneurial and investment activities, then a limit is set to the extent and nature of investment by their ability to manage a large and more diversified range of investment projects. The limited reliance on professional management, which is often a pronounced characteristic of these economics, depending mainly as they do on family management, further inhibits the growth and diversification of investment activity. The foregoing analysis indicates that import substitution strategy, as an aid to diversification and identification of profitable lines of investment in a developing economy, has its merits. But it should not be implemented without regard to the need for an efficient allocation of resources. It has sometimes been argued that the allocative inefficiency of the import substitution strategy, following from the use of inappropriate techniques, may be more than offset by the dynamic gains from modernization, such as the rise of entrepreneurship and accelerated capital accumulation. The above analysis attempts to suggest that these dynamic advantages may also be secured without the use of inefficient policies.

The experience of the last two decades indicates that the basic ingredients of an import substitution strategy should consist of: (a) movement towards a greater uniformity of exchange rates; and (b) elimination of discrimination against exports, implied in the multiple exchange rates built up as a part of the regime of exchange control. The basis for detailed commodity wise discrimination in terms of exchange rates, either for imports or exports, does not exist either in theory or in fact. It would be hard to suggest that the prevailing system of multiple exchange rates is based on a quantification of differential elasticities of demand and supply of exports and imports. For a developing country, generally speaking, one can postulate that industrialization would most probably require protection, for reasons discussed earlier, and therefore the manufacturing industries require a higher exchange rate. But the higher exchange rate should be available to the manufacturing sector, irrespective of whether it sells in the domestic or the export market. A system of dual exchange rates, i.e., a higher rate for manufactures, both for exports and imports, and a general rate for the rest of the economy, seems to be a justifiable pattern in the circumstances of developing countries. The general rate of exchange should take cognisance of the new analysis, which questions the universal justification of elasticity pessimism relating to the demand and supply elasticity of the agricultural commodities. Any suggestion for considerable and detailed discrimination in exchange rates for accelerating the pace of industrialization, or of diversification of the economy in general, must be based on a serious justification in terms of differential elasticity, associated external economies and 'infancy' of the particular industry or economic activity.

Chapter 7

Learning from East Asia: Lessons for South Asia – an Epilogue*

Introduction

The processes driving economic development are by no means fully understood. After years of analysis of development problems and policies, this is what the World Bank concluded in its *World Development Report 1991*. At the same time, it stressed that there was, however, a broad understanding that economic development is a multifaceted phenomenon requiring social, political economic and institutional changes.

In East Asia and in the high performing Southeast Asian countries in a particular historical context, these various factors acted in a synergistic and a mutually reinforcing manner to produce not only a high growth rate, but also a reduction in poverty and social deprivation. It is thus not surprising that other, less favoured, developing countries should be interested in diagnosing the East/Southeast Asian development process in order to understand what lessons may be assimilated from these countries into their own development agendas. Nowhere is the interest in East Asia more alive than in South Asia, their immediate neighbours in the Asian region, who constantly contrast their own backwardness with the economic dynamism of their fellow Asians to the East.

Several broad aspects of the development process in East and Southeast Asia have drawn particular attention, i.e., their high rates of growth and the pattern of transformation of their economic structure, openness vis-à-vis the world economy, poverty alleviation, participation and wealth sharing, the system of governance in the widest sense, i.e., political leadership and commitment, administration and implementation capacity, relative roles and

* First published in *Towards a Theory of Governance and Development* (Dhaka: The University Press Ltd, 1998). The chapter was specially prepared by Professor Nurul Islam, summing up the issues addressed in the CPD conference on *Learning from East Asia* is based on a review and analysis of the papers circulated, as well as the comments made and the views expressed by the participants, at a seminar on the above subject held by the Centre for Policy Dialogue at Dhaka in July 1996.

interaction of public and private sectors, etc. While seeking to analyse the East/Southeast Asian experience for purposes of comparison with and lessons to be learned by South Asia, one should, however, be careful to specify correctly the particular period in their development history that is relevant for or is roughly comparable to the current circumstances of South Asia. The experience of East Asia during 1950–70 and that of Southeast Asia during 1960–80, rather than the most recent periods when the latter already made significant socioeconomic progress, are likely to be of greater relevance for South Asia.

In order to interpret and understand the East/Southeast Asian experience, the *Centre for Policy Dialogue* convened an international seminar in Dhaka from 27–29 July 1996. This chapter attempts to draw some conclusions from these three days of intensive dialogue on the subject between a selection of senior policy makers and eminent scholars with expertise on East, Southeast and South Asia. The chapter is not a summary of the discussions, but draws upon the issues raised during the free-ranging dialogue in order to capture its essence. The sole responsibility for the interpretation and synthesis of the wide variety of views expressed at the seminar remains with the author, where his prejudices and bias may have inadvertently but understandably crept into the overview.

The paper is structured according to the main themes addressed in the course of the dialogue. These cover the following issues, deemed to be central to the positive outcome of the East/Southeast Asian development process:

1) high growth, savings and investment;
2) strategic intervention by the state to realize structural transformation;
3) quality of governance and the development orientation of the leadership;
4) reducing inequality and poverty;
5) lessons from China;
6) conclusions.

High Growth, Savings and Investment

The most outstanding achievement of the East and Southeast Asian countries was the high rates of growth, i.e., 7–8 per cent rates of growth sustained over 20 years or more. Very high rates of savings and investment contributed to the high rates of growth. The domestic savings rate as early as the mid-1960s, in these countries were higher than the current rates of domestic savings in South Asia (8–17 per cent), except in India (20 per cent). The Republic of

Korea[1] was the only country with a domestic saving rate during the mid-1960s that was increased to about 30 per cent over the next decade. In contrast, the domestic saving rates in the South Asian countries recorded a very modest increase over the last three decades.

What were the sources of such high rates as well as the significant increase over time in the rate of mobilization of domestic savings? In this respect, there were differences among the various countries in East and Southeast Asia. For example, in Korea, the predominant source of domestic savings was corporate savings. The spectacular increase in the savings rate in Korea, within a decade or so in the late 1960s and early 1970s, was attributed to the accumulation of savings out of the very high corporate profits that were created through a policy regime of high trade protection and targeted subsidies by the government. The market structure of the Korean manufacturing sector was oligopolistic and was nurtured as such by government policies. This structure, with protection from imports, contributed to the accumulation of high corporate profits. The prospects of earning high profits, in turn, stimulated savings and investment in the corporate sector.

Concurrent with the accumulation of corporate profits, several restraints were placed by the government on private luxury consumption. This was enforced through both prohibition of and high tariffs on luxury imports, on the one hand, and through excise duties on domestic consumption, on the other.

In other countries, the major source of savings traditionally originated in household savings, stimulated by the incentive of real positive rates of interest. These savings were captured through efficient intermediation by the banking system, both formal as well as informal in the form of credit institutions, which could successfully mobilize savings and channel them into productive investment. The capital market, i.e., the bond and equity markets for the mobilization of private savings, was not adequately developed in early years in East and Southeast Asia. These financial institutions were rather late in assuming any significant role and even today they are less developed than in other high income countries with comparable financial strength and economic structures. The banks thus had to play a catalytic role in mobilizing savings and financing private investment. Macroeconomic stability and low inflation rates, as well as the availability of profitable investment opportunities, assured by a favourable enabling climate created by the government, made an important contribution towards sustaining the high investment rates which remain an important feature of the East and Southeast Asian economies.

The East/Southeast Asian economies were characterized by a beneficent circle of high growth rates stimulating high domestic savings, which in turn financed the high rates of investment, which in turn stimulated the growth in income. To start with, the rates of saving in the Southeast Asian countries in the 1960s and 1970s were high partly because of the higher level of initial income in these countries (except Indonesia) compared to countries in South Asia today. Also, the initial spurt in investment was facilitated by external capital inflows including foreign aid. Moreover, the government provided tax incentives and concessions of various kinds to stimulate saving and promote investment. It is to be noted, however, that many of these fiscal and monetary incentives which proved serviceable in stimulating investment in East Asia are frowned upon today by many development specialists, who also question the extent and importance of such incentives in stimulating savings and investments in the East and Southeast Asian countries.

In several of these countries, the East Asians in particular, small and medium scale enterprises outside the corporate sector generated a considerable amount of savings and investment. The prospects of profitable investment in these enterprises, strengthened by the government's investment in infrastructure, both physical and human, helped to stimulate aggregate domestic savings. Macroeconomics stability, i.e., non-inflationary growth greatly facilitated this process. But this price stability was in no small measure helped by food price stability that was ensured through measures to stimulate food production, supplemented by necessary food imports.

The average tax to GNP ratio in the East/Southeast Asian economies remained around 20 per cent, which left resources for investment in the hands of both the state as well as with private individuals. Fiscal balance in the public sector was thus consistently maintained. The low population growth rate also helped to keep down the rate of private consumption and public expenditures needed to meet the current needs of a growing population.

South Asian countries not only generate low private savings but also have a poor record in mobilizing public savings in general. The exception to the South Asian norm, India, has reached a relatively higher rate of domestic savings and investment than its neighbours. Much of its higher rates of domestic savings originate in the private, mostly household, sector. Furthermore, higher levels of investment in India have traditionally yielded low returns, both because of organizational as well as managerial deficiencies and because of a predominant reliance on large-scale, capital-intensive industrialization, under a regime of persistently high trade restrictions.

In contrast to India, for years on end Bangladesh continued to register a single digit rate of domestic savings built around negative public savings with public investment largely financed by foreign aid. Other members of the region, bar India, also registered low levels of savings for reasons similar to Bangladesh's experience. However, on average, low inflation rates prevailed in most South Asian countries, except in years of food scarcities and rising food prices. But high population growth rates in most countries also depressed household savings. Importers did earn high profits from scarcity margins in view of the prevailing restrictions on imports. Furthermore, commissions earned by agents for foreign suppliers for contracts financed under various forms of external credit were also high, due to high trade restrictions as well as the non-transparency in the transactions of the public sector and private agents. In some countries, such as Pakistan, such rent-earning activities contributed to the emergence of some major private entrepreneurs. In Bangladesh, no corresponding growth of a group of major entrepreneurs has been possible because the investment climate was less favourable. But over the years a few, initially rent-oriented, business houses have developed a sizeable industrial stake with some export-orientation, though these corporate groups are still very small by Indian or even Pakistani standards.

In all countries in South Asia, higher levels of luxury consumption of the rich seemed more prevalent than was the case in East or Southeast Asia. It is argued that the explanation for such differential consumption behaviour of the rich originates in the differences in cultural traits between East and Southeast Asia, and South Asia. However, even if such an argument has some validity, it is of limited use as a guide for policy making since there is no feasible policy instrument to change cultural traits in the desired direction within a predetermined time period. Policy instruments to mobilize savings thus have to be designed to take these cultural traits into account rather than to assume that savings rates are culturally predetermined.

The financial institutions in South Asia, often dominated by publicly-owned institutions, unfortunately did not serve as effective channels for the mobilization of savings and investment. Loan defaults and poor management continued to plague these institutions for long periods, particularly in Bangladesh and Pakistan, where the concept of a 'default culture' severely compromised the viability of their financial institutions.

However, there are also positive features in the South Asian scene, where the poor and the not-so-rich have generated significant savings, whenever they faced or obtained profitable investment opportunities. Experience demonstrates that small and medium scale enterprises in South Asia have frequently relied on or utilized household savings for investment, often supplemented by informal sources of credit. Moreover, whenever financial intermediaries have proved to be efficient and provided appropriate outlets or financial instruments, they have been able to mobilize savings of the low or medium income groups. More than that, an institution such as Grameen Bank in Bangladesh has been able to mobilize savings from and to extend credit to the poorest of the poor – widely considered as non-creditworthy – for investment in rural micro enterprises that promote both equity and growth.

Strategic Interventions by the State to Realize Structural Transformation

Both the state and the market played their respective roles in East and Southeast Asia in the transformation of the economic structure as well as in keeping the economy relatively open in terms of foreign trade and investment. The state, in varying degrees, intervened not only to provide 'public goods', i.e., basic physical and human infrastructure (health, education, training, research and development as well as transportation, communication and power facilities, etc.), but also to guide, assist and stimulate private investment. Protection, export subsidies, directed subsidized credit and tax incentives were used to stimulate private investment as well as to promote specific private sector activities.

The magnitude of the above-mentioned state interventions, as well as their role or success in promoting the development of East and Southeast Asia, has been much debated. The greatest controversy has been about the role of the state in East Asia, specifically in Korea and Taiwan. Sometimes, one has the impression that the interpretation of the importance of and the nature of state intervention in these countries, depends upon the 'eyes of the beholder', in this case the eyes of the analyst. There are those who argue that state intervention helped to stimulate a particular set of industries or sectors that were considered strategic. This so-termed 'strategic intervention' was deemed to have played an important role in these countries in contributing to their long-term growth with dynamic efficiency. Moreover, interventions over the longer run were supposed to have more than offset the short run allocative inefficiencies which may have arisen out of such interventions. Others argue that interventions, such as they were, did not constitute a very important factor in East Asia's growth. There is still a third group who admit that strategic interventions did succeed in many cases in creating dynamism and growth in the East Asian economies, especially in promoting rapid and efficient industrialization. This success, however, was possible because of certain special circumstances that cannot be easily replicated in other countries. In fact, this school of thought argues that such interventionist measures or policies were tried elsewhere in other Asian countries or developing countries in other parts of the world, but without a comparable degree of success.

A broad consensus appears to prevail about two general characteristics which appear to have contributed to the success of strategic intervention in the East Asian countries. Firstly, these states worked with a relatively honest and competent bureaucracy under a committed political leadership, who interacted to formulate appropriate as well as workable policy interventions which could then be efficiently implemented. Secondly, the selection of industries as well as activities identified for active promotion or intervention were broadly in line with the country's comparative advantage, i.e., their dynamic rather than static comparative advantage. The departures from static comparative advantage that were dictated by a country is initial resource endowments, were kept within limits. In other words, the deviations from comparative advantage were not 'unduly stretched' beyond the managerial and technological capacities of these countries.

In addition, a number of special characteristics of the East Asian countries were emphasized: first, positive steps were taken by the government to upgrade skills and to promote technological innovations through public investment in R&D, as well as through licensing of foreign technology and foreign direct investment. The upgrading of skill and technology facilitated the gradual entry of these countries, over time, into the higher stages of industrialization, into more sophisticated branches or sectors of industry. The investment by the state in R&D was particularly relevant for the small and medium scale investors since, unlike the large scale corporate sector, they did not have the resources to organize or invest in R&D at a level where they could benefit from economies of scale.

Second, how did the countries in East Asia and later in Southeast Asia, select the specific industries appropriate for strategic interventions? Korea and Taiwan followed Japan in identifying the types of industry that were successfully developed by Japan in the early 1950s or 1960s, when her circumstances were comparable to those prevailing in other East Asian countries. The next round of latecomers from the Southeast Asian countries, in turn, followed the earlier examples of Korea and Taiwan. In fact, the mature industries of Japan became the nascent industries of Korea and Taiwan, and so was the case of the Southeast Asian countries vis-à-vis the East Asian countries. The latter's skill, and technology were continuously upgraded to take over the activities in which the former's comparative advantage declined over time, consequent on a change in resource endowment, manifested in the way of rising wages. This process of industrial mobility between countries at different stages of development has been defined as the 'flying geese' model of development. Learning from each other about the evolving pattern of industrialization was a salient characteristic of the industrial policy of the East and Southeast Asian countries. 'Looking East' was the slogan in Malaysia in the 1980s and early 1990s.

Third, it is suggested that the level of trade restrictions, both tariff and non-tariff, designed to promote import-substituting industrialization, was, on average, lower during the 1950s and 1960s in both East and Southeast Asian countries than in South Asia in the 1970s or 1980s. It is also argued that not only was the level of effective protection used in East Asia lower but also its dispersion was narrower than in South Asia, so that distortions in resource allocation among the import substitution sectors or subsections were kept low. The empirical evidence on this is inconclusive in the absence of detailed comparable data for the 1960s in these countries. Comparisons have been made between the levels and dispersion of rates of effective protection that prevailed in the 1980s in East Asia, with those prevailing in the 1980s or 1990s in South Asia. This is not the relevant comparison and may, in fact, be misleading because the level and dispersion of effective protection rates declined over time in East and Southeast Asia during the 1960–80 period.

Fourth, whatever the level of trade restrictions or export incentives in the form of tax concessions, export subsidy, or subsidized credit, they were all 'performance based' – i.e., failure to perform was sanctioned by the withdrawal of protection and other incentives on offer. This process or procedure was, however, more valid for the East Asian rather than for the Southeast Asian countries, who tended at a comparable stage to be more projectionist. The main criterion of performance used by the East Asian regions was the ability of the recipients of state assistance to meet the targets of growth in exports and to gain competitive advantage in the world market. This forced even the supported industries to increase efficiency and undertake product development, as well as to improve technology and marketing skills in order to be able to compete in export markets. All import-substituting industries were expected to enter the export market and eventually to compete successfully and expand their exports.

In general, import liberalization on a wide front followed rather than preceded or accompanied export expansion. This approach was influenced by the view that the import liberalization could have adverse balance of payments implications. Furthermore, it was argued that entry into export markets was the signal that the industry was efficient and could thereby withstand the onslaught of cheaper imports. Even though the industrial structure in these countries tended to be oligopolistic, where a few large conglomerates dominated the domestic market, in particular East Asian countries, they vigorously competed with each other to receive export performance-related incentives and assistance from the government to promote exports.

Moreover, the domestic cartels/oligopolies, both in the public and private sector, were always subject to the threat of withdrawal of import protection (in several cases a declining schedule of trade restrictions was announced ahead of time). This was much more the case with East Asia, and much less the case with the Southeast Asian countries.

Fifth, the latecomers among the high performing countries, i.e., the Southeast Asian countries (Indonesia, Malaysia, Thailand) during the late 1960s and 1970s had, in general, a higher level and dispersion of effective protection rates than those in the East Asian countries (Korea, Taiwan, Singapore, etc.), but these rates were, in turn, lower than those in South Asia during the 1980s. The Southeast Asians, therefore, exposed their economies to a greater degree of distortion of resource allocation than was the case with the East Asian countries at a comparable period. Moreover, the period of trade restrictions in any case persisted over a longer period of time compared to East Asia. During this projectionist phase, however, both East and Southeast Asian countries continued to promote their export industries through a wide range of fiscal and financial incentives.

Sixth, the Southeast Asian countries, in contrast with the East Asian countries, had a more liberal policy towards foreign direct investment. However, both groups of countries pursued policies designed to encourage direct foreign investment in selected sectors – not across the board in all sectors irrespective of their export orientation or ability to transfer technology and skills to local counterparts.

The Southeast Asian countries used fiscal incentives to attract foreign investment on a much more liberal scale than the East Asian countries such as Korea and Taiwan. The export push and performance in Southeast Asia, to a much larger extent than in East Asia, was dependent on foreign direct investment.

Seventh, the costs of inefficient policies could be sustained over a longer period in Southeast Asia than in East Asia. These costs were absorbed in respect of their impact on growth performance, since the Southeast Asian countries were cushioned by the relative abundance of their natural resource endowments in relation to population. This resource base included the availability of minerals and petroleum as in the cases of Indonesia and Malaysia, for example, and in some cases, was supplemented by abundant foreign development assistance. At the same time, in the Southeast Asian countries, the correction of distortionary policies was accelerated under the pressure of global recession in the late 1970s and 1980s which had led to the fall in their export earnings from oil and mineral as well as other commodity exports. At the same time, the structural adjustment programmes as well as reforms of trade and exchange rate policies were also put in place in various Southeast Asian countries, and tended to go further and faster than was the case in South Asia.

Quality of Governance and the Development Orientation of the Leadership

In order to appraise the varied outcomes from interventionist policies it is necessary to understand how some governments relative to others managed to minimize the costs of inefficiency over a prolonged period. There seemed to be a general agreement among analysts of the East Asian experience that the point of departure as between the more efficient and less efficient interveners lay in the varying capabilities of the respective bureaucracies to manage their national economies. It was thus agreed that bureaucracies in East Asia even more than in Southeast Asia tended to be more competent and honest than their South Asian counterparts in managing their economies.

Furthermore, the East/Southeast Asian governments did not stifle private enterprise but promoted, assisted and regulated it to meet the targets of growth and efficiency as well as to attain sectorial priorities set by the government. The governments in those countries evolved a systemic mechanism for interacting with the private sector. They established forums or so-called *deliberation councils* in some of these countries where the governments regularly consulted with private enterprises. They used these councils as 'sounding boards' for eliciting the reactions of the private sector to intended policy measures, prior to the formulation and implementation of national economic policies. Also, these government business consultation forums were useful in assessing the impact of policies on the private sector as well as for the identification of problems that may arise in the course of implementing such policies. The bureaucracy managed to establish a framework of discipline and to administer a regulatory system, without being in very close collusion with private enterprise. Such corruption and collusion did exist in East/

Southeast Asia, but was not deemed to be significant nor on a wide enough scale in the early years (1960s and 1970s) seriously to misuse the regulatory system for private gain and thus to heavily distort the allocation of resources.[2]

The growth-promoting role of the bureaucracy in East and Southeast Asia relates to the wider issue of governance. It is thus necessary to diagnose the circumstances whereby differential levels of integrity and competence are maintained as between the respective national bureaucracies of South as distinct from East/Southeast Asia, who were both exposed by state policy to access to large rents from the exercise of their respective regulatory powers. The Southeast Asian countries were not reputed to be free from the temptation of extensive rent-seeking on the part of the political leaders or the regulatory bureaucracy, and yet they seemed to have promoted development more effectively than their South Asian counterparts.

It was suggested that the nature of the East/Southeast Asian state provided, for regime continuity which in turn permitted for a more stable and predictable system of rent-seeking. Whilst such rent-seeking may have raised transaction costs for business enterprises and hence prices to the final consumers or users of goods and service, it did not seriously distort the allocation of resources or the incentives to save and invest. Also, there was a greater degree of stability and continuity of policies in the East and Southeast Asian countries, in spite of changes in government. This continuity in policy among the policy makers provided confidence to private enterprises for undertaking long-term investment.

Why did the governments or the political leadership and administrative machinery in these countries assign such a high priority to development? What were the circumstances that led to it? It is suggested that there were several ingredients in the system of governance in these countries. First, in most countries in East and Southeast Asia, political leadership had a vision or a framework of policy for the future development of their countries and had a commitment to achieve such a vision. Secondly, it is argued that commitment to development and to improvements in the standards of living of the people was due to the threat perception of the respective regimes or their desire to seek legitimacy or popular acquiescence.

The above proposition remains only partially valid. For example, it is argued that this compulsion to accelerate development in South Korea, Taiwan, Indonesia, Thailand and Malaysia was driven by the challenge of communism. To the extent that North Korea and mainland China succeeded in providing improved living standards to their populations it was a powerful challenge to South Korea and Taiwan, to prove that development could as effectively be delivered through a noncommunist economic system. The Suharto regime in Indonesia that came into power by overthrowing a Left-leaning government and suppressing, through a massive blood bath, the largest communist party in a noncommunist developing country, similarly took up the challenge of proving that it can deliver higher standards of living to the masses and that the attraction of the communist system was misplaced.

It was the external threat posed by the advance of communism in neighbouring former French Indo-China that guided the ruling classes in Thailand, i.e., the military, bureaucracy, and business enterprises through their shifting coalitions to make a commitment to development. Thus faster economic growth conferred benefits on all major interest groups of growth who could thereby work out a *modus operandi* for distributing the gains among themselves. The Thai king, with his commanding moral authority and the universally-accepted esteem reserved for the monarchy in Thai society, provided a symbol of national unity and of social stability which enforced the parameters within which conflicts between groups for sharing the gains from growth could be contained.

In Malaysia, newly independent, following the suppression of a communist insurgency, political leadership dominated by the Malays, in an economy dominated by the Chinese, took the political decision to expand the areas of economic opportunity for the economically disadvantaged Malays through a policy designed to promote rapid economic development. The Malays were given preferential access to the widening range of opportunities. This promoted growth and helped reduced ethnic disparity. In the absence of rapid economic growth, a redistribution of income and assets away from the Chinese, the principal source of enterprise and investment in Malaysia, could have brought short run gains to the Malays but might have led to a long-lasting damage to the economy and its future prospects as the redistributive process could have led to social and political destabilization. However, the consistently high growth rates achieved in Malaysia reduced the adverse effects of distortions caused by such interventions.

It is apparent from the historical experiences in Asia that the circumstances under which the political leadership becomes committed to development varies widely among countries. In some countries, the authoritarian regimes sought legitimacy through development and the diffusion of its benefits among the masses. In political democracies, such as those which now prevail in South Asia, the desire of the population for better living standards is reflected in voting into power those who commit and deliver the fruits of development However, if none of the principal political parties within the polities of South

Asia are so committed, or fail to honour their commitments, the voters have no real choice but to turn to alternative political parties who can better serve their aspirations. As a result, it is arguable that democratic competition in South Asia has failed to provide the desired push for development, as seemed to happen in East or Southeast Asia under a rather different political dispensation.

The role of the political leadership in promoting development remains a subject of debate. It has been accepted that the emergence of a political leader with a vision for development and the commitment as well as competence to realize this vision, is a rare phenomenon in most Third World countries. Thus the sociopolitical culture of a country would appear to be a far more relevant variable to influence the pace and quality of development. In countries such as in South Asia, widespread illiteracy and lack of awareness amongst the people as to their rights and potentials appear to have constrained them from putting sufficient pressure on the political parties to deliver on their promises of development. The organization and structure of the political parties and their vote-getting or -seeking strategies seldom tend to be strongly based on the realization of a clear development agenda.

In the absence of a sufficiently proactive commitment by the political parties of South Asia to promote development, in recent years non-government organizations (NGOs) have emerged to play an important role in raising consciousness among the populations and in articulating and voicing the demand for improvement in the conditions of living of the common people. The rising consciousness among the populations of South Asia may, in the near future, generate sufficient pressure from below to compel politicians to respond more effectively to the challenge of development. This process can be greatly strengthened by the spread of literacy, including primary basic education, but above all, it depends on the growth in the consciousness and political empowerment of the democratic majority of the poor in South Asia.

Given the political commitment to development, it is necessary to give concrete shape to development aspirations in terms of a framework or blueprint for the future of the economy. One can refer to such an exercise as an 'indicative plan' or an agenda for action, delineating the objectives of development as well as the projected structure of the economy in the future and the policies and programmes both in the public and private sectors that are envisaged to realize these objectives. The private sector cannot produce the broad perspectives for the future. Individual private entrepreneurs do not have adequate information to make informed guesses or projections regarding demand, production, or technological or supply possibilities in various interdependent or interacting sectors of the economy. The government is able to and can take an overall view, can analyse the interlinkages between different sectors and subsections and work out alternative scenarios based on different assumptions about various growth paths. The formulation of such a framework or an agenda for action assists in mobilizing public support and popular enthusiasm for its realization. It can help to promote a broad consensus around its realization, through debate, dialogue, persuasion and negotiation, so that the divergent views of various interest groups can be accommodated. The political leadership at a given time has the responsibility not only to initiate such a debate, but to also seek to build a consensus through negotiations with other concerned or contending groups.

The East/Southeast Asian leadership tended to be driven by a strategic vision for transforming their economies, sometimes incorporated in medium-term plans or development agendas. However, the East and Southeast Asian governments did not follow too assiduously the route of open debate, dialogue and negotiations among different interest groups and political parties in their attempts to evolve a consistent framework defining the future development path for the economy. The East and Southeast Asian countries were mostly run under an authoritarian system, where consensus tended to be established through far from transparent mechanisms. However, the dominant interest groups did share a common view that improved living standards for the masses were imperative if they were to survive in power. Hence, they kept their intragroup conflicts within limits and could thus reach a stable bargain among themselves.

There was little popular participation in debates on major issues on a national scale within East/Southeast Asia. The consensus and coalition building, it is alleged, was possible because the cultural and social systems in East and Southeast Asia enjoined a spirit of social responsibility and cohesion, so that individual freedom of choice was moderated or qualified by considerations of social interest or gains. The awareness of or consideration for the needs of others in society was combined with individual self-interest so that consensus building was relatively easy in such a 'top-down' process.

In contrast, in South Asia, though much was made of their Five Year Plans, the commitment to implement such plans remained weak and little effort was made to build a consensus behind the longer-term development agenda incorporated in these plans. In South Asia, the spirit of social cohesion tends to be weaker and the task of building a consensus for development thus becomes more difficult. Moreover, whilst the East and Southeast Asian countries had strong governments but a weak civil society, i.e., various non-government and civic groups or organizations the reverse was the case in South Asia, with a weak state confronted by a slowly emerging more active civil society.

The next aspect of governance is the competent implementation of programmes and policies put in place to realize the developmental vision of the leadership. It has been argued that the economic bureaucrats in East and Southeast Asia were more insulated from political pressure so that they could thereby implement, with relatively fewer constraints, the policies and programmes laid down by the leadership. This autonomy of the East/Southeast Asian economic bureaucrats was deemed to be of particular importance in explaining the more effective implementation of policies in these countries. This autonomy owed to the fact that, more often than not, the top bureaucrats were likely to be part of the political decision-making process. Because of the authoritarian nature of the governments, the political leadership freely inducted top technocrats/bureaucrats into the political arena to play a dominant role in both the formulation of national policies and in the implementation of their programmes. These politically powerful bureaucrats thus did not have to work their way through the political process to the top of the heap, accumulating political debts *en route*. This co-option of the bureaucracy into positions of political authority was itself facilitated by the top-down nature of the political system in these countries.

In a more open democratic political system, especially in parliamentary democracies of the type prevailing in South Asia, the important economic decision-making is more often than not dominated by the politicians. This was not always the case and, at least in Pakistan, under a very limited or constricted parliamentary system, the bureaucracy was once an important part of the policy making group. However, in more recent years, in a few instances in South Asia, bureaucrats or experts have been drafted into the political process and are reincarnated into practising politicians. Therefore, those bureaucrats turned politicians are not in any sense insulated from political pressures and compromises impinging upon or interfering with economic policy making and its implementation.

In East and Southeast Asian countries, on the other hand, the technocrats could play a significant role in economic policy making and in its implementation, without having to politicize themselves and to be constantly subject to political compromises. They were often called techno-pols, i.e., the technicians who played political roles without having to actually become politicians. These techno-pols or technocrats turned political decision-makers were protected by the highest authority from being buffeted by the competing interest groups and thus they could enjoy a degree of relative autonomy in decision making and its implementation.

In a multiparty parliamentary democracy, such autonomy can only be exercised by the bureaucracy if the political leadership stands firm in protecting its non-political bureaucrats and technocrats from pressure from the various interest groups which have infiltrated the political system. However, if transparency, openness and accountability are to be enforced in a multiparty parliamentary democracy, the technocrats/bureaucrats can seldom be totally protected from the shifting compromises constantly made by political leaders under whose authority they work. They cannot, for any length of time, be insulated from the political process. In most plural political systems, the political leadership has to engage itself in accommodating various interest groups, which may require changes of policy and/or flexibility in its application over relatively short periods of time. Such political dispensations lend themselves to instability in policy making in contrast to the more stable policy regimes realisable within a more authoritarian system. Unless political leaders can stave off and neutralize the competing interest groups seeking to subvert the stability of policy or its effective implementation, there is no way that economic bureaucrats in such a system can be invested with the autonomy necessary to effectively implement the development goals of the regime.

Reducing Inequality and Poverty

The East and Southeast Asian countries not only attained a high rate of growth of per capita income, but also managed to reduce both inequality and poverty during the period of rapid development during the 1960s–80s period. They succeeded in not only reducing the percentage of the population below the poverty line but also the absolute numbers of poor people. In other words, reduction in the percentage of the population below the poverty line was not offset by high population growth, which was the case in other countries exposed to rapid population growth. Throughout this period, the level of income inequality in the Southeast Asian countries, as indicated by the Gini coefficient or in terms of the percentage of national income accruing to the highest or lowest 20 per cent of the population, was generally higher than in East Asia. However, East Asia also realized lower levels of income inequality than South Asia. In contrast, in the Southeast Asian countries income inequalities tended to be higher than was evident in South Asia, though absolute and relative poverty levels were significantly lower. The East and Southeast Asian countries seemed to have refuted the Kuznets hypothesis that rapid economic growth was associated with an increase in income inequality.

The reduction in poverty in both the East and Southeast Asian countries was, in part, attributed to their high rates of growth. However it was also observed that in these countries the initial emphasis on employment generating, labour-intensive industrialization, strengthened by high rates of growth in agriculture, was no less important in reducing poverty and inequality. Thus, both the pace and the pattern of growth remain important for the reduction of poverty. Moreover, given a low level of income inequality in East Asia, a high rate of growth could be widely shared and served to reduce poverty. Low income inequality enhances the poverty-reducing impact of high growth rates. Moreover, in all countries, low population growth rates also helped to strengthen the poverty-reducing impact of growth by raising the per capita income of all, including the poor, who tend to have larger families.

The rapid expansion of labour-intensive, employment-generating industrialization helped improve the conditions of the poor, whose most important asset was their labour. At the same time, high agricultural growth, where most of the poor tend to be located, helped raise their incomes. The rapid expansion of labour-intensive industries was facilitated by the emphasis on export markets, since this removed the constraint that might otherwise have been exercised by the narrow size of the domestic market due to low per capita income. This export orientation also ensured that the industries would remain competitive by exploiting their global comparative advantage in labour intensive industries.

Relative emphasis on agricultural growth varied among individual countries. For example, Taiwan attached higher priority to agriculture than did South Korea. In the resource-abundant countries in Southeast Asia, agriculture played a leading role. In all cases, agricultural growth, through intersectional linkages, stimulated rural non-farm employment and income generating activities, including rural industrialization, all of which were relatively labour-intensive and provided higher incomes for the poor. Investment in the rural infrastructure, transportation, and communication facilities as well as in rural electrification helped the expansion of the non-agricultural sector in the rural areas.

Apart from agricultural growth and expansion of the rural non-farm sector, reduction in rural poverty was facilitated by land reforms in some countries and by liberal access of the poor to surplus land in the relatively land-abundant countries. Since inequality of land distribution was the dominant cause of inequality in the rural areas, and since, in the early years, the economies of

East and Southeast Asia were mostly rural, a reduction in rural income inequality through land redistribution or increased access of the landless or small farmers to land contributed to the reduction in overall income inequality. In Korea and Taiwan, there was an effective and radical land redistribution. In the Southeast Asian countries, the initial land distribution was less unequal than in East Asia. For example, in the most populous region of Indonesia, i.e., Java, the small farmers were traditionally the dominant class. Moreover, increasing population pressure in Java was relieved through migration and resettlement in other less populated islands. Similarly, in both Thailand and Malaysia, with excess land and a low density of population, their increasing population was provided access to land through the settlement projects in the land surplus areas of the country.

South Asia has not gone through any comparable experience in land reforms or redistribution even in countries (Pakistan) or subregions (northern states in India) where land distribution was highly unequal. The various legislative attempts at the imposition of moderate land ceilings and their weak enforcement did not produce any significant results. In Bangladesh and in several states in India, the heavy pressure of population on land provided little scope for land redistribution to meet the requirements of a rapidly growing rural population operating small farms, which were becoming progressively less capable of sustaining their operators above the poverty line. Neither could much progress be made to promote security of tenure or a more equitable system of crop sharing between owners and tenants. Over the years, the high growth of population in South Asia has resulted in a process of subdivision and fragmentation of land, enforced through laws of inheritance, which has contributed to the growth in the ranks of marginal farmers and in landlessness, as well as to the increase in rural poverty.

This diminishing capacity of the land to provide an adequate livelihood to its cultivators was partially relieved by the growth in non-farm rural employment – in some cases located in low productivity activities yielding lower average incomes than in agriculture, but which nonetheless supplemented agricultural income and relieved poverty. At the same time, employment in a few non-farm dynamic sectors such as trade and services, provided higher incomes – higher than in agriculture – and thus contributed to a growth in average per capita rural income. In many instances, the combined result of landlessness and poverty in rural areas and expectations of higher income in the urban areas led to the migration to urban areas, thereby adding to urban poverty. However, in the East and Southeast Asian countries, dynamic, fast-growing agriculture promoted a rapid growth of non-farm activities including industries, both urban and rural, through intersectional consumption and production linkages. In contrast, in South Asia, because of a sustained slow growth in per capita income in agriculture, there was no corresponding stimulus to rural or urban non-farm employment growth. Moreover, the prevailing pattern of inefficient import-substituting urban industrialization, compounded by high import restrictions and narrow domestic markets, based on inappropriate technology, provided limited scope for absorbing the increasing number of the rural poor into the modem industrial sector. Furthermore, the bias against exports in trade and exchange rate policies limited the expansion possibilities of the urban industries of South Asia in export markets.

The rapid expansion of primary and secondary education on a wide scale in East and Southeast Asia enabled the poor to participate in the rapid growth of income and employment opportunities provided by the prevailing development strategies of these countries. The expansion of education facilitated skill formation in both agriculture and industry, as well as in assimilating the technology being transferred from abroad into the new industries, and thus contributed, to a more efficient use of resources. Unlike in South Asia (except Sri Lanka), investment in health, in education and in related social sectors received high priority in East and Southeast Asia. Not only public investment but also private investment in health and education was significantly higher in these countries. In South Asia, investment in education and health, even as late as the 1980s, was lower (in terms of percentage of GNP or overall public expenditure) than what it was in many East and Southeast Asian countries in the early 1960s and 1970s. Moreover, the relative emphasis on primary and secondary education was much weaker in South Asia compared to East and Southeast Asia. Sri Lanka was the only exception in South Asia with its intersectoral educational priorities more closely reflecting those of East and Southeast Asia.

The distorted priorities in educational investment created a serious imbalance between supply and demand for different types of skills and educational attainments in South Asia. For example, the imbalance resulted in a large number of graduates from universities and colleges with a form of higher education that could not equip them for entering their domestic employment market. In East and Southeast Asia, the increasing number of primary and secondary school graduates was readily absorbed in a growing economy that generated demand for such skills. The pattern of sectoral economic growth required, and thus rapidly absorbed, the kind of skill that was generated by their respective education systems. In East Asia, in recent years, emphasis shifted to demand for higher levels of engineering, scientific, technical skills and educational attainments as their economies became more diversified with a growing demand for skills from more sophisticated industrial and related activities. The imbalance between supply and demand for higher levels of skills in these countries was met by an initial inflow of foreign technicians, but went in tandem with accelerated domestic programmes for skill upgrading. Their success in developing at a rapid pace the requisite level of advanced and higher levels of skill needed to keep the East/Southeast Asian countries globally competitive will be critical in determining the rate of their future economic progress.

The lessons of the East and Southeast Asian experience with regard to skill formation are that heavy prior investment in primary and secondary education is a *sine qua non* of economic development, and indeed holds the key to the diversification of a rural economy based on primary production. Agricultural development, after an initial period, becomes more skill- and management-intensive. The impact of extension and training programmes for farmers depends on their literacy and, in fact, primary and sometimes even secondary education become vital for the acquisition of necessary skills and technology for further agricultural intensification.

Heavy emphasis on higher levels of education (tertiary) at early stages of development, as happened in South Asia, given the limited resources available for investment, creates a shortage of skilled labour with a capacity for gradual upgrading through on-the-job training. This hampers the process of economic diversification and creates an excess supply of highly-trained individuals leading to either a 'brain drain' abroad or social and political instability at home due to unemployment-induced discontent and frustration amongst the 'educated' classes. At the same time, as economic diversification proceeds apace and the economic structure becomes more sophisticated, requiring higher levels of education and skills, the supply of these skills must rapidly expand or else growth will be constrained. Markets, left alone, will not necessarily balance supply and demand for skills in the right proportions over time. There are external economies in this sector which create a divergence between social and private costs and benefits, which justifies the need for public action. The role of the state thus becomes critical in guiding the market to overcome the delays and time lags through which market signals operate in the labour market, to bring forth a supply of workers needed to service the areas of the economy with a potential for growth.

The East and Southeast Asian countries relied on economic growth – high rates of labour intensive, employment generating patterns of growth – as well as on investment in human capital formation which, together with land reforms and/or access to surplus land, led to a reduction in poverty and income inequality. However, in recent years, in the high performing Southeast Asian countries, income inequality is increasing quite significantly. There is increasing evidence of growth in regional as well as interpersonal inequalities of income. While absolute poverty is low in these countries, as earlier stated, having diminished over time significantly, the general perception among the respective populations of these countries is that the increase in inequality and the concentration of economic power in a few families at the top in countries such as Indonesia is breeding social and political discontent. The perception that economic opportunities are concentrated and that inequalities of income and opportunities are increasing over time tends to have a destabilizing influence on the sociopolitical environment which could prejudice the sustainability of the growth process. This instability is already manifest in the growing militancy of the labour and student movements in South Korea. The perception of economic inequity is thus no less important than the actual fact of inequality and is often strengthened by public demonstrations of the extravagant lifestyles of a few at the top.

In some of these countries where an authoritarian political system persists in all its primitive rigour, economic prosperity is building up pressure for the political opening-up of the system to permit for greater popular participation in the political decision-making process, especially on the part of the rising middle class. This seems to corroborate the hypothesis that, whilst economic growth is not always dependent on the nature of the political system, be it democratic or authoritarian, the level of income is positively correlated with a more open political system.

The East and Southeast countries did not adopt, on any significant scale, direct poverty-alleviating measures involving transfers of income to or subsidization of consumption of the poor, Given the circumstances of slow growth and limited employment opportunities generated by the rate and pattern of economic growth, the South Asian governments undertook a wide variety of measures to initiate a direct assault on alleviating the poverty of the poorest. These measures ranged from public employment programmes to self-employment schemes and subsidized food distribution for the poor, including the free feeding programmes for the most vulnerable groups. In many cases, these were supplemented by subsidies for agricultural inputs such as fertilizer, irrigation, seeds, etc., as well as subsidized credit intended for the rural poor and, in recent times, also for the urban poor.

The programme for food or input subsidies did not work well and the benefits were often captured by the rich and the non-poor. However, the public employment and self-employment schemes, financed by either public funds or by liberal credit facilities, did provide a significant amount of employment for the poor, especially in Bangladesh and India. Both NGOs and government agencies have played an active role in these targeted programmes. In the past, it was the government that played a leading role in poverty alleviating measures, in providing employment or delivering credit and services for the poor. Increasingly, the non-governmental, non-profit organizations, both national and international, have come to play an important role. The NGOs have taken on the role of organizing and raising the consciousness among the poor regarding their constraints and possibilities. These organizations helped the poor to make the best use of services provided by the government and have increasingly acted as agents for delivering services on behalf of the government. They also acted as watchdogs to monitor the performance of various measures undertaken to help the poor.

In the very recent years, the Southeast Asian governments have also felt the need for more direct interventions to assist those social groups left behind in the process of rapid economic growth, even though the poor in these countries are at a higher absolute level of living and constitute a much smaller percentage of the population than their counterparts in South Asia. This is indicated by the recent interest amongst some Southeast Asian governments in self employment creating schemes for the poor as, for example, similar to those of the Grameen Bank in Bangladesh.

South Asia, however, can learn from the experience of the East and Southeast Asian countries that direct poverty alleviating measures, even if they are successfully targeted and efficiently implemented, can supplement and not substitute the indirect effect of rapid economic growth on raising the levels of living of the poor. Given the scale and intensity of poverty, such direct poverty alleviating measures cannot alone meet the challenge or be adequate to alleviate the conditions of the large numbers of people that live in poverty. More importantly, poverty-alleviating measures cannot be financed and sustained for long unless economic growth is accelerated and resources are generated out of increased income which can finance such poverty-alleviating programmes. At the same time, experience in South Asia does demonstrate that non-subsidized financing of self-employment-generating schemes as well as efficient public employment schemes designed to build the physical infrastructure of the rural areas, along with investments in education and health can expand income, employment as well as investment. Such measures do not constitute pure income transfers, but serve to promote growth-generating investment.

Lessons from China

It may be mentioned that the above analysis of the experience of the East and Southeast Asian countries does not include China, a country in this region that has enjoyed very high rates of growth, in some periods at double digit rates, since the economic reforms started there in the late 1970s and continued into the 1980s. China is a unique case, not comparable to the rest of Asia. It is a country in transition from a fully-planned economy to a market economy, under the guidance and control of a fully authoritarian state run by the Communist Party. Its liberalization and reforms started with a gradual restoration of market prices in the agricultural production system, where communes were replaced by the *household responsibility system* and the privatization of some productive activities. The use of market-based price incentives and the freedom to respond to such incentives available to private producers have led to phenomenal rates of growth that may now decelerate as technical progress is almost stagnant. While in the urban industrial sector, state enterprises dominate and are maintained by heavy state subsidies, there has been a rapid growth outside the state sector of small industries, especially village and township enterprises, that are allowed to operate in a free market system, sometimes linked to urban industry in the state sector and sometimes in joint collaboration with foreign enterprises linked to export markets.

The South Asian countries have lessons to learn from the experience of village and township industries in China, as a way of promoting labour-intensive rural industrialization as well as evolving innovative patterns of ownership in relation to local government institutions. At the same time, the South Asian countries may learn from the problems and difficulties China faces with the state enterprises, burdened with surplus labour and exposed to substantial losses. South Asians as well as the Chinese have to learn how to improve the efficiency of state enterprises. South Asia, in particular has to learn how to manage a process of privatization, backed by a viable system of financial institutions and banks, which is open, accountable and results in the establishment of a competitive private enterprise system rather than in inefficient monopolies. This again takes one back to the issue of a strong state with an efficient and honest system of governance.

Conclusions

In attempting to learn from East and Southeast Asia there are at least two factors that may be kept in mind by the policy makers of South Asia. First, one cannot totally divorce the economic development experience from historical accidents and fortuitous circumstances. As was pointed out at the outset of this overview, one cannot completely understand or explain all the factors that led to development in a particular context. Each case is, to some extent, unique. At the same time there are certain features that can be isolated and can be followed elsewhere with suitable adjustment but particular sociopolitical circumstances. The role of the overseas Chinese community in Southeast Asia is, for example, unique in providing both capital and entrepreneurship in most Southeast Asian countries. The nonresident Asians of South Asia do not play anywhere near as decisive a role within their respective countries of origin as has been exercised by the Chinese communities of East and Southeast Asia in the land of their ancestors.

In the most recent years, a new factor, aside from the role of governance, has been gaining prominence among the development specialists as a possible factor in explaining economic development: the role of social capital, broadly comprising a network of social associations and groups that promote social responsibility as well as stimulate interest and participation in activities that promote social welfare or development in the broadest sense. Such stocks of social capital are seen to promote growth and contribute towards sharing the fruits of growth. The role of mutual trust and confidence which reduces substantially the costs of economic transactions, i.e., the formulation and enforcement of contracts, is also seen as a positive return from social capital. The East and Southeast Asian countries are stated to have accumulated higher levels of social capital including mutual trust and confidence, which serves to promote economic growth. Also these societies demonstrate a sense of social responsibility, built around a concern for both family and community, which transcends the exclusive focus on individual interest. This social consciousness helps in the enforcement of discipline, and makes the task of governance relatively easier than in societies where a more exclusive concentration on individual or self-interest provides for a very limited recognition of any social or community responsibility.

Second, on the narrower issue of economic factors, the East Asian experience is straightforward, even if one does not subscribe to Krugman's oversimplification. Policies, institutions, including a development-oriented governance system and technology, including an emphasis on human capital formation, all contributed to the process of growth and poverty alleviation.

An attempt has been made to assess these factors in the foregoing pages. It is, however, clear that debate on the relative role of various factors will continue in the future. Hopefully, debate will lead to further clarification of unsettled questions. For South Asian countries, while the experience of East Asian countries is relevant, the Southeast Asian experience may be even more relevant, not only because the latter experience is more recent but also because some of their policies, especially for trade and exchange rate management as well as their industrial policies, may have more relevance to the contemporary circumstances of South Asia. Similarly, Southeast Asia's much greater reliance or openness to foreign direct investment may be worth examining. Unlike the experience of East Asia that has been subject to relatively extensive investigation for some time, even though not yet fully understood, the experiences of the high-performing Southeast Asian countries have not received such detailed scrutiny and analysis. The need for such an analysis may be overdue.

Third, on the much-debated role of the state versus the market there has been a growing consensus on the basis of what Marshall once referred to as 'introspection, observation, and analysis', that the scope of state intervention in many developing countries may have been greatly over-extended in the past and a correction in the other direction was needed. While the need for public intervention to correct market factors is not questioned, consensus breaks down on the question of what constitutes market failures in a particular instance, whether public intervention can correct the particular market failures and whether government failures may not exceed market failures.

In the resolution of such an argument, judgement plays a crucial role. The strategic interventions by the state in East Asia, in particular, were time-based, transparent and reversible. The nature and extent of such interventions evolved over time in different countries at a different pace, adjusting their role to changing circumstances. Most of the incentives were performance based. In fact, whilst an import substitution bias was greater than their export promotion bias at the initial stages of their development boom, later on this bias was offset by the strengthening of export promotion incentives. In fact, export bias or incentives might have been stronger in the later stages so that incentives were no longer neutral between production for the local market and for exports but clearly favoured the export market.

Fourth, a relevant factor in learning from past experience is the change in the external economic environment facing South Asia compared to what the East and Southeast Asian countries had to deal with in the past. The South Asian countries have a more limited room for manoeuvre in terms of economic policy, especially external economic policy. There are two sets of restrictions on the choice of economic policies. These restrictions did not affect the East and Southeast Asian countries to any significant degree in the early stages of their development history. One is the new international trade regime ushered in by the World Trade Organization (WTO) and the other, so-called 'donor regime' implemented by the international donor community, i.e., both bilateral and multilateral. The kind of trade restrictions, i.e., import restrictions and export subsidies, including domestic subsidies trade implications, that were permissible practices up to the early 1980s, are no longer permissible to the same extent. The developing countries are under obligations in the WTO not only to convert quantitative trade restrictions into tariffs and to reduce their tariffs but also to eschew, apart from a few exceptional circumstances, export subsidies and domestic subsidies that lead to significant divergence between world prices and domestic prices. The new anti-dumping regulations and dispute settlement procedures also reduce the freedom of manoeuvre in trade policy.

It is true that even within the new trade regime the developing countries have some leeway insofar as they are allowed a longer period of adjustment to new rules and can invoke trade restrictions on balance of payments grounds, even though not on grounds of the need to protect domestic industry. They have also been given some freedom of manoeuvre in the use of tariffs since they have replaced quotas by very high tariffs for purposes of binding a ceiling on tariffs and to the extent that what is called 'dirty tariffication' is practised by some developed countries, the ceiling binding is far above the current level of applied tariffs. There is no leeway on export subsidies and domestic subsidies are mainly permissible for the promotion of rural development, research and extension, etc. In respect of the new WTO regime, the least-developed countries of South Asia such as Bangladesh and Nepal enjoy a much wider range of exemptions than other South Asian countries.

More importantly, the liberalization of the trade and exchange rate regime in most developing countries in recent years has predated the Uruguay Round as a part of structural and stabilization lending programmes of both the bilateral and multilateral donors. Most countries liberalized their trade regime to a much greater extent than was required under the new WTO rules. Moreover, apart from the WTO regulations, and until the latter comes into full play, the threat of unilateral sanctions by the major trading blocs such as the United States and European Union still exists. Furthermore, there are a number of new aspects of economic policy, in fact, domestic policy, that are increasingly being brought within the ambit of international negotiations. The freedom of manoeuvre which the East and Southeast Asian countries enjoyed in respect of labour standards – including trade union and wage regulations, environmental standards regarding both products and processes, rules governing or regulating foreign direct investment as well as freedom of copyrights and patents for access to foreign technology (protected by intellectual property rights), domestic procurement policies of governments for indirectly protecting domestic industry, etc. – are no longer available to South Asian or other developing countries. Some of these are already under negotiation in the WTO and others, such as labour and environmental standards, etc. are sought to be put on the WTO agenda by the developed countries. The resort to any of these measures for promoting domestic industries or for obtaining cheap access to foreign technology on the part of the present-day developing countries may well be defined as unfair competition in international trade and hence subject to multilateral or unilateral pressure by developed trading partners, buttressed, wherever feasible, by the negotiating or bargaining power that the donor community can exercise on aid-dependent developing countries.

South Asia, along with other developing countries, would therefore face a narrower range of options in economic policy relating to road or strategic government interventions than was available to East and Southeast Asian countries in the dynamic phase of their development. This constraint should be recognized, quite apart from the merit or otherwise of the case for such state interventions that have been examined in the earlier sections of this overview, when we seek to learn from the experience of East Asia.

Notes

1 The Republic of Korea refers to what is popularly known as South Korea. In the remainder of this text in all further references, 'Republic of Korea' will be abbreviated to 'Korea'.

2 The recent disclosure of very large-scale corruption in the highest political circles in Korea, fed by the big industrialists/heads of conglomerates and senior members of the bureaucracy, seems to suggest that not enough is known about the impact of corruption on economic development.

Chapter 8

Reflections on Development Perspectives since the 1950s*

The decades since the 1950s have witnessed significant evolution in the approach towards and perspectives on development processes, patterns and policies. This is particularly so with respect to growth and equity issues. There was a time when economic growth was considered to be a sufficient objective of development policy. But during the last two decades, it has been increasingly recognized that economic growth, while necessary, is not a sufficient objective, and that equity, stability and peoples' participation are equally important objectives to pursue.

Those who propounded economic growth as a sole objective of development argued that it was possible to attain other objectives such as equity through growth itself. Research, analysis and empirical experience, however, subsequently demonstrated that economic growth unassisted by supplementary policy measures may, in fact, aggravate inequality. Simon Kuznets offered the hypothesis that growth, unaccompanied by any corrective measures, would lead to an increasing inequality of income until a level of per capita income characterizing present day middle income countries – say $800–1000 per capita in 1970 prices – is reached; afterwards, as growth continued, its cumulative effects would help reduce inequality. It must be noted here that this hypothesis of a changing relationship between growth and inequality over time did not comment on the subject of the impact of growth on absolute poverty.

In the early stages of growth, there are various factors that tend to increase inequality, both between groups within a country or within a region, as well as between regions within a country. It is most unlikely that the various regions within a country would enjoy the same rates of growth, or equally enjoy the benefits of growth; they have dissimilar endowments, and differential human and physical infrastructure. Moreover, as the process of growth acquires momentum, increasing demands are made on technical, professional and

* First published in K. Ahooga-Patel, A.G. Drabek and M. Nerfin, *World Economy in Transition* (Oxford: Pergamon Press, 1986).

managerial skills; there is a mounting demand on entrepreneurship, innovative spirit and ability, capacity and willingness to take risks, to explore and exploit new opportunities for financial rewards for pride of achievement. The supply of such skills and abilities is scarce in the initial stages in development; they have to be acquired and learned through experience and training over time. Hence, in view of scarcity of such skills and abilities, their financial rewards are high compared to the average income level. Also, economic growth is associated with a transfer of population from the agricultural to the non-agricultural and from the rural to the urban sectors. The non-agricultural and urban sectors have higher per capita incomes so that the increasing weight of the high income sector would tend to increase inequality.

The various factors, which cause increasing inequality in the early stages of growth, become less important in the later stages. The lagging regions develop infrastructure or receive an inflow of resources, human and material; simultaneously, migration out of them increases the income of those left behind. Their weight in the national economy declines. Similarly, urban and rural, agricultural and non-agricultural disparities decline; scarce resources (such as skilled labour or managerial capacity) increase in supply, partly through 'learning by doing', and partly through institutional development, training and education.

This is what in general has happened in the developed countries of today in their earlier stages of growth. Over time, inequality of income has ceased to increase and, indeed, has started to decline. There is a political consensus in the developed countries that corrective measures to redress extreme income inequalities are not only socially desirable, but also politically necessary.

So much for the inter-relationship between growth and inequality; what about the impact of growth on absolute poverty? Doubts have been raised as to whether absolute poverty declines with growth. However, a decline in absolute poverty is consistent with increasing inequality. A distinction needs to be made between: (a) the percentage of population who are poor; and (b) the absolute numbers of the poor. With growth it is conceivable that the percentage of the population in absolute poverty may decline; however, their absolute numbers may increase because of an increase in population.

Research and empirical evidence seem to suggest that if the growth rate stays high enough and is sustained for a sufficiently long period, it is most likely that the percentage of the population living in absolute poverty will decline. Whether the absolute number of people in poverty declines will depend on the interaction between population growth and economic growth. The dynamics of population growth and its interaction with economic growth are not fully known. There are a number of available empirical studies demonstrating that high growth has been associated with an increase in the absolute number of the poor. However, an interpretation as to why it is so is not easy; one encounters so many contributing factors that it is difficult, if not impossible, to ascribe it to a unique set of factors. Furthermore, there are controversies as to the nature of data and their correct interpretation. At the same time, however, very few countries in the last three decades or so have been able to maintain very high rates of growth and for sufficiently long enough time.

However, even if the growth rate is high and sustained over a long time – long enough to make a dent on poverty – there are forces at work in a market economy that may delay the process of alleviating poverty. This may be due to interventions in the market mechanism and pricing process by governments or organized economic pressure groups. Attempts to force wages above what the supply and demand for labour would otherwise determine, or to push interest rates down below the opportunity-cost of capital, will reduce employment as well as encourage the substitution of capital for labour. High wages for those who are employed keep the rest of the labour force out of employment. Monopolies tend to restrict output and employment and hence contribute to inequality and poverty. For the process of economic growth to prevent the emergence of forces tending to aggravate inequality, the free market mechanism must be allowed to function freely, and prices must reflect scarcity. But free market forces do not pre-empt the growth of monopolies and market imperfections are not necessarily due to state intervention. To suggest that market forces must not be interfered with is to deny the existence of vested interests, as well as unequal opportunities arising out of the unequal ownership of assets. Those who advocate economic growth first and distribution later, aver that poverty can be alleviated through redistribution only when the size of national wealth and the level of income have grown to high enough levels; otherwise, there will be a redistribution of poverty rather than an alleviation of poverty.

It is contended that premature attempts to redistribute income or assets or to undertake poverty-alleviating measures would divert resources away from growth, i.e., to consumption away from investment and to current expenditures away from savings, and would in the long run hurt efforts to redress poverty. To have growth, savings and investment must be accelerated; in a private enterprise system, it is high private incomes and thus inequalities of personal income as well as high corporate profits that are needed to generate high savings. This implies a slow growth or stagnation in the income and the consumption of the poor. In a socialist system, it would imply a constraint on the consumption of all, including the poor, unless at the early stages of development, income and assets are redistributed so that growth builds itself on an already egalitarian base of assets and income distribution.

The protagonists of growth first and distribution later are unable to demonstrate that those who secure the lion's share of the benefits of growth and major control of wealth and income – hence political and economic power – will volunteer to surrender their wealth, privileges and power for the benefit of the poor and that they willingly encourage or readily sponsor the adoption of poverty-alleviating or redistributive measures. Furthermore, in a context of rising expectations of the poor, a long period of waiting for the process of 'trickle down' to work itself out is neither feasible nor politically acceptable. Sustained poverty is likely to destabilize a sociopolitical system through the aggravation of tensions and conflicts, and can endanger growth itself.

Social expenditures for the benefit of the poor or the provision of assets or income-earning opportunities for the poor can in many ways and within limits, help the process of overall growth and capital accumulation as well. Not only do social expenditures improve their health and enhance their contribution to the national product, but investment or assets accumulated by the poor for the poor also yield high returns that are no less than those obtained on investment by the rest. This is evidenced by the productivity of small farms or small-scale industries.

Small farmers respond to incentives, undertake innovations, save and invest; they generate returns on investment and produce output per hectare no less than the big farmers. Redistribution of land, therefore, helps achieve growth, equity and poverty alleviation. There are ways of taking advantage of the economies of scale in farming operations, to the extent that they exist, as example, in the marketing and processing operations or in the use of specific capital equipments for irrigation or land development. The institutions, such as the cooperatives or mutual aid arrangements, help combine the advantages of scale with those of individual incentives. Under the conditions of heavy pressure of population on land, not enough land is available to provide all with viable farms; some form of joint or communal production arrangement, combined with expansion in non-agricultural employment, promote equity without sacrificing output.

Along with equity and growth, stability of income and consumption has become an equally important goal. Growth, equity and stability are interrelated. Important sources of instability in developing countries are instability in food and agricultural production caused by variations in weather or in the availability of purchased inputs, or by pests and diseases, etc. Instability in trade characterized by fluctuations in price and volume of trade in turn causes instability in consumption. Trade-related instability affects income and employment in export sectors and, via fluctuations in availability of imports, affects the import-competing or import-dependent sectors.

Instability of employment and income has its worst effects on those who are already at a low level of income; at times of economic downturn, the consumption and welfare of the poor are affected most severely. Food shortages cause starvation among the poor; they do not have adequate purchasing power to offset the fall in income or in self-production of food, nor have they the means to hold stocks during periods of scarcity. Instability of prices and supplies adversely affects expectations of the future and slows down savings and investment. Stability in prices and supplies of food, adumbrated in the concept of food security, has come to be accepted as a very important objective of development policy.

People's participation, both as a means as well as an end of socioeconomic development, has been in the forefront of development thinking and dialogue in recent years. People's participation is independent of the objectives of equity and alleviation of poverty – the latter defined as the poor benefiting from the fruits of development. People's participation is intended to secure their involvement in the decision-making process, in the determination of development objectives, targets and policies, as well as in the formulation, implementation, monitoring and evaluation of programmes and projects. Two prerequisites for people's participation are: 1) representative and democratic organizations; and 2) decentralization at the grassroots or the local level of the decision-making process in which people's organizations can effectively participate.

Both these prerequisites are linked with political processes, systems and structures of a country; basically, they relate to the way in which political power is organized, mobilized and exercised. Democratic, representative and self-governing institutions and people's organizations are not universally accepted in all societies. Administrative decentralization by itself does not ensure people's participation. The structure of power is often centralized; those who are in control of political power at the centre, be they politicians or bureaucrats, are afraid or loth to give up or delegate authority or functions to lower levels at the grassroots level.

Many governments are reluctant to encourage people's organizations, lest they weaken the control of the governments. Participatory democracy is the antithesis of centralized control. Issues of efficiency versus centralized political processes continue to arouse controversy; they involve appropriate criteria for judging the efficiency of political organizations in the context of society's goals and objectives.

People's organizations that are not in tune with or do not work in conformity with the political power structures of the day can seldom play a developmental role without creating tensions and conflicts. How to generate a stable political system in which voluntary, freely-functioning people's organizations can play a developmental role without creating 'unmanageable tensions' is a challenge to the practitioners of development policy. Development without change is not conceivable. It is not without risks to the existing political and social order; successful development depends upon unleashing people's energies and motivating them towards the attainment of growth and equity in a participatory process. The appropriate balance between conformity and dissent, control and freedom, harmony and conflict is to be sought by each society in its own way; the process of adjustment and conciliation may change over time. Economic development is not an entirely economic process; it is to an important extent a political and social process.

Chapter 9

Lessons of Experience: Development Policy and Practice*

Analysis and empirical investigation of factors which promote or retard growth and development in the poor countries have proliferated since the 1950s. A consensus has emerged on a number of important aspects of development patterns and policies which deserve recapitulation so that one can assess the state of our understanding and identify areas which require further exploration.

This note focuses attention on a few limited aspects of development experience on which a growing consensus is discernible. Firstly, the role of organization and management in promoting development; secondly, political elements in economic planning and policy making; and thirdly, interrelations between growth and poverty.

Organization and Management

Experience of two and a half decades of development planning and policy making teaches us that in the task of promoting socioeconomic development, the role of organization and management is equally, if not more, important than that of capital investment. I use the term, organization and management, in this context in its broadest possible sense. In this broad concept are included institutions and administrative machinery for policy making and planning, ability to examine issues and options, capacity to formulate as well as to implement policies projects and programmes.

A development plan which works out neat quantitative forecasts using sophisticated mathematical exercises in the optimum allocation of resources is at best an aid to a clearer perception of issues and options in development policies, and at worst a vehicle for the diversion of highly scarce trained manpower away from the urgent tasks of policy making based on a simpler analysis. The critical components in any plan are not the targets and forecasts, but a detailed analysis and prescription of policies and institutions which would

---

* First published in J.R. Parkinson (ed.), *Poverty and Aid* (Oxford: Basil Blackwell, 1983).

mobilize resources and ensure their effective utilization in order to generate equitable growth.

It is now increasingly appreciated that the greater the shortage of resources for investment, the greater is the need for efficient organization and management throughout the economy, not only in government but also in relation to projects. The poorer a country is, the greater is the sacrifice in current consumption that it has to sustain in order to accelerate the rate of saving and investment. There is correspondingly a greater need to ensure that savings are efficiently managed and productively utilized and that investment yields returns which amply justify the sacrifice in current consumption. The spectacle of inefficient and wasteful use of resources in many countries discourages sacrifice in current consumption and reduces public support for the mobilization of additional resources, or increases in taxes. A vicious circle is thus created. There are many instances where investments in new capacity are made although existing capacity is partially inefficiently utilized. New development projects are built without making adequate provision for their operation and maintenance, including repair of capital equipment.

It is not only technical and professional training and education, but also skills, attitudes and incentives which are crucial for building efficient organization and management. Moreover, a wide range of economic, pricing or fiscal policies and incentive structures are relevant for the efficient use of investment resources. Concern about appropriate incentive structures and policies is increasing amongst policy makers and analysts in the developing countries, but their actual implementation falls far short of what is needed.

It is a difficult and complex task to improve organization and management and to increase the efficient use of resources. If asked to show better results or produce higher output, those in charge of development projects would rather press for additional investment than concentrate on producing more from a given capital stock. Improvement in organization and management is a slow and time-consuming process. It involves changes in attitudes and institutions, in established habits of work and thinking, and in traditional cultural patterns which are not easy to shake off. There are, undoubtedly, signs in many developing countries of improvement in this direction; frequently, development projects incorporate components for training and institutional development. This is especially so for externally financed projects as the donors become increasingly aware of the need for improved management and institutional frameworks.

Often international and donor agencies throw up their hands in despair when confronted with inefficiency in decision-making, management, organization and implementation. Unfortunately, these inefficiencies are characteristics of underdevelopment itself. It would be unrealistic to assume that efforts extended over only a few years would be sufficient to overcome these shortcomings. In this, as in many other aspects of the development process, there is no short-cut. It is counterproductive to proceed from a recognition of current inefficiencies to a counsel of despair and frustration. Development has to be conceived not in terms of decades but in terms of generations. Both the international community and the developing countries themselves have to assign high priority to the task of improving organization and management and persevere patiently with it as well as improving the institutional and policy making framework.

While the need for improvement of organization and management is as strong as ever, there has been a declining trend in technical assistance for institution building and training. In the early 1950s donor communities, both multilateral and bilateral, including private foundations, placed a great deal of emphasis on manpower training and institution building. The declining interest in technical assistance of this kind may partly be traced to impatience or disappointment with the long time-lag involved or with the inadequacy of perceptible results of technical assistance efforts. Manpower training and institution building is not only a time-consuming process; the risks of slippages and failures are high. The accomplishments are not always quantifiable due to the institutional, political and social constraints or factors affecting the process and the output of training efforts. The results accordingly are uncertain. Misallocations of trained manpower resources within a developing country are not infrequent. The phenomenon of brain drain is widespread even under circumstances of acute scarcity of trained personnel. However, experience reconfirms that there is no escape from intensified efforts, both national and international, for building up indigenous capacity for organization and management. It is often preferable to over-invest in training and management and to supply an excess of trained and managerial personnel than to under-invest with resultant scarcity.

Political Factors in Economic Development

This brings us to the role of political elements in economic policy making. Development activities benefit different groups and regions differently. It is not only that some gain more than others, but also that while some may gain, others lose. Economists often assume that when the gainers gain more than the losses suffered by the losers, and hence could overcompensate the latter, a decision will somehow be taken as to whether to compensate and by how much. But whether or not to compensate and how much to compensate is a matter of political decision and of the relative strength of the contending parties. This is a part of the received, contemporary economic wisdom applicable to both developed and developing countries.

What is probably peculiar to the process of development and policy making in a developing country is that the process of change, involving the distribution of gains and losses, is rapid and substantial with marked differences between individuals. Given the prevailing levels of income and welfare, the magnitude and distribution of gains and losses has important implications for the relative structure of economic and political power. And they raise great expectations as well as possibilities of severe disappointments and frustrations. This is exacerbated because changes in the distribution of gains and losses are such that they are often concentrated into a period of time which is much shorter than comparable periods in the history of the economic development of the developed countries of today.

There is another distinctive feature in the historical experience of the rich and poor countries. In many Western developed countries the growth of a pluralistic political system often accompanied the process of economic development. Various interest groups had recourse to a process of bargaining and a peaceful process of determining how to share in gains and losses thrown up by development.

In the developing countries such a political system will require years to develop. In many countries there are only one or two dominant groups which control a more or less authoritarian political system. They guide and determine the process of change which necessarily carries differential consequences for the various social groups. Tensions are generated by a process of change and modernization. But a consensus as to how to diffuse the tensions or how to mitigate the short-term adverse consequences or dislocations does not easily or quickly emerge under these authoritarian systems. Instead tensions and conflicts are often suppressed until, after an interval of time, they generate an explosion.

An authoritarian system in which diverse interest groups do not participate freely in the decision-making process may sometimes achieve rapid overall progress. It may succeed in taking quick, essential decisions and in implementing necessary changes expeditiously. In the postwar period, such systems have often been welcomed by developed countries as being in line with their priorities on investment and growth as well as their emphasis on efficient and quick decision-making and implementation. The absence of wide participation in the decision-making process or in the benefits of development did not worry the international community in the 1950s and 1960s. Their complacent attitude was shared by many in the developing world who were anxious to press forward for a rapid pace of development, defined mainly in terms of rate of overall growth, irrespective of its composition or distribution.

There are indeed alternative scenarios: one in which decisions about the speed and pattern of economic development including capital accumulation and income distribution are taken by consensus and negotiation between the contending interest groups; or one in which a dominant group within an authoritative framework takes such decisions in the light of its perception of the requirements of modernization and economic development. The alternative process of consensus through negotiation and wide participation of the affected parties is time-consuming and untidy; it is often full of compromises and shifting patterns of intergroup alliances leading to inefficiencies. The result is not always the most effective method for accelerating economic growth; it is often impeded by protracted and frustrating discussions and debates among contending interest groups. The disagreements and debates regarding the pros and the cons are all the time kept in the public eye and one has the feeling of not always being sure that the right things are being done. In the closed, oligarchic system the pros and cons are not debated in public; differences of opinion and approach regarding the patterns and the process of change are hidden or submerged; the process of reconciliation goes on only within a small dominant group or person. One man, or small group, takes quick decisions and accomplishes a few big things often efficiently and competently. History shows, however, that after a time such suppression gives way to social upheavals. This is fed and caused by rising expectations amongst the newly emerging economic groups, such as middle-class professionals as well as traders and industrialists, for a share in political power. Faced with these two alternatives, one is confronted with a dilemma. The authoritarian approach promises quick results; the democratic one is plodding and indecisive, but necessarily so, if democratic political institutions are to be forged.

What is the appropriate rate of modernization; how do the various interest groups, new and old, participate in determining the rate and pattern of modernization and development; how and by what process can a consensus be reached? These are the questions which lie at the heart of the problem of development policy and planning in the poor countries.

Experience indicates that it is probably better to allow a process of consensus to emerge through negotiations and bargaining between different interest groups. We cannot totally avoid a process of uncertainty and slowness, nor can we avoid inefficiencies of a decision-making and implementation process which involves many interest groups but which is relatively peaceful. Even if there are periodic minor eruptions, in such an open system we hope that they will be contained within limits and dislocations will be minimized.

How does one create a viable, open, pluralistic democratic system in developing countries which have only recently emerged from colonial rule or from a feudal, autocratic regime? How can it be ensured that those who seize power in the aftermath of independence or after the overthrow of feudal and autocratic regimes initiate a democratic process? For one thing, it seems that it is necessary to foster a coalition of interest groups, none of whom is strong enough to impose its will on the rest and, therefore, all of them compromise to achieve a stable equilibrium in the sharing of power. But the process by which such a stable regime of distribution of power is highly disturbing and disrupting. In industrialized countries, the process of economic and political development has taken 100 years or more. Also, it has often been painful, created tensions and instability and caused great dislocation and distress. There is no clear cut and ready-made solution; each individual case is unique.

There is often a tendency in advanced countries, especially amongst the aid agencies and the development specialists, to advocate show or model cases or development. The rest of the developing world is urged or exhorted to emulate their examples. The case of Taiwan and South Korea in Asia, and that of Kenya and recently, the Ivory Coast and Malawi in Africa come readily to mind. The models which are advocated change over time as the perceptions of the processes and objectives of development, as well as the predilections or the prejudices of the observers change. While, for some time, a few growth models are applauded for having achieved high rates of growth, they soon go out of fashion and fascination grows about countries like Sri Lanka and China, with radically different sociopolitical systems, which have succeeded in achieving higher levels of satisfaction of basic needs in spite of lower rates of growth. No matter which models are exhorted in a given period, in the course of time they develop symptoms of tensions and start faltering. Erstwhile sympathizers soon start raising doubts about the validity of these approaches and emphasize their limitations.

The lesson of past experience seems to be that one must not be too quick in arriving at a judgement on the models of success or on the identification of success stories for others to follow. Development is not only a multidimensional process often with conflicting objectives; but also each nation has to evolve its objectives, priorities and processes through a process of experience and experiments. The example of even a highly organized and ideologically-oriented society such as China demonstrates that the path of economic progress is neither smooth nor stable nor peaceful. Rather, it is bumpy and often follows changes of direction. There may not be one or a very limited number of very clearly defined ways of achieving the goals of growth, equity and welfare. So much depends on the relative emphasis a nation places on its multiple objectives, not often even fully articulated, on its historical circumstances, on its political and social characteristics, as well as on the specific environment – political, social and economic – which a nation faces in its relationship with the rest of the world.

This sounds like rather trite pragmatism, not ideologically pure and analytically neat. This is very true but it is what seems to be the case with human or social development.

Growth and Poverty

The question of growth versus equity or, more appropriately, how to make growth consistent with the alleviation of poverty, is one of the challenges of our times. There is no simple answer. There are countries which have grown fast without diminishing inequity or poverty; in some cases poverty may have even increased. And there are also countries which have grown fast while greatly reducing poverty. Poverty persists and deepens in most of the countries that have grown at a slow pace, but a few slowly growing countries have made progress towards reducing poverty by specially designed policy measures.

In the light of the above, would it be fair to conclude that while the conventional wisdom of a trickle-down process is not valid, growth is still and continues to be a necessary but not a sufficient condition for the alleviation of poverty? There are those who argue that high growth, free competition and efficient pricing policies would lead to a reduction of poverty. How high should the growth rate be in order to achieve this objective? Under the prevailing circumstances of the developing world and given the present level of poverty, are such high rates of growth feasible in the majority of the countries? At the same time, are the institutional barriers to free competition easily removed? Does contemporary experience have any lessons about how effective free competition can be in performing this role? It remains doubtful whether without positive redistributive measures in favour of the poor, such as the public provision of social services for them, direct transfer of income and special investment efforts to increase the employment and income of the poor, poverty can be really relieved.

There is a second set of questions which remain unresolved. They relate to whether supporting or redistributive measures undertaken to alleviate poverty would not reduce the rate of growth itself. Undoubtedly, there are policy measures or programmes for poverty alleviation which, to a certain extent, would not militate against growth. Land reforms are a case in point. Under certain circumstances a highly unequal distribution of land acts as a constraint on growth in agriculture. Land distribution policy can achieve equity and higher employment, without producing negative effects on efficiency and growth, provided supporting measures are undertaken to supply inputs and resources to those who gain access to land. This is true especially in labour-abundant poor countries where unemployment is a serious problem and non-agricultural employment possibilities are severely limited.

Similarly, a more equitable distribution of public services, such as health, education, nutrition and social welfare, etc., is likely to increase the productivity of the poor and contribute to increased output. Would redistributive measures reduce private saving on the part of the rich? Perhaps, but it is argued that this can be offset by providing incentives and institutions for mobilizing the savings and investment of the majority of the not-so-rich and poor. And, of course, there is always the instrument of public saving which, within limits, can serve to offset adverse effects on private saving.

Depending upon the level of income, the magnitude of poverty and the time horizon over which poverty is sought to be reduced, it may be necessary under certain circumstances to sacrifice growth for the alleviation of poverty in the short tun. This is a choice which is preferably made through a process of negotiation and bargaining and by consensus. This process would not be without tensions and frictions.

This brings us back to the political power structure and decision-making process. It is very rare for a dominant group in an authoritarian regime to take a long-term view. How does one really persuade the dominant groups to adopt measures for alleviating poverty? The usual concept of the state or government serving as a neutral agent that mediates among the contending interest groups in favour of reducing poverty is not very realistic. Those who run governments or make policies can scarcely divorce themselves from the dominant groups whose support is essential for their political survival. Enlightened self-interest is slow to develop and its acceptance requires pressures from below.

What is frequently necessary is the existence of countervailing power amongst the disadvantaged and the poor, who have to be organized to exercise such power. The first prerequisite for the organization of the poor is, of course, a greater awareness amongst them of their rights and responsibilities and also a realistic appraisal of the potentials or prospects for success. The critical element in this process is undoubtedly widespread education. But once we are on the subject of the organization of the poor, we are back to the nature of the political system which facilitates the organization of interest groups, especially of the poor, and accommodates the risks and uncertainties caused by tensions amongst them.

It is often suggested that progress towards equity in the poor countries can be accelerated or initiated by directing the use of external assistance toward poverty groups and betterment of their conditions. This assumes that external assistance directly used for alleviation of the conditions of the poor would not substitute for the domestic resources which otherwise could have been so used. If such a substitution does take place, this would result in a paradoxical role of external assistance in economic development, in that the domestic policy makers would appear to be less interested than the aid givers in the alleviation of poverty. Secondly, even otherwise, the role of external assistance in alleviating poverty is likely to be limited. It is often contributes a very small proportion – no more than 15 to 20 per cent – of total resources for development within a country. Moreover, foreign-aided projects cannot always be so devised in order that most of their benefits go to the poor. The domestic and foreign resources have to be used jointly, combined with appropriate institutional and policy changes, towards the alleviation of poverty.

It is sometimes said that foreign aid should employ leverage so as to assist a redistribution of income or assets in a recipient country; for example, under a programme of land redistribution, aid might be used to compensate the landlords so that they can invest in alternative channels. Besides being a source of income and wealth, landownership confers economic and political power, which in turn secures access to resources. How are the landlords going to be compensated for the loss of political and economic power? Are they to be compensated by an equivalent, dominant economic power in non-agricultural sectors? If this is so, the process of removing inequity or constraints to growth in the agricultural sector is going to promote inequity in the rest of the economy.

However, foreign aid can facilitate redistributive growth in selective ways, especially in the agricultural sector. When a developing country musters sufficient political will to redistribute land and is in favour of promoting equitable growth, foreign aid can provide funds for compensation which will be used not for creating alternative sources for the concentration of economic power but for contributing to a wider diffusion of investment opportunities in the non-agricultural sector. Secondly, foreign aid, especially food aid, can compensate or reduce the effects of possible short-term dislocation in food production, following land redistribution; it can also help provide resources, inputs and credit to the new, small-scale owners of land to help them to improve output and efficiency.

The foregoing remarks strongly suggest that there is a considerable gap in our knowledge about the process of development and about the appropriate combination of policies and institutions which promote growth with equity. Increasingly, it is obvious that the discipline of economics by itself, concentrating as it does on the analysis of investment and its allocation between alternative uses, without taking into account political and institutional factors, is quite inadequate from the point of view of policy makers or analysts in developing countries.

PART IV

FOOD SECURITY: AGRICULTURAL PROGRESS: PAST PERFORMANCE

Chapter 10

Green Revolution in Asia: a Few Aspects and Some Lessons*

Introduction

Since the 1960s food and agricultural production in Asia has increased at a faster rate than that of population, with variations among individual countries; in many instances, the growth rate of production was higher than that of demand of cereals and resulted in reducing the food gap, i.e., a reduced dependence on cereal imports in the face of an increasing domestic food demand. The rates of growth of production of food in Asia (excluding China) were 2.5 per cent, 3.2 per cent and 4.0 per cent during 1961–70, 1970–80 and 1980–85 respectively.[1] The rates of growth in per capita food production in Asia (excluding China) were 0.1 per cent, 0.91 per cent and 1.8 per cent during these three sub-periods. If China was included in the Asian aggregate figures, all the corresponding rates of growth were even higher. India reached self-sufficiency in cereals whereas Indonesia and Philippines achieved self-sufficiency in their predominant cereal crop, i.e., rice. Pakistan and Thailand expanded their rice exports.

What were the main sources of growth in the Asian agriculture? What were the consequences in terms of equity and consumption of the poor? What were the policies, institutions and the nature of available technology, which contributed to growth and affected its distributional consequences? Are there any lessons which are relevant for other regions such as Africa?[2]

Sources of Growth in Asian Agriculture

In view of a high density of population on arable land and the limited opportunity for the further expansion of land area, the major source of increase

---

* First pubished in *Proceedings of the Symposium on the Sustainability of Agricultural Production Systems in Sub-Saharan Africa*, Occasional Papers Series C, Norwegian Centre for International Agricultural Development (Washington, DC: International Food Policy Research Institute, 1990), Reprint No. 181.

in agricultural production in Asia was the increase in productivity yield per hectare rather than expansion in land under cultivation. For example, the rate of growth in yield per hectare in the crop sector as a whole increased from 1.6 per cent per year during 1961–70 to 2 per cent per year during 1970–80 and 2.8 per cent per year during 1980–85. During the corresponding periods, the rate of expansion of area slowed down from 0.93 per cent during 1961–70 to 0.85 per cent, 1980–85. More than two-thirds of the increase in cereal production were accounted for by increase in yield per hectare even though there were inter-country variations. In South and Southeast Asia, contribution of the increase in yield to the increase in cereal production was 60 per cent during 1961–70 and 72 per cent during 1970–80 (Grigg, 1985, pp. 220–40). Associated with increase in yield was the increase in cropping intensity, i.e., multiple cropping on the same land within a year.

The major factors which were responsible for the rise in yield and cropping intensity were the increase in: (a) irrigation; (b) the availability and rapid diffusion of high yielding varieties of seed; and (c) the use of fertilizer. Seventy-six per cent increase in rice production in Asia was due to a combined effect of increased use of fertilizer, HYV (high yielding varieties) and irrigation, the latter leading the way; there was a synergistic interaction between the three contributing factors. The extension of irrigation was the necessary condition; the returns from the use of fertilizer and HYV were highly positively related with the extension of irrigation (ibid.).

The total land area under irrigation expanded at the rate of 2.3 per cent per year during 1961–80.[3] Between 1966 and 1980 there was 39 per cent increase in the irrigated land. By 1980 27 per cent of the arable land was irrigated. In a few countries such as Pakistan, Sri Lanka, and Indonesia, the percentages of irrigated land were as high as 71 per cent, 51 per cent and 38 per cent respectively (Vyas, 1983a). The extension of irrigation facilitated multiple cropping which went up to 129 per cent by 1982/84.

The use of fertilizer increased at a rate of 15 per cent per annum during 1961–70 and at 10 per cent during 1971–80. It is not only the rate of growth which was high but the absolute amount of fertilizer use per hectare, which was 1.6 kg per hectare of arable land in 1950, increased to 11.8 kg in 1961–65 and 49 kg in 1983–84.

Along with an increase in irrigation facilities and in the use of fertilizer, there was a considerable acceleration in the diffusion of high yielding varieties of seed of the two principal crops, rice and wheat. In the region as a whole (excluding China) by 1976–77, the percentage of the rice output under high yielding variety was 46 per cent and in the case of wheat the percentage was

48 per cent (Grigg, op. cit.). In a few countries such as Sri Lanka and the Philippines, HYV accounted for 63–68 per cent of land area in rice by the early 1980s. During the same period the percentage of land under HYV in wheat varied from 72 to 76 per cent in South Asian countries (Vyas, op. cit.). The new high yielding varieties were developed by IRRI for rice and for wheat by CIMMYT; they were, however, adapted to the specific conditions of each country, by national agricultural research systems. The role of national and international agricultural research system was crucial. The expenditure on agricultural research increased three times between 1959 and 1968/71 and 2.5 times between 1968/71 and 1980. During every decade from 1960 to 1980, the man-years of scientists engaged in agricultural research more than doubled (Jain, 1988).

Irrigation

What were the factors and policies which contributed to the growth of irrigation, fertilizer consumption and diffusion of high yielding seeds?

Take, for example, the case of expansion of irrigation; undoubtedly there was a considerable potential for irrigation in Asia based on both groundwater and surface water. In India, a few of the irrigation works based on surface or gravity flow irrigation were established hundreds of years ago; however, a rapid expansion took place in the 1950s and 1960s in groundwater irrigation.

In many cases, detailed hydrological surveys were undertaken to map the irrigation potential, both ground and surface water; alternative strategies for exploiting such water resources were evaluated. A balance was sought in some countries between large and small-scale irrigation projects, depending on the institutions and capacity to manage irrigation works of different types, and favourable results were achieved. An important contribution of irrigation in Asia was not only to provide water in areas or in seasons where there was no rainfall, but also to provide supplemental irrigation to rainfed agriculture. The irrigation system thus provided both an adequate and assured supply of water.

Given the high cost of investment involved in large-scale projects, and long time-horizons over which they had to yield high returns in order to make the initial investment worthwhile, the Asian countries soon learned that preference should be given to small-scale projects, which had a shorter life and required smaller investment. This was found cost-effective whenever there was an uncertainty about the quantity of long runwater availability. The determination of the irrigation potential, especially the quantity of available ground water, required detailed and time-consuming hydrological surveys and assessment. At the same time, there was an urgent pressure to increase the supply of irrigation water in the short run or in the immediate future on the basis of inadequate or uncertain knowledge. Under these circumstances, irrigation projects which were relatively inexpensive with a short gestation lag and a limited life span had an advantage; adjustment to changes in the irrigation technology in the medium or long run was easier.

Fertilizer

The expansion of fertilizer use in Asia was principally due to the following factors: (a) high return from the use of fertilizer in irrigated land and in improved varieties of rice and wheat; (b) favourable price ratio between fertilizer and crop; (c) expanding and assured supply from domestic production plus imports; and (d) education and the training of farmers in the efficient use of fertilizer. In Bangladesh, for example, research and extension services were coordinated in joint programmes for organizing demonstration plots in farmer's fields. Again, the period of the most rapid growth in the use of fertilizer in Bangladesh was also the period when the supply of institutional credit almost doubled in real terms, i.e., the late 1970s. Farmers who began to use fertilizer for irrigated rice also gradually spread its use to local, traditional varieties as well, as they accumulated experience in the appropriate application of fertilizer for the new varieties (Ahmed, 1988a). In order to provide an assured supply of fertilizer, it was necessary not only to supplement domestic production with imports on a well programmed basis, but also to maintain stocks in order to compensate for fluctuations in domestic production and in imports. For getting the best results from the use of fertilizer, it was very important to supply appropriate quantities and varieties of fertilizers at the appropriate time. Otherwise returns on the use of fertilizer were found to be low or uncertain, and this tended to discourage the farmers from expanding the use of fertilizer. In this regard, the role of the internal marketing and distribution system in ensuring an assured supply to the farmers at the right time and in the right quantity was very critical. As the amount of fertilizer used per hectare increased over time, the need for an efficient management of fertilizer use, including appropriate timing and a balance between different types of fertilizer such as nitrogen, phosphate and potash, as well as a complementary supply of micro nutrients, becomes critical. This, in turn, required a more sophisticated education and extension or training facilities for the farmer.

Two issues were of particular significance in the development of irrigation and extension of fertilizer use in Asia. First was the role of the state and second was the pricing of water and fertilizer. As far as irrigation was concerned, the government played an important role not only in surveying the irrigation potential, but also in investing in irrigation projects, especially in the large-scale irrigation projects. At the early stages, the government took the lead in developing or introducing the new technology for medium- or small-scale irrigation projects such as tubewell or lift irrigation systems. The role of private investment, which was in any case important in the old traditional system of irrigation, depending primarily on animal power or human labour, was expanded in the modern irrigation system as well. This happened only after the state initiated or sponsored small-scale irrigation methods were popularized and their profitability was demonstrated.

Similarly, at these initial stages there was some subsidy involved in both large- and small-scale projects; in the former, it was due to inadequate water charges and in the latter, irrigation equipment was rented out to the private farmers or farmers' groups, at less than its cost. In many countries, private investment in small-scale irrigation projects such as ground water irrigation through tubewells or low lift surface water irrigation was facilitated by provision of low interest credit to private farmers.

As in the case of irrigation development, the relative role of private and public agencies in the production, import, marketing and distribution of fertilizer has come under critical review in recent years. The fast growth in the supply and use of fertilizer was in the past affected by the pattern of ownership, i.e., whether the public or private sectors were involved in production and/or in distribution. The Asian countries had wide-ranging experience in this regard; both public or private sectors played roles in varying degrees of importance in different countries.[4] In recent years, in most Asian countries, there has been an increasing tendency towards the privatization of the marketing and distribution functions especially. This was part of a general policy of reducing burden on the limited administrative and implementation capacity of the government; it was partly in response to the past experience in several instances of limited efficiency of the public sector enterprises. The policy of privatization was also adopted with a view to encouraging competition in marketing, not only among the private sector enterprises but also between the private and the public sectors.

In many Asian countries, subsidy played an important part in stimulating the demand and expanding the use of fertilizer in the early years, as the bottlenecks in the way of supply of fertilizer were removed. The introduction of fertilizer as a new input involved risks of failure and of low returns, unless (a) adequate water supply and (b) correct timing and method of application of fertilizers were ensured; also new seed varieties were often associated with an increased variability of yields; returns from fertilizer use under these circumstances must be high enough to offset not only the risks of variability in yield but also the high cash costs of fertilizer input. To risk averse farmers, fertilizer subsidy at this stage played an important part in offsetting or reducing their perceptions of risk. In the early years the volume of sales was also low; the cost of distribution per unit of fertilizer was high. A subsidy helped to offset the high cost of distribution. As the volume of sales expanded over time and costs of distribution fell, the need for subsidy to offset high costs disappeared. As experience with the new fertilizer technology accumulated, the additional stimulus through subsidy was not needed, except to overcome the high cost of transportation and distribution in remote regions. In some cases, fertilizer subsidy compensated for the lack of access or high cost of credit to the small farmers who, more often than not, operated in a non-competitive, imperfect credit market.

Asian experience re-emphasizes that sustained growth in the use of modern agricultural inputs, including fertilizer, depends on: (a) the accumulation of information through experimentation and field trials about the relative yields and profitability of fertilizer use by crops and regions; (b) dissemination of this information to the farmer; (c) provision of credit to enable the purchase of the costly inputs; and (d) an efficient input delivery system.

Public Expenditures on Agriculture

As stated earlier, the technological breakthrough in rice and wheat varieties which provided the main spurt to the growth of food production in Asia required concentrated and sustained research efforts over a large number of years in both international and national research systems. At the same time, for the technological potential to be actually realized once it was available, investments in irrigation, fertilizer supply and production and multiplication of new varieties of seeds, education and extension of the farmers, as well as rural infrastructure such as roads and transportation and credit, had to be significantly expanded.

Substantial public expenditures on the various elements involved in agricultural modernizations were undertaken in Asia. This in turn stimulated complementary investment by farmers and follow-up investment by the private sectors in the provision and distribution of inputs. Public expenditure in agriculture per hectare of arable land under permanent crops was $3.95 in Asia in 1980, as against $2.38 in Africa, whereas the amount of scientific man years in agriculture per hectare of arable land was 0.10 in Asia, and 0.04 in Africa (FAO, 1986a, Annex III). However, the cost per man-year of scientific manpower in agriculture was higher in Africa varying between $52,000 and $68,000 as compared to $38,000 and $30,000 in Asia (ibid., see also Jain, op. cit.).

It is to be noted that national agricultural research systems in Asia had a long history prior to the technological breakthrough originating primarily in the international research centres which occurred in the 1960s. Thus, indigenous capacity had been in existence for some years before the 1960s, even though it was not adequate to meet the challenge; this capacity was greatly expanded in subsequent years in order to undertake adaptive research in cooperation with international research centres to adjust/modify the new varieties to suit the local agro-ecological conditions.

Foreign assistance played an important role in promoting national agricultural research in Asia; also associated with it was a rapid expansion in agricultural education at all levels, including extension and training.

At about the same time that the technological breakthrough took place, a higher priority was accorded to agriculture by the governments of the Asian countries. This was occasioned by a growing perception about the need for increasing food production in several Asian countries like India, Pakistan, Indonesia and the Philippines, etc., where increasing food imports were causing an unsustainable burden on the balance of payments. Also, in some countries the growing industrial sector was beginning to face an increasing demand constraint in view of the low rate of growth in agricultural income.

The most significant manifestation of the higher priority accorded to agriculture was an increase in the proportion of government expenditures allocated to the agricultural sector. The shares of total (current and capital) expenditures on agriculture in total government expenditures were 10.4 per cent and 10.1 per cent respectively during 1974–79 and during the entire period, 1974–83; they were higher than the corresponding figures for the entire developing world, i.e. 8.4 per cent and 8.3 per cent respectively for the two periods. The share of agricultural development expenditures in total development expenditures was 22–24 per cent in Asia as against about 18 per cent for the developing world as a whole (FAO, 1986b, pp. 94–5). Also the rate of growth during 1974–83 in real government expenditures (in 1980 prices) on agriculture was the highest in Asia (7.2 per cent for total expenditures and 8.6 per cent for development expenditures); the corresponding figures for the developing world as a whole were 4.0 per cent for total expenditures and 5.2 per cent for development expenditure. In addition, the year to year variation in the government expenditures on agriculture was the least in Asia. For example, the average percentage deviation from the trend level of agriculture expenditures in Asia was 13 per cent during 1974–83, as compared with 17 per cent for the developing world as a whole. Furthermore, faced with the need to adjust government expenditures in response to variations in available resources, they attempted to protect the development expenditures from declines and to let the current expenditures bear a greater burden of instability. During the period 1973–84, the average percentage deviation from the trend level was 15 per cent for development expenditures whereas it was 20 per cent for current expenditures (ibid.).

The Asian experience confirms that along with an expansion in research and educational expenditures, the availability of infrastructure, i.e., rural roads, transportation and communication facilitates, financial institutions, electricity and markets had a very important role to play in stimulating agricultural production, employment and household income, including non-agricultural income. It promoted the diffusion of modern technology through an improved and efficient marketing and distribution system for both outputs and inputs; also costs of marketing were reduced, output prices improved and input prices were lowered. It facilitated specialization and division of labour between regions by expanding the size of the market. For example, in Bangladesh the percentage of irrigated land, that of cropped area under HYV and the kg of fertilizer per hectare of cropped land in the infrastructurally developed villages were twice as high as those in underdeveloped villages.

Furthermore, infrastructural development, along with technical education and training and supply of credit, helped stimulate non-farm activities closely tied to agricultural production, i.e., processing of agricultural output and supply of inputs/consumption goods to the agricultural sector. Increased non-farm income in turn stimulated demand for food and expanded food production, income and employment in the rural areas. In the infrastructurally developed villages in Bangladesh, for example, agricultural income per acre, as well as non-farm income per household, were 20 per cent higher than those in the underdeveloped villages. There was increased employment in both farm and non-farm activities, more so in the latter than in the former, with wage income from agriculture being 88 per cent higher and that from non-farm income 108 per cent higher in the developed villages than in the underdeveloped villages (Ahmed, 1988b, pp. 33–5).

It is noteworthy that in Asia the acceleration in the rate of growth in the production of basic food crops was associated with an increase in the rate of growth of cash crops. For example, the rate of growth of production of basic food crops accelerated from 2.5 per cent during 1961–70 to 3.1 per cent during 1970–85 and at the same time, the rate of growth in the production of cash crop, accelerated from 2.2 per cent during 1961–70 to 3.8 per cent during

1970–85. Pakistan, Bangladesh, Indonesia and the Philippines expanded their export or cash crops while they were striving for increased self-sufficiency in food production. There was no serious discrimination against cash crops in respect of macroeconomic policies or investment programmes excepting: (a) that unfavourable external market conditions affected their production and export earnings; and (b) that technological progress in food crops, especially rice and wheat, increased their relative profitability as against export/cash crops as well as other food crops such as pulses and edible, oil, etc.

Incentives and Government Policies

In their drive towards industrialization in the early 1950s, the macroeconomic policies relating specifically to agriculture in Asia did discriminate against agriculture in relation to industry. There were, however, considerable divergences among individual countries. Several countries intervened in the marketing and distribution of food grains with a view to: (a) stabilizing inter-year variations in food prices; (b) ensuring incentive prices to the farmers in respect of government procurement of food grains; and (c) providing subsidized food grains to the selected consumer groups. A few countries taxed their agricultural exports, either to raise world prices or to extract income from the exporters or producers of export commodities.

However, over the years, discriminating policies against agriculture took their toll on agricultural growth and they were gradually reduced as a result of this realization. The biases against agriculture originating from overvalued exchange rates and the industrial protectionism were sought to be offset by subsidies on inputs, credit and export incentives specially to non-traditional exports, including the minor agricultural exports. But even then the compensation was only partial. The net effect was that import-competing products were subject to a lower degree of discrimination than the export commodities (Bautista, 1990; Krueger, Schiff and Valdez, 1988).

Direct government intervention in the marketing or distribution of agricultural commodities in Asia was limited to one or two cereals (i.e., rice and wheat) only; in a few countries, there was direct government participation in the marketing of export commodities through export marketing boards. However, the private sector, within the framework of export/import taxes or sometimes quantitative controls, was allowed a role, even though there was a variation among individual countries. There were virtually no instances of significant state farms outside the centrally planned/socialist economies in Asia.

An important second generation problem for Asian agriculture in the aftermath of the Green Revolution was the need for diversification of agriculture. With an increase in income, demand was increasingly diversified, i.e., away from cereals, specially in high income regions or countries, towards non-cereals such as edible oils, pulses, fruits and vegetables, sugar/livestock and dairy products. In some countries, an increasing self-sufficiency in principal cereals was matched by rising import bills or rising prices for non-cereal foods such as pulses or edible oils, meat and dairy products, etc. Further increases in the production of cereals, especially rice, were beginning to face demand constraint which originated partly from poverty and limited purchasing power of the poor on the one hand, and partly from limited prospects in the export markets aggravated by large surpluses of cereals in developed countries. The thinness of export market in rice makes it especially difficult to absorb additional supplies. Therefore both supply and demand considerations contributed to bring the issue of diversification to the forefront of policy concerns in Asia.

Equity and Power

Did the Green Revolution aggravate or alleviate inequity or poverty in the agricultural sector?

Neither farm size nor tenure was a serious constraint to the adoption of new high yielding grain varieties. The rates of adoption of new technology did vary, depending on the size of farm or tenurial status. However, evidence seems to indicate that within a relatively few years after the introduction of new technology, lags in adoption rates due to differences in size and tenure disappeared (Vyas, op. cit.; Hosain, 1988, ch. 6, pp. 71–88). In some cases, it was found that not only the percentage of adopters among the small farmers was no less than that among the large farmers, but also the per hectare use of modern inputs such as fertilizer, and irrigation water, etc., was higher among the small farmers than among the large farmers. The small farmers, faced with the pressing need to obtain the maximum increase in output from a limited land area, made a more intensive use of land through the maximum utilization of family labour, which had a low opportunity cost; this enabled a more intensive use of modern inputs per unit of land since many of the required additional work was labour intensive. The new technology increased the demand for labour not only in the agricultural operations but also in the post-harvest operations in view of a larger output resulting from new technology. As a result not only was the small farmer's family labour more utilized, but also there was an increase in wage employment or in the employment of hired labour.

There was an increased absorption of labour in agriculture, due to the fact that new technology increased the demand for labour, for greater application of fertilizer, including other related operations, as well as for enabling multiple cropping which resulted from short maturing new cereal varieties.

The Green Revolution technology, which by now is widely recognized, was scale neutral; in other words, both the large and small farmers achieved increased productivity or efficiency from the new technology. There were no significant differences in efficiency between the large and small farmers. However, the new technology was not resource neutral. The small farmers did not have equal access to credit, inputs and services as the big farmers did (Lipton and Langhurst, 1985, ch. 3). The small farmers paid higher prices per unit of inputs and services and hence the percentage increase in their net income was smaller than that of the larger farmers. Thus, though there was an absolute increase in the income of the small farmers as a result of the adoption of the new technology, the big farmers gained proportionately more than the small farmers.

Thus, in Asia landowners gained from the Green Revolution relatively more than the tenants and labourers. Given the initial disparity in income and rural assets, both land and non-land assets like irrigation equipment or animal power, the Green Revolution tended to aggravate inequality. Given the magnitude of the problem of landless labour and marginal farmers in Asia, the increase in employment or the labour absorption in agriculture resulting from new technology was not large enough to make a significant dent in unemployment or underemployment of the landless labours or on the income of the small farmers. Even with an increased income and productivity of the small or marginal farmers, the absolute increase in their income derived from their tiny farms was not large enough to make an impact on poverty in most of the densely populated countries of Asia.

One way in which the Green Revolution contributed towards the alleviation of poverty was to dampen the rate of increase in food grain prices at the consumer level. An increase in supply resulting from the Green Revolution exerted a downward pressure on prices, other things being equal, provided an increased domestic production did not merely replace imports but resulted in a net increase in supply, and provided countries were closed economies, in the sense that changes in domestic prices did not correspond to changes in world price. Since the poor allocated a very large proportion of their income and expenditure to food, the low price of food was of direct benefit to the poor.

An important reason why the Green Revolution did not make a significant impact on poverty or unemployment is that it did not spread widely enough among geographical regions and crops. The Green Revolution did not spread over time to all cereal crops, let alone non-cereal crops, because technological break through was yet to take place in respect of a wide range of other cereal and non-cereal crops. Nor was appropriate technology developed for the rainfed or arid land agriculture. Attempts are under way but no significant breakthrough is yet in sight (Ruttan, 1977).

Hence the overall growth rate was not high enough to make a significant impact on unemployment and poverty, given the high level of prevailing landlessness and poverty. On an aggregative basis, it was found that the employment elasticity of agricultural production in Asia ranged between 0.40 and 0.75, the most common figure being 0.5. With an increasing population and labour force, the impact on the alleviation of unemployment and poverty would be even less. For example, even with a 4 per cent rate of growth in production and 2 per cent rate of growth in labour force, no dent can be made on overall underemployment or unemployment, given the employment-output elasticity (Vyas, 1983a, pp. 52–8; Maunder and Ohkawa, 1983, pp. 52–8). At the same time, unless incomes and purchasing power of the rural masses are raised, growth in agriculture cannot be sustained for long. This is because, in most Asian developing countries, over 50 per cent of the land holdings are less than 0.5 hectare and account only for 10–12 per cent of the total cultivated area. This implies that 50 per cent of the agricultural households account for 10–12 per cent of agricultural output and receive at most 10–12 per cent of agricultural income.

In one respect, however, the Green Revolution tended to aggravate inequity or disparity, i.e., inter-regional disparity. Since the Green Revolution was confined to rice and wheat and mainly spread to the irrigated areas, the impact of the Green Revolution was uneven among regions. The rainfed and irrigated areas with favourable agro-ecological circumstances and necessary physical infrastructure such as roads and communication, transportation and marketing facilities principally benefited the Green Revolution. To the extent that in the less favoured regions rural populations were also poorer, the better-off farmers in well endowed regions got richer, while the poor producers in the disadvantaged regions stayed poor.

Was the Green Revolution associated with an increased concentration of landownership, through increased sales of land by the marginal farmers to the medium or large farmers or through the eviction of tenants or sharecroppers so that the owners acquired land for self-cultivation? Empirical evidence to date indicates that these consequences did not follow from the Green Revolution to any noticeable extent, firstly, because there were no scale economies in new technology and secondly, the status of tenure or the size of farms did not prevent the diffusion of new technology.[5] In fact, the landowners could extract high surplus from letting or even facilitating the adoption of new technology by the tenants and share croppers. They stood to gain out of increased output and income through either rent or sharecropping arrangements. The pressure of population and landlessness, coupled with the lack of alternative income and employment opportunities in the non-farm sector, greatly reduced the bargaining power of the tenants/sharecroppers in their contractual arrangements for a share in the absolute or in the incremental output resulting from technological progress.

The attempts at land reforms in Asia either for imposing a ceiling on landownership or for protecting the rights of the tenants were not very successful. The laws setting limits on land ownership in some countries, including India, might have helped to stem the trend towards a growing inequality in landownership; this encouraged the big landowners to sell their land rather than to face the uncertainty and risk of having to surrender it to the government sometime in the future if a shift in the balance of political forces led to a strict enforcement of the land reform legislation. The legislations protecting the tenants' rights had some positive impact only in those Asian countries where the social gap between the landlords and the tenants was not very wide and where the tenants could organize themselves for the achievement of their legal rights (Vyas, 1983b).

The Asian experience also confirmed that the most important factor strengthening the relative position of the labourers/tenants was the expansion of non-farm employment in both the urban and rural areas. Whenever the rate of growth in agricultural production was substantial and was broad-based and shared by widely among the small and medium farmers, rather than being restricted only to the large farmers, the growth linkages with the non-agricultural sector were strong. The increase in the expenditures by the farmers on the non-farm goods and services, particularly on the labour intensive goods and services, provided a strong stimulus to the expansion of non-farm activities and hence to the employment of the poor. This in turn stimulated an increase in the demand for food and led to increased food production and so on in a cumulative process of expansion of income and employment.

In most countries, non-agricultural income constitutes an important proportion of the total income of the farmers; an improvement in the levels of living of the farmers thus requires a simultaneous improvement in farm as well as non-farm incomes. At the low levels of landholdings, i.e., for small and marginal farmers, non-farm income in fact constitutes as high as 44 per cent of their total income; the relative importance of non-farm income declines as farm size increases because larger farms provide a higher level of income from farm sources. At the same time, at the upper end of landholdings, i.e., for big farmers, non-agricultural income progressively constitutes a higher proportion of the total income. This is because with higher income or landholdings, they are able to devote their resources to non-farm activities.

While at the lower level of income the predominant form of non-farm income is wage labour, the non-wage component of income becomes more important at higher levels of income because the big farmers tend to widen the range of their activities in the non-farm sector by engaging or investing in trade and services, etc. In India, for example, for low income groups, the share of non-agricultural incomes was 60 per cent of the farmers' total income; this proportion decreased to 25 per cent as the level of aggregate income rose, but then again, at the higher level of income, the proposed share of non-farm income was as high as 60 per cent; at the same time, the share of wage income decreased significantly for the higher income groups (National Council of Applied Economic Research, 1980).

Attempts have been made in various Asian countries to increase the income and productivity of small farmers by expanding and improving the delivery of inputs, credit and services to them. New experiments in and approaches towards a more efficient delivery system have been and are being carried out with varying degrees of success, especially in two important areas which are crucial of the development of smaller holder agriculture, i.e., supply of credit and extension and training.

Many Asian countries have adopted supplementary measures with a view to making a more direct assault on poverty rather than relying exclusively on the 'trickle down' effect of the process of growth. These measures in particular include public distribution of food grains, with varying degrees of subsidy to the selected consumer groups as well as employment-generating programme for the poor such as rural food for work projects. Experience with food distribution system has demonstrated that unless they are narrowly targeted they are very expensive, often diverting resources away from essential development expenditures, which generate income and employment of the poor. Also, targeting may involve high administrative costs and detracts from the very resources which otherwise would accrue to the poor as a transfer either of income or of food. The choice of an appropriate method of targeting on the basis of either income or occupation, geographical location or type of food consumed is very relevant. There is a trade-off between the high administrative costs of narrow targeting, on the one hand, and, on the other hand, the coverage of a wide range of beneficiaries, including those who may not strictly qualify on the basis of poverty, especially in countries where the large majority are in any case very poor.

Asia has considerable experience with rural public works and employment programmes, a large majority of which were financed by food aid in the early years and some of which were subsequently financed from domestic reserves. They involve self-targeting; the poor and the unemployed seek employment in such projects, often at wage rates below the prevailing market rates; these projects not only generate income and employment for the poor and enhance their access to food but also build productive infrastructural projects such as roads, irrigation and drainage system etc. which contribute to the long run growth. However, the Asian experience confirms that both these types of poverty-alleviating measures are very demanding in the use of management and administrative resources. In addition, the rural employment or works programme impose two additional requirements. One is the need for decentralized local governments in the rural areas which are able to undertake such public works projects with the active participation of rural people in their identification, design and implementation.

Second is the problem of financing the local resource component of such projects; for example, if the food component is derived from the central government, the non-food components, i.e., materials and labour may partly be raised locally. The possibility of such a mobilization of local resources is greater, the greater is the perception by the local population that locally financed projects are useful to them, and the greater is their participation in the decision making process. Equally important is the role of the local institutions in the repair and maintenance of such rural works projects.

Possible Lessons for Africa from the Asian Experience

As already stated, Africa's food production has been suffering from relative stagnation in the last two decades, with increasing food imports meeting their rising demand. Technological progress, in terms of development of high yielding varieties of seeds for Africa's traditional crops, coupled with the extension of irrigation and use of fertilizer, was very slow. By 1983, Africa did not have more than 2 per cent of her arable land under irrigation and fertilizer consumption was no more than 9–10 kg per hectare. The comparable figures for Asia were several times higher. In Africa, the fertilizer that was used on food corps, i.e., wheat, rice and coarse grains was about 34 per cent of the total fertilizer use, whereas 66 per cent of the fertilizer use in Asia was on food crops.

However, in view of the rapid increase in food demand in recent years, caused mainly by increasing population as reinforced by increasing per capita income in at least a few countries, the recognition of the need for accelerated rate of growth in food production is high on the agenda of the African development strategy. As the Asian experience demonstrates, institutions, infrastructure and public policies will determine the pace and pattern of agricultural growth in Africa. There are at least three questions which deserve priority attention. First, how far the growth rate can be accelerated under the available technology, especially: (a) in the relatively well-endowed regions, i.e., regions reasonably well supplied with water either from rainfall or irrigation; and (b) for traditional crop varieties or currently available improved varieties which respond to the use of modern inputs such as fertilizers. An improved system of marketing and distribution of fertilizer, and a rapid pace of extension of the irrigation facilities seem to be essential. Second, appropriate technology needs to be developed for the arid/semi-arid and the less well-endowed regions, suited to diverse agro-ecological characteristics, and for coarse grains, including maize, sorghum and millet as well as for root crops for which technological breakthrough to date has been very limited. This will require not only a strengthening of the agricultural research system but also its close interaction with the extension and training system and the development of infrastructure. Third, simultaneously with the search for technological breakthrough, macroeconomic and sector-specific public policies and the system of incentives facing the farmers, including public expenditures on the agricultural sector, need to be substantially improved.

Africa in general is very different from Asia in terms of agro-ecological features. Also there is a much greater diversity in terms of agro-ecological characteristics between different subregions/regions within a country in Africa than is the case in Asia. However, there are some agro-ecological zones in Africa, which are similar to those in Asia.

About 30 per cent of the land area is either marginally or eminently suitable for cultivation. The land which is suitable for cultivation from the agro-climatic point of view is highly unevenly distributed. Three-quarters of the reserves of unused cultivable land are located in two climatic zones, i.e., hot, humid forest zones of the Central Region and the cereal growing areas of the Southern Region (Vyas and Casley, 1988).[6]

Constraints to Higher Productivity of Agricultural Labour

The key to accelerated agricultural growth in Africa is increased productivity of labour. The relative resource endowments of different regions in terms of land/labour ratio, especially the density of population and labour in relation to arable land or good rainfall or good quality land, are widely different in the different regions of Africa. The evolving labour intensity in different regions of Africa has to be dynamically evaluated in the context of appropriate technology – i.e., relatively labour-using or labour-saving technology – which needs to be developed in Africa. Even in those areas where labour intensity is currently low, the rate of growth of population and labour force is not only currently very high but also the rate of growth is accelerating

In areas where labour/land intensity is low, the problem of labour shortage is aggravated by a very high seasonality of labour input requirements, which is much greater than in Asia. The peak season labour requirements cannot be met by hired labour, because, in view of relatively abundant supply of land, landless or idle labour is not available for wage employment as in Asia; moreover, the opportunity cost of working in a non-farm occupations, specially in the urban areas, is high; the rural–urban real wage differential is much higher in Africa than in Asia, if account is taken of the very high seasonality of labour demand in the rural areas.

Technologies which increase average yield per hectare and marginal product of labour, but at the same time increase labour requirements, are not found profitable in view of high seasonal labour requirements, especially if, at the same time, new technologies also involve large expenditure on cash/purchased inputs, while seasonal or annual variability in the crop yield or outputs remains very high. In view of the highly seasonal nature of labour requirements, and the quality and nature of soil, including the topography of land, the substitution of labour by mechanical equipment is not always feasible or cost effective. In most regions of Africa, the use of draught animals is still very rare.

The above agro-ecological characteristics or peculiarities in several regions of Africa constrain the very rapid spread of the seed/fertilizer/irrigation technology along the lines of Asia without necessary modifications or adjustments. If the problem of peak season labour requirements can be resolved through the use of the draught animal, for example, and if high seasonal nature of agriculture operations can be mitigated, let us say, through multiple cropping, then labour intensive technology of the Asian type would be worthwhile in several regions in Africa. It is clear that the problems of low overall labour productivity, compounded by seasonal bottlenecks, require concerted interdisciplinary effort. How to alleviate peak season labour constraints, and how to increase returns to seasonal labour input to compete with high returns to labour outside of agriculture consistently with increasing yield per hectare, are challenges confronting the agricultural researchers in Africa.

This is not to suggest that there are not regions or areas in Africa where technologies based on fertilizer, irrigation and high yielding seeds are not appropriate. There are such areas and they include high altitude upland areas and low altitude irrigated areas; admittedly they represent a small share of total cropped area in Africa. It should be noted, however, that the currently available appropriate technology in these areas is underutilized and with prospects for substantial progress if appropriate measures are adopted.

Irrigation Potential and Problems in Africa

Admittedly the irrigation potential is more limited in Africa than in Asia. For example, according to an FAO estimate, whereas in Asia 42 per cent of the total potential land area can be irrigated, the corresponding ratio is no more than 3.4 per cent in Africa. This could be an underestimate as considerable uncertainty exists regarding the potential for irrigation in Africa in the absence of extensive hydrological surveys.

The irrigation potential in Africa, limited as it is, has not been fully exploited in a cost-effective manner – much less than in Asia. About 51 per cent of the irrigation potential in Asia has already been exploited, as against only 10 per cent in Africa (FAO, 1987a). There has been relatively greater attention paid in Africa to large-scale gravity flow irrigation projects, which are capital intensive and costly; furthermore, they often strain severely the limited administrative and management capacity in Africa. As Asia learned through experience during the 1950s and 1960s, much greater attention needs to be paid in Africa to small-scale irrigation projects than in the past. Secondly, investments in irrigation projects in Africa have been partly rendered unattractive by including in the cost of irrigation projects such extraneous overhead costs as infrastructure including road, housing and public services; in Asia these overhead costs were kept distinct from or were developed independently and ahead of investment in irrigation projects as they contribute to overall agricultural and rural development. Thirdly, it has often been inadequately appreciated in Africa, according to the specialists in water control, that an assured water supply and efficient management is frequently the more crucial or important requirement rather than a mere increase in the supply of water. Traditional types of micro-irrigation practices, prevalent in Africa, often on small farms, have not received much recognition in the discussions of the future irrigation systems/structures in Africa (Svendsen and Meinzen-Dick, 1989). It has been suggested that there are possibilities of improvement and adaptation of the traditional systems with a view to improving the quality and expanding the quantity of assured water supply for the development of agriculture in Africa (ibid.).

Asian experience indicates that before substantial resources are committed to irrigation projects, especially the large-scale capital intensive projects, a three-pronged approach towards irrigation development in Africa maybe appropriate: first, emphasis in the immediate future on small-scale irrigation projects, which do not have long gestation lags and on investments which would pay off in relatively short run; second, better management and control of water on existing irrigation projects/schemes following a diversity of approaches to irrigation and water control, including traditional schemes and their improvement; and third, properly organized hydrological surveys, to be launched as soon as possible, in different regions/subregions in Africa for the assessment of the surface and ground water potential. The third component is critical for formulating the long-term programme of irrigation development, keeping in view a great diversity of conditions in Africa.

At the same time insofar as the existing irrigation projects are concerned, intensified efforts are needed to reduce the risks associated with: (a) waterlogging and salinity; and (b) waterborne diseases such as malaria, schistosomiasis, etc. in the different irrigated areas of Africa.

Along with an improvement in, and expansion of, the irrigation system is the urgent need to remedy the deficiencies in the system of input delivery and input marketing. For example, the full potential for fertilizer use is far from being realized even in areas of good rainfall, high yield and fertilizer-responsive crops. This requires adequate supplies of fertilizer, principally through imports, as well as maintenance and management of stocks in various geographical regions, reinforced by an improvement in the distribution and marketing facilities coupled with a systematic monitoring of fertilizer requirements on commodity/regional basis.

Two issues have emerged in this connection. First, what is the relative role of public and private sectors in the import, marketing or distribution of fertilizer? Second, is there a need for public subsidy on fertilizer?

Asian experience in this respect indicates that choice of policies would depend upon specific country circumstances. There are a few factors which are peculiar to Africa and require public intervention in the marketing and processing of an input like fertilizer. For example, Africa's dependence on imports is much greater than in Asia; hence management of imports to ensure adequate supply at the right time and at the right place is very crucial. Secondly, in view of small amounts involved, as compared to Asia, at least in the early stages, the domestic market is thin, with very little prospect for the realization of economies of scale. Thirdly, transportation cost is very high. Fourthly, demand is highly variable in response to variability of rainfall. Above all, it needs to be noted that the use of fertilizer in Africa is in its infancy.

Therefore, a rapid dismantling of the role of the public sector agencies either in the marketing and distribution, in the absence of a viable private sector, or in the pricing of fertilizer, in view of high initial costs and risks, may result in a dislocation of the distribution and marketing system or in the expansion of fertilizer use.

Agricultural Research: Improved Priorities and Management

The balance between public expenditures on extension and training on the one hand, and research on the other, needs to be seriously evaluated. To spend a large amount of resources on extension in order to implement a succession of very small technical improvements may be less cost effective than concentrating significantly considerable resources on agricultural research with a view to achieving breakthroughs. What is needed is a large increase in productivity, rather than isolated marginal increments, in view of the low income of African farmers and a large gap between urban and rural wages.

In Asia, breakthroughs in rice and wheat in the irrigated areas were primarily responsible for large growth in production; it is reasonable to expect that acceleration in growth rates in Africa in the foreseeable future will occur in no more than two or three key food staples in the relatively better watered areas. Given the scarcity of the resources and the urgent need for technical breakthrough, a limited number of crops or areas may be selected for high priority attention, in more responsive areas and for crops in which breakthroughs are more likely; rice, maize and sorghum may provide good opportunities for growth through the development of drought-tolerant varieties and improvement of soil and water management, etc. The choice of priority areas/crops depends not only on the biological potential but also on the long-run demand prospects (Mellor, Delgado and Blackie, 1987, especially ch. 28, 'Priorities for Accelerating Food Production Growth in Sub-Saharan Africa').

It is believed that the complexity of African agro-ecological environment, and a relative lack of knowledge about its characteristics such as fragile soils with low nutrients, limited water holding capacity and unique disease problems etc., require that substantial basic scientific-cum-strategic research is undertaken not only on plant breeding and genetics but also, to much larger extent than in Asia, on soil science and plant physiology/pathology. This provides an additional argument for high priority to be accorded, at least in the short run, to a limited set of commodities than were necessary in generally less complex and difficult environments such as in Asia. In the long run, the expansion of trained manpower and institutional capacity will widen the range of choice in research activities.

In recent years agricultural research expenditures in Africa have greatly expanded. A very substantial part of this increase is due to external financial assistance without a corresponding commitment of domestic resources to support external efforts. This also led to a lack of balance between capital expenditures, predominantly externally aided, and current expenditures, mainly from domestic resources. However, increasing concerns have been raised about the quality of output emanating from the agricultural research efforts in Africa. The competent observers of the African agricultural research system conclude that the small size of countries and of research stations, the dispersion of research institutions and high turnover of researchers make it hard to attain a 'crucial' mass of national agricultural research scientists (Lipton, 1988). The lesson of Asian experience is that successful research system depends on: (a) long-term commitment of national resources to selected priority areas; (b) adequate status and incentives accorded to the agricultural research scientists within the hierarchy of professional and administrative services of the country; and (c) limited bureaucratic control with autonomy, flexibility and initiative allowed to the research institutions and their leadership.

Institutions and Infrastructure

Simultaneously with the development of new, appropriate technology for arid/semi-arid zones, the exploitation of available technology, suited to irrigated areas or areas of good rainfall should be expedited. This requires appropriate institutions, educated and trained manpower at all levels, appropriate incentive policies, as well as adequate investment on both capital and current inputs including rural infrastructure.

The debate in Africa over large-scale or state farming versus private farming appears now to be at least partially resolved with possibly less than the degree of conviction and certainty than is generally observed in. Asia; the lingering predilection towards large-scale farms or state farming may be partly due to: (a) the past colonial experience, carried over to the post-independence period, with efficient large-scale estate-farming run by the European settlers, including large-scale plantations managed by expatriates; and (b) socialist penchant in a few countries for large-scale state farms. However, the lessons of Asian experience with regard to smallholder farming appear valid for Africa. This is reinforced by experience in Africa in the food sector and, in a number of cases, in the export sector as well, which confirms the efficiency of the small farmers as well as its responsiveness to incentives and supporting services. More perhaps than in Asia, the role of women in food production as well as in the marketing and distribution of food in Africa is important. In some regions of Africa, an increasing number of women are *de facto* if not *de jure* heads of the households responsible for farming decisions, because of the migration of the male heads of households. However, the institutional arrangements for the supply of inputs and credit are not oriented towards and in fact biased against the women farmers.

The resource requirements for agricultural and rural development in Africa are considerable; they are larger in Africa than in Asia, in view of more sparse populations, poor initial infrastructures, lower level of trained manpower and more difficult soils, as well as requirement for larger variety of inputs for the improvement and management of soils, etc. Public expenditures on agriculture were relatively low in the past even though there has been an increase in recent years. It increased at an annual rate of 6.3 per cent during 1974–83 with development expenditure increasing at 6.5 per cent per year and current expenditures increasing at 4.4 per cent per year. Even though Africa started from a lower base than Asia, even then the rate of increase was lower in Africa than in Asia during this period. The percentage of public expenditures on agriculture (development and current) was 9.2 per cent in Africa (as against 9.6 per cent in Asia) during 1979–83, compared to 7.7 per cent in 1974–79, (as against 10.4 per cent in Asia). However, year to year instability in public expenditures in real terms in Africa was considerably higher than in other regions, including Asia. The average percentage deviation from the trend expenditure real terms was 23 per cent for Africa as against 17 per cent for the rest of the developing world; the percentage deviation from the trend was even higher in the case of development expenditures. Faced with a variation in public resources from year to year, the African countries protected the current expenditures rather than development expenditures (FAO, 1986b, pp. 94–7).

African agriculture has been the recipient of a considerable inflow of external assistance; per capita, official commitments of external assistance to agriculture more than doubled between 1974–76 and 1982–84; furthermore, it was much higher than in Asia, i.e., almost twice as high as in Asia in 1982–84. Average annual commitment of external loans (official and officially guaranteed private credits) during 1980–84 as a percentage of both total merchandise exports (3.7 per cent) and total central government expenditure (2.6 per cent) were the highest in Africa among all the developing regions; the rate disbursement was, however, slower than in the other regions (ibid., pp. 102–5).

This highlights the importance not only of appropriate selection of projects and priorities but also of trained manpower and implementation capacity. In Asia, external assistance in the early days of the Green Revolution played a vital role in: (a) promoting agricultural research systems; (b) training both scientific and administrative manpower; and (c) building up institutions not only for policy making, planning and formulation of projects but also for implementation. These are long-term undertakings as the Asian experience amply demonstrates. Unless commitments of external and domestic resources are made to promote these preconditions of accelerated agricultural development on a sustained and an adequate scale, the pace of agricultural development would suffer in Africa.

Africa needs a vast increase in trained manpower, not only in scientific manpower for agricultural research, training and extension, including the development of training institutions, both national and regional but also social scientists, policy analysts and administrators to undertake agricultural data analysis, policy research, policy advice and implementation. The Asian experience demonstrates that the development of trained manpower took place over a long period.

While the importance of infrastructure in agricultural and rural development is recognized, the relative allocation of investment resources between infrastructure on the one hand, and development of technology on the other, has been a subject of some controversy in Africa. The state of underdevelopment of physical infrastructure such as roads, transportation and communication system in Africa relative to Asia is substantial. The more sparse populations of African countries compared with those in most Asian countries and the general underdeveloped state of physical infrastructure facilities imply a higher marketing and distribution costs and general backwardness of agricultural marketing in Africa. The density of developed road networks is of 0.01 to 0.11 km per square km of land area, compared with 0.30 to 0.40 km per square km in Asia. Moreover, only about 10 per cent of the roads in Africa are paved compared with 35 per cent in Asia. The consequences in terms of high marketing margins in Africa are obvious. The average producers' price in Africa, expressed as a percentage of the terminal market price, is 35 per cent to 60 per cent, compared to 75 per cent to 90 per cent in Asia; similarly, the regional price differentials are higher in Africa than in Asia implying a poor or inadequate integration of markets in Africa. Similar is the extent of seasonal price differentials. Transport and associated marketing costs explain 39 per cent of the differences in the marketing margins between African and Asian countries (Ahmed and Rustagi, 1987, p. 107).

What is debated in Africa is whether investment in infrastructure should precede technological progress, i.e., new, high yielding technology, or should follow it. Investment in infrastructure should receive high priority where there is existing technology to be exploited or where technological breakthrough is in the offing. In areas where potentials for technological breakthrough or growth prospects on the basis of existing technology or agro-ecological characteristics are meagre, large investment in infrastructure in the short run may not deserve very high urgency. In some countries investment in infrastructure facilities has been biased for the regions with large-scale farming, to the detriment of the small farmer. However, for the sake of efficiency in allocating infrastructural resources, small-scale and peasant farmer should be organized or restructured to generate enough demand for the services of infrastructure facilities.

Adequacy of investment in rural infrastructure, however, cannot be judged only in relation to the implementation or facilitation of technological progress in agriculture. Infrastructure plays a wider role than that of merely facilitating or quickening the pace of technological progress in agriculture. It stimulates rural non-farm employment and income generating activities; it contributes to the diversification of the demand pattern and strengthens the interlinkages, between farm and non-farm activities through the consumption and production interrelationships. These are the lessons from the Asian experience.

Incentives and the Role of Public Sector

The adverse impact of macroeconomic policies, including output/input pricing policies, exchange rate overvaluation and industrial protectionism on agriculture has already been mentioned in the context of Asia. The discrimination against agriculture in Africa arising from the macro and sectoral policies had been more severe than in Asia. Also, the direct intervention by and participation of the public sector in the production, marketing and distribution of both input and output was on a larger scale than in Asia. The role of the public sector in the marketing and distribution of agricultural products, especially export crops, was inherited from colonial times. In the post-independence period they were extended to other crops and given additional functions such as input distribution. Furthermore, the dualistic nature of the African agriculture, divided between large-scale commercial/estate farming, on the one hand, and smallholder agriculture, on the other, had implications for the relative importance of public marketing agencies. Large farmers tended to integrate marketing and production vertically. This left a very thin market for the small, peasant sectors that were not large enough for efficient, private marketing intermediaries. The thin market was generally an unstable market.

The Asian experience about the role of both public and private enterprises in these activities has widely varied across countries and over time. The balance was not always correct as evident from recent attempts at privatization in Asia which seek to redress the balance. The crucial considerations in reaching an appropriate balance is the extent of: (a) administrative and managerial capacity at the disposal of the government (which, it is often not appreciated, is the scarcest resource in developing countries, especially in Africa); (b) the prospects of and time lag involved in the development of the private sector and in its ability to take over functions previously entrusted to the public sector; (c) availability of adequate credit to the private marketing agencies; (d) extent of infrastructure development which affects the profitability of private trading activities, etc. It is often found, paradoxically enough, that such social overhead facilities as research, education, extension and development of infrastructure, etc., which can only be provided by the public sector, suffer from inadequate public sector participation, while at the same time the public sector undertakes directly productive activities which are able to attract and can be performed well by the private sector. This occurred in several cases in spite of the severe limitation of administrative capacity and trained manpower in the public sector.

Instead of carefully selecting priority areas, where the public sector has a critical and dominant role to play such as human, social and physical infrastructure, it was thinly spread over a wide range of activities, some of which could be done by the private sector, given appropriate incentives and opportunities; this wide dispersion detracted from efficiency, in some cases, in a significant manner.

At the same time, the very recent experience of rapidly dismantling the public sector enterprises without first taking steps to nurture and develop the private enterprise goes to the other extreme, resulting in several cases in a serious dislocation and decline of services and inputs previously provided by the public sector, albeit not always very efficiently. Africa needs to follow a path of gradual transition to a progressively larger role for private enterprise through selected public intervention. The main requirement is that private sector should be allowed to function freely – assisted by the provision of adequate market intelligence on prices and supplies through the government agencies. Direct public marketing may still be needed in areas with backward infrastructure, low and uncertain agricultural production and marketable surplus and in areas where the maintenance of secure stocks of food grains, for example, is a priority (ibid., p. 115).

The impact of public policies on the relative prices of agricultural commodities, both food crops and export crops, over time and in relation to world prices, has been unfavorable in Africa more so than in Asia. For example, whereas the longer term cereal prices (1967–83) in other regions averaged 15 per cent and export crops averaged 13 per cent above 1969–71, the corresponding prices in Africa were barely above the 1969–71 level. While the international real prices increased by 27 per cent to 29 per cent during 1967–83 above the level of 1969–71 for cereals and export crops respectively, their real domestic prices in developing countries rose by one half of the increase in international prices, excepting in Africa, where there was no increase in real domestic prices of export crops and only 4 per cent increase in cereal prices. In fact, in a number of African countries real export prices fell, whereas the corresponding world prices rose by 43 per cent and world cereal prices rose by 20 per cent. During this period, the bias against agriculture originating from the overvaluation of exchange rate was the highest in Africa, considerably more than in other regions, especially Asia. Six countries in Africa during this period had overvaluation of 20 per cent and above as against one in Asia (FAO, 1987b, pp. 26–8).[7]

In very recent years in several African countries, price incentives have improved in favour of agriculture partly as a result of the structural adjustment process under way. In countries, where price distortions have been very severe, the reduction of distortion has encouraged production of both export and food crops. At the same time, as the Asian experience indicated, these improvements would be short run unless infrastructure, technology and institutions are built up. Also, the availability of technology and infrastructure will greatly strengthen the response of agricultural production to price incentives.

An issue which often has come up in Africa is the relative emphasis on food crops against cash/export crops. It has been a matter of greater concern in Africa than in Asia. As stated earlier, the complementary nature of the cropping pattern developed in Asia on the whole without raising serious concern among analysts and policy makers. In Africa, a much greater preoccupation with this issue may have been partly due to the severity of food crisis in recent years leading to the conclusion that emphasis must shift away from export crops. Over the years, in fact, priority in research and related public expenditures, including macro policies, have shifted in favour of food crops. However, Africa, no more than other regions, cannot ignore comparative cost considerations and prospects of world market for its products in terms of a choice of appropriate cropping patterns.

Food and export/cash crops are frequently complementary in various ways. Investment in infrastructure and in education and training benefits both sectors. Both food crops, to the extent that they save on imports and foreign exchange, and export crops, to the extent that they increase exchange earnings, have favourable impact on the balance of payments. Farmers everywhere faced with risks of crop failures or market uncertainties or imperfections, tend to diversify their cropping pattern, i.e., produce both export/cash and food crops.

Equity in the Agricultural Sector

How far do the present patterns of development and the future trends in African agriculture tend to promote or detract from equity? Unlike in Asia, landlessness in African agriculture is much less of a problem, excepting in a few countries, at least in the foreseeable future. The problem of poverty relates principally to the small farmers/herdsman. Two characteristics of the smallholder agriculture need to be emphasized. First, while under a communal system of landholding, especially within the context of abundant land, there is not much inequality in landholdings, this phenomenon is in the process of change quite substantially in some areas, in view of *de facto* privatization, if not always legally enforced, giving rise to inequalities of landholdings; second, in view of low input technology, labour's share in output is very high and hence the rate of accumulation in agriculture for investment by farmers is low; this can only be accelerated through a much higher level of labour productivity. Third, contrary to the commonly-held view, a large and substantial percentage of the small farmers are net purchasers of food from the market and not self-sufficient in food consumption. This tendency is on the increase. This is the familiar situation in Asia. Fourth, the small farmers in Africa, though varying widely between regions, draw a large part of income from non-farming occupations, specially in view of high seasonality of agricultural operations. The reliance on animal husbandry and non-farm sources of income is a very essential strategy for them to meet the risks of agriculture and inadequacy of farm income for subsistence.

These characteristics have implications for equity in African agriculture. The role of technological progress is critical in alleviating poverty in two ways. First, to the extent that the cost reducing innovations will keep food prices from rising under pressure from increasing population, the poor/small farmers will benefit; second, the development of non-farm activities is crucial to the alleviation of rural poverty. As technological progress proceeds and labour productivity and income go up, demand for non-farm foods and services will also go up; at the same time, the additional labour will be released from agriculture for non-agricultural occupations.

Like in Asia, the impact of new technology in agriculture, given increasing inequality in land holdings and limited access of the small farmers, including the poor women farmers, to input, credits and services will aggravate inequality; the poor and small farmers will absolutely gain in income but proportionately less than the larger farmers. In this respect, the recent innovations in Asia relating to the delivery of credit inputs and services to the small farmers including women may be of relevance to the African policy makers.

The cost of delivery of inputs and services to widely scattered small or poor households/farmers, specially in sparsely populated countries in Africa is high and consequently limit the access of the latter. This puts the farmers in remote regions in a position of disadvantage. Furthermore, as in Asia, the administrative and delivery costs of credit to the small farmers is very high. How to efficiently provide credit and ensure its timely repayment and encourage its utilization for essential consumption and productive purposes on the part of the small farmers, remains a challenge to the policy makers in Africa. The role of institutions or organizations of farmers, be it cooperatives or traditional forms of village/rural associations or groups, in order to receive and utilize credit is very critical.

In view of the wide diversity in agro-ecological potentials and possibilities of technological progress between different regions of Africa, agricultural progress is going to be highly uneven as among different regions. Growth is likely to be faster through a concentration on high potential areas and this would aggravate regional inequalities in Africa that has ethnic and tribal connotations leading to greater social tensions than in Asia. This may require that regional priorities based on efficiency considerations may need to be modified to accommodate political and welfare concerns to a greater degree than in Asia.

Notes

1 The corresponding figures for cereals production were 2.9 per cent, 2.8 per cent and 3.9 per cent and those for total agricultural production were 2.5 per cent, 3. 1 per cent and 4.0 per cent respectively.
2 During all the three periods, per capita cereal production in Sub-Saharan Africa declined consistently. In fact, during 1970s this rate of decline was accelerated, as a result of stagnating or slow rate of growth of production (with 2.0 per cent, 2.1 per cent, and 1.4 per cent rates of growth in cereal production during the three periods) on the one hand, and increasingly accelerating population growth, on the other (FAO, various years; also FAO, 1987b).
3 FAO database at 2000.
4 Private marketing agencies often performed better in output marketing than in the distribution and marketing of new, non-traditional inputs, especially in the early stages of development or spread of new technology. For example, informal markets did not develop as rapidly for modern agricultural inputs as for agricultural outputs. Private sector involvement was concentrated where the turnover was high; hence private sector often catered effectively to the needs of the large commercial farmer, rather than small ones with low demand. In India, for example, appreciable fertilizer use outside the plantation sector did not begin until the government got directly involved in supply and distribution. Again private sector, even after it extended its participation, was not able to displace cooperatives or improve upon performance of the latter or to expand geographically the fertilizer distribution network (Mellor, 1987, pp. 199–202).
5 There were other factors in the rural Asian scene which aggravated inequality, arising from the pressure of expanding population leading to increased subdivision and fragmentation of land holdings; also in many instances, land size through this process became too small, to provide any reasonable return to the tenants/sharecropper/owner so that very small farms were frequently sold out to the medium or the not so marginal farmers.
6 The ecology of the humid central region of Africa is considered under the present state of knowledge more appropriate for roots and tubers than for cereal production. Moreover, substantial investment in soil improvement, infrastructure including transportation and communication facilities is needed for the development of this region. Potentially this area is suitable for the crops which are suitable for export markets.
7 Real domestic price = domestic nominal price index deflated by domestic consumer price index. Real international price = nominal dollar price in international markets deflated by deflator in the USA for consumption goods; exchange rate overvaluation bias = ratio of official over real exchange rate, where real exchange rate is the official rate (1970) adjusted for differential inflation overtime as measured by rate of domestic consumer price index over the US deflator for consumption goods.

References

Abernethy, C.L. and Berthy, D., *An African Strategy for IIMI* (Sri Lanka: IIMI, 1986)
Ahmed, R., *Agricultural Price Policies under Complex Socio-Economic and Natural Constraints: the Case of Bangladesh*, Research Report No. 27 (Washington, DC: IFPRI, 1981).

Ahmed, R. *Fertilizer in Asia: Lessons from Selected Country Experiences*, unpublished report, March (Washington, DC: IFPRI, 1988a).
Ahmed, R., *Employment and Income Growth in Asia: Some Strategic Issues*, unpublished report, August (Washington, DC: IFPRI, 1988b).
Ahmed, R. and N. Rustagi, *Marketing and Price Incentives in African and Asian Countries: A Comparison*, reprint (Washington, DC: IFPRI, 1987).
Bautista, R.M., *Development Strategies. Foreign Trade Regimes and Agricultural Incentives in Asia*, unpublished manuscript (Washington, DC: IFPRI, 1989).
Bautista, R.M., *Regional Survey: Asia n Trade and Macro Economic Policies: Impact on Agriculture in Developing Countries* (Washington, DC: IFPRI, 1990).
Bayliss-Smith, T.P. and Wanmali, S. (eds), *Understanding Green Revolution* (Cambridge: Cambridge University Press, 1984).
Dantwala, M.L. et al. (eds), *Asian Seminar on Rural Development* (New Delhi: Oxford and IBM Publishing Co., 1986).
Desai, G.M., *Development of Fertilizer Markets in South Asia: A Comparative Overview*, unpublished manuscript (Washington, DC: IFPRI, 1988).
FAO, food production year books, various years.
FAO, *African Agriculture: the Next 25 Years* (FAO: Rome, 1986a).
FAO, *The State of Food and Agriculture* (FAO: Rome, 1986b).
FAO, *Agriculture Towards 2000* (FAO: Rome, 1987a).
FAO, *Agricultural Price Policy: Issues and Proposals* (FAO: Rome, 1987b).
Grigg, D., *The World Food Problem* (London: Blackwell 1985).
Hossain, M., *Nature and Impact of the Green Revolution in Bangladesh*, IFPRI Research Report, July 1988.
Jain, H.K., *Role of Research in Transforming Traditional Agriculture: an Emerging Perspective*, ISNAR, Reprint series Amsterdam, 4 October 1988.
Krueger, A.O., Schiff, M. and Valdez, A., 'Agricultural Incentives in Developing Countries: Meaning the Effect of Sectoral and Economy Wide Policies', *The World Bank Economic Review*, Vol. 2–3, 1988.
La-Anyane, S., *Economics of Agricultural Development in Tropical Africa* (New York: John Wiley and Sons, 1985).
Lipton, M., *New Seeds and Poor People* (London: Unwin Hyman, 1989).
Lipton, M., 'The Place of Agricultural Research in the Development of Sub-Saharan Africa', *World Development*, Vol. 16, No. 10, 1988.
Mellor, J.W., Delgado, C. and Blackle, J.M., *Accelerating Food Production in Sub-Saharan Africa* (New York: IFPRI/Johns Hopkins Press, 1987).
National Council of Applied Economic Research, *Household Income and Distribution* (New Delhi, India: National Council of Applied Economic Research, 1980).
Paulino, L. and von Braun, J., *The Food Needs of African Countries and the Potential and Desirability of Meeting Them Through Domestic Production and Intraregional Trade*, unpublished manuscript (Washington, DC: IFPRI, 1989).
Ruttan, V.W., 'The Green Revolution: Seven Generalizations', *International Development Review*, Vol. XIX, No. 4, 1977.

Svendsen, M. and Meinzen-Dick, R., *Garden Irrigation: the Invisible Sector*, unpublished manuscript (Washington, DC: IFPRI, 1989).

Vyas, V.J., 'Asian Agriculture: Achievements and Challenges', *Asian Development Review*, Vol. I, No. 2, 1983a.

Vyas, V.J., 'Growth and Equity in Asian Agriculture: A Synoptic View', in A. Maunder and K. Ohkawa (eds), *Growth and Equity in Agricultural Development* (London: Gower, 1983b).

Vyas, V.J. and Casley, D., *Stimulating Agricultural Growth and Rural Development in Sub-Saharan Africa*, Agriculture and Rural Development Working Paper, World Bank, June 1988.

Chapter 11

Overview to *Population and Food in the Early Twenty-first Century**

The experts, researchers and policy makers, as well as the public in general, seem to alternate between moods of pessimism and optimism, anxiety and complacency, about the world food situation and outlook. Frequently, short-term food shortages or surpluses, or movements in food prices, influence the perception of the future world food situation. During the world food crisis of 1973, marked by a sudden and substantial rise in food prices that resulted in acute distress – and even starvation and famine in several low-income developing countries – there was great apprehension about the future food supply. The year 1973 was seen by many as a harbinger of things to come, signalling the emergence of a long-run world food shortage. During the late 1970s and 1980s, as the food situation improved, food prices fell. Faced with an abundance, the food-exporting developed countries resorted to measures to curtail production or at least to reduce price support and subsidy programmes. Optimism about the future food situation was revived. This assessment was further reinforced by the declining trend in the real prices of wheat and rice over the past several decades.

As the 1980s rolled into the 1990s, concerns arose that the increasing scarcity of land and water resources and environmental degradation, including erosion, pollution and loss of biodiversity, might constrain the expansion of food production in both developed and developing countries. Furthermore, climate change, including global warming, might adversely affect the prospects of production growth. The late 1980s had witnessed a fall in per capita food production in developed countries and a slowdown in the rate of growth of production and yield in developing countries. It was argued that low cereal prices throughout the 1980s were not so much a sign of abundance as of inadequate demand due to a lack of income and employment opportunities.

Optimists countered these concerns by asserting that technological breakthroughs will continue to stimulate production growth, that the impact

* First published in *Population and Food in the Early Twenty-first Century: Meeting Future Food Demand of an Increasing Population* (Washington, DC: International Food Policy Research Institute, 1995).

of global warming is highly uncertain and that the dangers of resource scarcity and environmental degradation are exaggerated and can be mitigated by improved policies, institutions, and technological progress.

As IFPRI was planning its roundtable, 'Population and Food in the Early Twenty-first Century: Meeting Future Food Demand of an Increasing Population', the preparations were under way for the International Conference on Population and Development to be held in late 1994, with a focus on the consequences of increasing population and measures to meet them. The anticipated pressure of increasing population on future food supplies contributed to a new awareness about the need to examine the long-term demand and supply outlook for food. This was an occasion to look beyond the short term to examine whether the developments of the 1980s and early 1990s were transient phenomena or had in them seeds of long term change.

The papers in this volume examine food demand and supply prospects up to 2010 and explore the following questions: what is the projected increase in world population and in the different regions; what is the likely demand for food, especially cereals; will per capita food production and consumption continue to increase; what are the constraints, such as land and water, on increasing food supply overall and in different regions; will the increased food supply be available at constant or rising prices; what are the prospects of technological progress, either intensification of known technology or breakthroughs in conventional plant breeding or biotechnology; what are the implications of global warming for future food supply?

At the roundtable, participants generally agreed that the world food supply in 2010 would probably meet global demand, but that regional problems could occur. South Asia and Sub-Saharan Africa were recognized as the most vulnerable regions. The key to future food supplies was seen as increased productivity, that is, yields must continue to rise; to accomplish this, sustained support for investment in agriculture, including research expenditures, would be needed. The participants also identified areas where knowledge is inadequate. For example, uncertainty continues to surround the possibilities of technological progress as well as the environmental consequences of agricultural intensification. Sources of growth in yield are inadequately understood, as are the effects of changes in income distribution and urbanization on food demand. The need for further analysis and research was obvious.

In this context, an effort is made in this volume to take a step forward in furthering analysis of and insight into the evolving world food situation. The following provides an overview of the major issues raised and conclusions reached.

- For the past several decades, the rate of growth in world food production exceeded the population growth rate. This was true not only for developed and developing countries, as two separate groups, but also for developing regions, except Sub-Saharan Africa. Moreover, the trend over the last three decades shows an increase in average per capita food consumption for all developing regions except Sub-Saharan Africa.
- The annual increase in world population has continually risen and is expected to peak during the decade of the 1990s, remaining high for a couple of decades before declining. Additions to the world population will be around 933 million and 921 million, respectively, during the two decades up to 2010. Nearly all of this will occur in the developing world. The growth rate is expected to decline from 2.1 per cent a year during 1985–90 to 1.5 percent during 2005–2010.
- About 90 per cent of the rate of increase in aggregate food (cereal) demand from now to 2010 will be due to population increase. During this period, per capita growth in income is expected to be around 3.4 per cent. The result of a slowdown in the rate of population growth combined with a modest increase in per capita income will be a slow growth in per capita cereal consumption. For the developing countries as a group, per capita consumption is expected to increase at an annual rate ranging from 2.2 to 2.4 per cent during 1990–2010, that is, from 237 kilograms a year during 1989–91 to around 250–255 kilograms in 2010.
- The main driving force behind projected growth in per capita cereal consumption is growth in the indirect demand for cereals, that is, use of cereals as a feed for livestock. Average per capita direct (human) consumption of cereals for the world as a whole registers a very small or no increase. However, indirect per capita consumption of cereals is projected to increase by about 80 per cent, albeit from a low level of 38 kilograms per capita to 57 kilograms.
- The annual growth rate of food production in developing countries is expected to range from 1.8 to 2.1 per cent against a rate of increase of 2.2–2.4 per cent in domestic consumption. This will lead to an increase in imports from about 90 million tons during 1989–91 to more than double this volume by 2010. A given change in growth rate of production or demand causes a much larger change in growth of net import volumes, since the latter constitutes a relatively small proportion (about 10 per cent) of domestic production. By 2010, both the absolute volume of net imports and the ratio of imports to domestic production are projected to rise in developing countries.

- The projected changes in food supply, demand and prices are very sensitive to the assumptions made about trends in population growth and agricultural productivity. A small increase in population growth rate above the projected rate or a small decline in the yield growth rate below the projection raises the price of cereals in world markets, often significantly, and hence adversely affects the per capita consumption in developing countries. For example, a 25 per cent decline in the growth rate in yield per hectare leads to a rise in the world price by 70 per cent for wheat, 40 per cent for rice, 50 per cent for maize and 58 per cent for other coarse grains; as a result, per capita consumption falls by 4 per cent for all developing countries. A 20 per cent increase in population growth rate causes increased demand and a rise in world price of 30 per cent for wheat, 18 per cent for rice, 11 per cent for maize and 18 per cent for other coarse grains.
- South Asia and Sub-Saharan Africa, with the highest rates of increase in their net cereal imports, are the two regions that face the greatest challenge in meeting food demand. Net imports in Sub-Saharan Africa could increase as much as four times and in South Asia as much as 10 times. In spite of increased cereal imports by 2010, per capita cereal consumption is expected to rise only slightly in these two regions; in Sub-Saharan Africa, it may even decline, according to one projection. The two regions' current per capita consumption is the lowest among developing countries – lower in Sub-Saharan Africa than in South Asia – and will remain so even to 2010.
- The prospects of financing the rising cereal import costs of developing countries depend on an increase in their export earnings or 'import capacity' and in food aid. The export earnings are linked to prospects of economic growth and trade liberalization in developed countries, their principal export markets. The future of food aid depends on the availability of food surpluses in food-exporting developed countries and on the latter's willingness – in the face of competing demands for development assistance – to provide food aid under the new rules of the Uruguay Round regime.
- The growth in production and net exports of developed countries is expected to be adequate to meet the rising imports of developing countries. This will be facilitated, first, by a very small increase in per capita cereal consumption in developed countries, given their already high level of direct and indirect consumption, and, second, by growth in their production, which, though lower than in the past, is nonetheless expected to exceed consumption growth.
- A major change in the net trade position of Eastern Europe and the former Soviet Union is projected. By 2010, their per capita domestic absorption

of cereals will decline because of more efficient use of cereals as animal feed and a reduction in the prevailing high level of post-harvest losses. Moreover, the recent substantial decline in their consumption of cereals and livestock, following the elimination of subsidies, is expected to recover very slowly up to 2010 and only to the average level of 1989–91. As a consequence, the formerly centrally planned economies will no longer be net importers and are likely to generate net export surpluses that will partly offset the increase in cereal imports of developing countries. At the same time, research and investment in developed countries must continue in order to ensure the necessary increase in their production and exports in a manner that meets their rising concerns about the environmental effects of agricultural intensification.

- The two most important factors contributing to future increase in food supply in developing countries are expansion of land under cultivation (including an increase in irrigation) and increase in yield.
- The amount of land currently under cultivation for all agricultural crops is about 750 million hectares. The land area in the same countries, not in agricultural use but with rainfed crop production potential, is more than twice this amount.
- Not all this potential agricultural land will be available for crop production because of increased competition for land for human settlements, industrialization and urbanization. The land with potential for crop production that is available is unevenly distributed among developing countries and regions. Also, bringing land under cultivation can entail high economic and environmental costs, requires infrastructure development and in many cases requires eradication or alleviation of livestock and human diseases.
- Water may become more of a constraint than land. The area available for additional irrigation is estimated to be 50 per cent above the currently irrigated area, and 80 per cent of that will be in developing countries. Increasing competition for water in non-agricultural uses, rising costs of irrigation projects and the likely adverse environmental effects of these projects all combine to limit the rate of expansion of irrigation in developing countries.
- Because of constraints on the expansion of land and water resources, growth in future food supply depends predominantly on growth in yields. At the same time, research and development efforts to expand the quantity and improve the quality of land and water resources deserve high priority no less than yield-enhancing technologies.

- Because of a multiplicity of factors, yield growth in cereal production in developing countries has been declining in recent years. While the rate of growth in yield is on the decline, absolute increases in yield over time have been on the whole positive. The future rate of growth in yield, while generally of lower magnitude than in the past, will not suffer a continuous decline. All projections confirm that absolute yields will increase over time.
- A recent disturbing development relates to the decline in per capita world cereal production from the mid-1980s onward. In the net cereal-exporting countries, which are mostly developed countries, there has been a fall in both total and per capita cereal production. This is most likely a reflection of a demand constraint. Many developed-country exporters resorted to supply management, but food aid flows continued to rise – an unlikely occurrence in the event of supply constraint. In the importing countries, the per capita growth rate in cereal production has declined during this period but has not turned negative. The reasons for this decline, which is not necessarily a reflection of long-term trends, require deeper analysis and watchful monitoring. In earlier decades, there were instances of a pause or decline in growth rate in per capita production (temporarily during the 1960s and 1970s), but recovery soon followed.
- How to achieve a 13–2.0 percent annual rate of growth in yield of major cereals in the next 20 years or so constitutes a major challenge in developing countries. This requires a wide diffusion and improved efficiency of use of existing technology. Adaptive and strategic as well as maintenance research, with a greater emphasis on location-specific research, will play a major role not only in maintaining yields as new regions and farmers adopt new techniques but also in improving potential yields. A constant stream of plant varieties, even with similar yield levels, needs to be generated for the breeding of improved varieties. As food production expands to 'less favourable' or 'low potential' areas with poor soil or limited water availability, the research needs of such areas assume increasing importance, even though returns on research in less favourable areas may be lower, slower to be realized, and more uncertain. Future genetic improvement will rest on conventional plant breeding, including hybrid seeds, and on biotechnology. Prospects for biotechnology to provide a significant breakthrough in yield in the next 10–15 years are limited; its major near-term contribution will be to provide greater resistance to pests and diseases as well as enhanced stability by reducing periodic decline in yields.
- The wide diffusion and efficient use of existing technology and the generation of new technology require a significant upgrading of education,

training, and extension as well as rural infrastructure – a difficult and time-consuming task. It is not certain that required increases in food and agricultural production can be secured without any environmental costs; trade-offs between environmental costs and increased food production in specific regions cannot be entirely ruled out. This emphasizes the critical need for designing appropriate environmentally friendly technology.

- In view of the challenges to expand and exploit the world's food production potential, especially in developing countries, and a long gestation lag of 10–15 years between the initiation of research and the achievement of its results, the priority given to agricultural research and investment needs to be raised, not lowered. The low cereal prices currently prevailing in the world market have generated a sense of complacency leading to a stagnation or even a decline in public-sector agricultural investment, including research expenditures, both national and international.
- Food security in developing countries cannot be viewed with reference solely to cereals or even to all food crops. Nor does it involve the pursuit of national food self-sufficiency, irrespective of comparable cost considerations. The agriculture sector as a whole provides income and employment and hence entitlements or access to food, including access to imports financed by agricultural export earnings.
- The projected food demand and supply balance refers to effective 'market' demand, not to nutrition or food needs based on energy and protein requirements. The projections imply a doubling of per capita income by 2010 and, provided the distribution of income remains unchanged, a doubling of the per capita income of the poor. In fact, in poor countries, no significant increase in per capita consumption is expected. Meeting nutritional needs would require not only adequate food supply but also a significant increase in effective demand or entitlements for the poor.

Chapters 2–10 in this volume are based on papers presented by the chapters' authors at the IFPRI roundtable. These chapters are organized into three parts: Part 1, Global and Regional Population Projections (chapter 2); Part II, Global and Regional Food Demand and Supply Prospects (chapters 3–5); and Part III, Selected Issues Affecting Future Food Supply (chapters 6-10). Chapters 2, 6, and 7 are each followed by a comment by another expert on the issue addressed in the particular chapter. Each of the six comments in Part II address all three chapters in this section. An appendix to Part II that compares the alternative demand and supply projections offered by the authors of chapters 3–5 is also included.

Chapter 12

Asia's Food and Agriculture: Future Perspectives and Policy Influences*

Introduction

Agriculture accounts for about 30 per cent of total income and from 60 to 70 per cent of total employment in many Asian developing countries. Future agricultural progress therefore will play a crucial role in the economic development of these countries.

Agricultural growth can also have a major positive influence on poverty alleviation, because the majority of the poor live in rural areas and are mostly engaged in agriculture and agriculture-related activities.

In most Asian developing countries, agricultural growth in the past two decades was, to a significant extent, led by food production. Much emphasis was placed on achieving self-sufficiency in food staples with a view to promoting national food security. In these countries, per capita food availability during the 1970s and 1980s rose, more so in Southeast and East Asia than in South Asia. In 1990, based on FAO data, average per capita daily food availability reached 2,500 kilocalories for Asia as a whole (2,550 in Southeast Asia and 2,297 for South Asia). Despite growing per capita income and food availability, the incidence of undernutrition continued to be high in some parts of Asia. The average percentage of malnourished children in 1990 varied from 59 per cent in South Asia to 24 per cent in South Asia to 22 per cent in China and much less in the rest of East Asia. The objectives of improving food availability and reducing undernutrition are high in the policy agenda of Asian developing countries.

This chapter presents first a perspective on Asia's food and agriculture, based on a recent assessment of likely future developments to the year 2000 in the world economy by the International Food Policy Research Institute

* Co-author R.M. Bautista. First published in Johnston, A.E. and Syers, J.K. (eds), *Nutrient Management for Sustainable Crop Production in Asia* (UK: CAB International, 1998), papers presented at the International Conference on Nutrient Management for Sustainable Food Production in Asia held in Indonesia, 1996.

(IFPRI). It then discusses two of IFPRI's alternative global scenarios, in which: (i) public agricultural investment and social expenditure are significantly increased; and (ii) agricultural markets worldwide are fully liberalized; focusing on their impact on per capita food availability and nutrition status in Asia. The implications of these future perspectives for formulating national policy in low-income, heavily agriculture-dependent Asian countries are outlined in part using the 'lessons of experience' from existing studies. Finally there are some concluding comments.

A Future Perspective of Asia's Food and Agriculture

Under the 2020 Vision project, IFPRI (1995) has undertaken a detailed analysis of the future prospects of food security in Asia in the context of a global, agriculture-focused model designed to project world food demand and supply balances in the year 2020.

The IFPRI model also makes projections of the number of malnourished preschool children in developing countries, using estimated relationships that explain the percentage of malnourished children by country in terms of per capita calorie availability (determined within the model) and growth in social expenditures, female education and access to clean water. The last three growth variables are projected on the basis of recent trends.

A major aspect of the IFPRI model is the endogenous determination of cereal output by the product of area and yield. Growth of cropping area is a function of changes in the relative price of output, taking account of the prices of competing crops, as well as in non-price factors such as irrigation and technology. Similarly, yield growth is not only a function of changes in output price relative to the prices of inputs, but also of changes in non-price factors such as public investment in agricultural research, irrigation, infrastructure, marketing and distribution, extension and training, etc.

Among the baseline assumptions of the model, Asia's population is expected to increase from about 2.76 billion in 1990 to 4.17 billion by 2020. South Asia will have a population of 1.90 billion by 2020, compared with 1.11 billion in 1990, an increase of 1.8 per cent per year, and China will have 1.48 billion as compared with 1.13 billion in 1990. Both India and China will grow at a slower rate in the next decades relative to the past

Along with population growth, which is projected to contribute more than three-quarters of the increase in future food demand, there will be an increase in per capita income. The highest growth rate is projected for East Asia, including China, at about 6 per cent per year. Per capita income in South Asia is projected to grow at a slower rate of 5 per cent but much depends on India's growth performance in the coming decades, assumed to be 5.5 per cent annually. At the same time, accelerated urbanization in the Asian countries will have consequences for the pattern of food demand. It will result in a dietary transition from coarse grain and root crops to cereals such as rice and wheat, as well as to livestock, horticultural and processed foods.

As shown in Table 12.1, growth in Asia's food demand is projected in the baseline scenario to slow down to an average 1.88 per cent per annum between 1990 and 2020, ranging from 2.34 per cent in South Asia to 1.56 per cent in East Asia. These projections take account, on the one hand, of the slowdown in population growth, the dominant source of growth in food demand and the already high average per capita cereal consumption in most Asian countries, and on the other hand, the low income-elasticity of demand for cereals.

Table 12.1 Baseline scenario: food prospects up to year 2020 supply and demand balance for cereals

	Net trade[1] (Mt)		Growth (1990–2020) (%)	
	1990	2020	Demand	Production
South Asia	-1.17	-22.71	2.34	2.17
Southeast Asia	-0.57	-5.86	2.01	1.89
East Asia (excluding Japan)	-25.10	-49.96	1.56	1.50
Asia (excluding Japan)	-26.84	-78.53 [2]	1.88	1.79

Notes

1 - for imports; + for exports.
2 Deficit for Asia, excluding Japan, increases by 250 per cent.

The IFPRI model works out the food supply prospects under certain assumptions concerning the determinants of growth in area and yield. The projected growth in yield is derived from the analysis of recent past trends, examination of future yield potential for crops and livestock, and in-depth assessment of the future trends in the sources of growth, including public investments in research, extension, irrigation and infrastructure. It is assumed that public investment in these sectors will continue at levels prevailing in the late 1980s and early 1990s.

The resultant growth in production projected until 2020, as shown in Table 12.1, is 1.79 per cent per annum in Asia, ranging from 2.17 per cent in South Asia to 1.50 per cent in East Asia. This has to be put in the context of the projected rates of growth in production and demand in developed countries which at 0.97 per cent year^{-1} and 0.75 per cent year^{-1} respectively, will lead to a 23-fold increase in net surplus, excess of supply over demand, by 2020.

The IFPRI model projects the declining past trend in world food prices to continue through 2020. The rise in food surpluses in food-exporting countries is projected to match the increase in deficit in food-importing countries at real prices that are lower than those prevailing in recent years. Real cereal prices are projected to fall by 11 per cent below the 1990 level, including a fall in price of 23 per cent for maize and rice and 15 per cent for wheat. The overall global supply and demand balances, however, conceal considerable differences among regions and among countries.

The rate of growth in demand for cereals in developing countries as a whole is projected at 1.88 per cent year^{-1}, which exceeds the 1.79 per cent increase in supply (Table 12.1). This will result in an increase in net cereal imports by more than a factor of 2.5, i.e., 78.53 Mt by 2020. The largest increase in cereal imports will be in South Asia, from about 1.2 to 22.7 Mt. However, more than half of Asia's cereal imports, 50 Mt, will be by East Asia, because of its relatively high level of income and high income-elasticities of demand for livestock products which greatly boost the indirect demand for cereals (Table 12.2). Among the South Asian countries, the largest increase in cereal imports is projected in Pakistan, i.e., 16 Mt out of a total increase of 23 Mt for the entire region; India emerges as a small net exporter, about 2 Mt.

For Asia as a whole, in terms of cereal composition, increases in demand for and production of wheat and maize will be larger than that of rice and other cereals. An increase is projected in the net export surplus of rice, whereas imports of wheat and maize will increase by more than twice and five times, respectively. The indirect demand for cereals for livestock production more than doubles for the whole of Asia. The largest absolute volume of per capita indirect demand is in East Asia, the lowest in South Asia.

IFPRI's baseline projections for 2020 of per capita food availability and undernourishment are given in Table 12.3. There is a 20 per cent increase in per capita food availability for the whole of Asia, with the largest increase, 35 per cent, in China and smallest increase, 11 per cent, in South Asia. South Asia had the lowest per capita food availability in 1990 and by 2020 this would still be so despite an improvement in the relative position vis-à-vis the other Asian countries. South Asia would continue to have the highest

Table 12.2 Total demand for food and feed cereals, kg per capita annually

	1990			2020		
	Total	Food	Feed	Total	Food	Feed
South Asia	186.7	171.8	3.1	219.8	201.1	5.3
Southeast Asia	203.5	173.5	12.5	237.1	188.9	25.9
East Asia (excluding Japan)	304.1	217.7	24.1	367.2	230.3	52.3
Asia (excluding Japan)	240.5	.192.1	13.7	278.7	210.2	26.6

Note: Southeast Asia consists of Indonesia, Thailand, Malaysia, the Philippines, Singapore, Hong Kong and Myanmar. East Asia consists of South Korea, North Korea, Taiwan and China. South Asia consists of India, Pakistan, Nepal, Bangladesh, Bhutan, Sri Lanka and Maldives.

Table 12.3 Per capita food availability, kilocalories per day, and percentage of malnourished children, in parentheses, 1990 and 2020, from various scenarios

	1990	2020 Baseline	2020 High investment and rapid growth	2020 Trade liberalization
China	2667	3408	3616	3420
	(21.80)	(13.78)	(7.18)	(13.65)
South Asia	2297	2640	2831	2711
	(58.50)	(41.37)	(31.77)	(38.97)
Southeast Asia	2555	2840	2953	2853
	(23.97)	(16.58)	(10.61)	(16.32)
Asia (excl. Japan)	2500	3034	3225	3083

Sources: Rosegrant et al. (1995) and Agcaoili-Sombilla and Rosegrant (1995).

percentage of undernourished children, though it is projected to decline from about 59 per cent in 1990 to 42 per cent in 2020. The average percentage of undernourished children for the whole of Asia would decline to about 17 per cent, with China, 14 per cent, having least.

Alternative Scenarios

The baseline results are based upon the best assessment of future developments in the world food situation. The results are, however, sensitive to rates of growth in agricultural production that result from different assumptions about the level of investment and growth in income as well as changes in policies affecting input and output prices that may result from trade liberalization, for example. The consequences for world food supply and demand as well as the state of undernutrition based upon two alternative scenarios are described below.

The first scenario examines the consequences of a higher level of agricultural investment and income growth. It incorporates the following assumptions:

- increased agricultural investment, particularly national and international agricultural research, resulting in a 6 per cent increase above the baseline scenario in crop yields by 2020;
- non-agricultural income growth in developing countries will increase by 25 per cent;
- the proportion of public expenditures in health, education, and other social sectors; the percentage of the population with access to safe drinking water; and the proportion of females enrolled in secondary schools will rise by 20 per cent in 2020 above the baseline scenario.

As a result of high investment and income growth (Table 12.3), per capita food availability in 2020 increases to between 2,831 and 3,616 kilocalories day^{-1} compared with between 2,640 and 3,408 in the baseline scenario (Table 12.3). In addition, the percentage of malnourished children decreases to between 7.18 and 31.77 per cent compared with between 13.78 and 41.37 per cent in the baseline scenario; this decline of almost a half is a significant improvement in the nutritional situation. In India, the number of malnourished children is projected to decrease by almost a half and in South Asia as a whole by 42 per cent from the 1990 level.

Food demand in almost all countries increases above the baseline level; the level of production also increases (Table 12.4). Annual cereal production in the developing world goes up by 95 Mt above the baseline scenario. Significant improvement takes place in per capita demand for food, in particular, high value products such as meat. For Asia as a whole, excluding Japan, there is a slight decline in net cereal imports (cf. Table 12.1, -78.53 Mt

Table 12.4 Impact of high investment and income growth on cereals (Mt) in the year 2020

	Production	Demand	Net trade
South Asia	419.79	443.45	-23.66
Southeast Asia	163.36	170.89	-7.53
East Asia (excl. Japan)	567.17	613.42	-46.25
Asia (excl. Japan)	1150.31	1227.75	-77.44
India	327.66	323.12	4.54
Pakistan	38.57	55.17	-16.60
Bangladesh	41.59	48.05	-6.46
Other South Asian countries	11.97	17.11	-5.14

Source: Agcaoili-Sombilia and Rosegrant (1995).

and Table 12.4, -77.44 Mt). There is a decline in imports in East Asia, excluding Japan, and a rise in imports in Southeast Asia and South Asia (compare data in Tables 12.1 and 12.4), proportionately less in South Asia than in Southeast Asia. Within South Asia, imports by all countries, except India, will increase, a large proportion of the increase being met by increased exports from India. Compared with the baseline scenario, the price of wheat goes up and that of rice goes down as demand shifts from rice to wheat; prices of maize and other coarse grains will remain the same.

The second, alternative scenario is based around trade liberalization. It considers the impact of price policy reforms insofar as they affect appropriate choice of inputs and optimum combination of outputs, i.e., allocative efficiency. This scenario is based on the elimination of public interventions in markets, i.e., both domestic price support and subsidy programmes, import restrictions, and export subsidies that drive a wedge between domestic and world prices of cereals. This excludes the consideration of public interventions that distort domestic prices and incentive structures such as subsidies on credit, extension and education, infrastructure and marketing and irrigation. The trade liberalization scenario is based on the assumption of a worldwide liberalization, i.e., it embraces all countries of the world, including Asia. Accordingly, this scenario has implications for production, consumption and trade, i.e., exports and imports, for all countries. While the currently concluded Uruguay Round has reduced protectionism in agriculture only to a very limited extent, the IFPRI model assumes total liberalization, i.e., elimination of all restrictions on agriculture but does not include liberalization in non-agricultural sectors.

As a result of trade liberalization, there is a reallocation of production and trade among individual countries and regions in the world. Also, real prices of most cereals will be a little higher than in the baseline scenario. This is the result of the combined effects of supply and demand, changes resulting from elimination of subsidies in countries that protect agriculture and removal of taxes in countries that tax agriculture. The effects of trade liberalization compared with the baseline and the high investment, rapid growth scenarios on the per capita food availability and percentage of malnourished children are shown in Table 12.3. The effect of trade liberalization compared with the baseline scenario is very small.

However, trade liberalization has a major effect on production and demand (Table 12.5). South Asia as a whole will reduce import demand significantly, i.e., from 23 Mt in the baseline scenario (Table 12.1) to about 8 Mt under complete trade liberalization (Table 12.5). There will not be much change in Southeast and East Asian countries taken together but there will be reallocation of production between individual countries (Table 12.4). More efficient producers of rice, such as Thailand, Vietnam, Laos and Cambodia, will gain, whereas Indonesia, the Philippines and Malaysia will restrict inefficient production and would rely on cheaper imports from abroad.

The above projections do not incorporate all the policy variables affecting incentive structure; for example, exchange rate policies that discriminate against agriculture. Moreover, the gains recorded in the projections are static gains, rather than dynamic gains, which result from improved incentives and increased competition in domestic and world markets, and consequential improvement in the efficiency of investment over time.

Policy Influences

The benefits to food security and nutrition status arising from increased public investment in agriculture and trade liberalization as quantified in the IFPRI projections are based on a multilateral implementation of the policy measure. However, similar benefits, albeit at lower levels, will also accrue to developing countries that undertake these policy improvements unilaterally, i.e., even if other countries do not follow suit.

As indicated earlier, the poor are the most vulnerable to food insecurity and undernutrition. In order to increase per capita food availability and food access among the poor, governments need to promote economic growth that is sustainable and equitable. For predominantly agricultural economies, broadly

Table 12.5 Impact of trade liberalization on cereals (Mt) in the year 2020

	Production	Demand	Net trade
South Asia	409.76	417.50	-7.73
	(394.02)	(416.73)	(-22.71)
Southeast Asia	157.85	163.22	-5.37
	(157.03)	(162.89)	(-5.86)
East Asia (excl. Japan)	535.95	583.55	-47.61
	(531.22)	(581.18)	(-49.96)
Asia (excl. Japan)	1104.00	1164.71	-60.71
	(1082.27)	(1160.80)	(-78.53)
India	320.61	297.27	+23.34
	(306.57)	(304.32)	(+2.25)
Pakistan	36.34	52.32	-15.98
	(36.29)	(52.20)	(-15.91)
Bangladesh	41.23	52.38	-11.15
	(39.77)	(44.44)	(-4.67)
Other South Asian countries	11.58	15.52	-3.95
	(11.39)	(15.77)	(-4.38)

Note: The figures in parentheses represent baseline projections.

based agricultural growth is of paramount importance (see, for example, Mellor, 1995). What do past experiences of Asian developing countries indicate concerning the impact of public investment and trade policy?

There is wide agreement in the empirical literature (see, for example, Vyas, 1983) that in virtually all Asian countries with high rates of agricultural growth in the 1960s and 1970s, the national government invested heavily in irrigation, research, extension, communications, rural roads, marketing and other infrastructure facilities. However, there were some differences across countries in the distribution of benefits from these public investments. To illustrate, both the Philippines and Taiwan recorded high agricultural growth rates for much of those two decades. Agricultural income growth in the Philippines, however, was heavily concentrated among large-farm households, arising in part from the limited access of small and upland farms to infrastructure investments (Bautista, 1995). The effect on the pattern and growth of rural consumption demand was to favour capital-intensive products of urban industry and imported goods rather than locally produced goods and services. This served to weaken the demand stimulus generated by the country's rapid agricultural growth to the growth of rural non-agricultural production and to the overall growth of the national economy. Moreover, there was no discernible impact on poverty and income inequality.

On the other hand, earlier, Taiwan had implemented an effective land reform and the income gains from agricultural growth were widely shared. It contributed to the rural-based, small-scale and labour-intensive character of early industrial development in Taiwan (Ranis and Stewart, 1987). Over time, with the accumulation of human and physical capital, the production structure and composition of exports shifted towards more skill- and capital-intensive products. Also, a remarkable aspect of Taiwan's development was the continuous decline in income inequality and poverty (Kuo, 1983).

The policy inference that can be drawn from the foregoing discussion is that public investment in agriculture can lead to rapid agricultural growth. However, increased participation of the poor in that growth is needed to promote the twin objectives of improving food security and reducing undernutrition.

With respect to trade policy, it may be recalled that the IFPRI trade liberalization scenario involves only agricultural commodities. The continuing heavy protection of domestic industry presents the more potent source of economic inefficiency in most low-income Asian countries. In fact, there has been a recent trend towards the lowering of industrial trade barriers among South Asian countries. Competition with foreign suppliers can be expected to induce cost and quality consciousness among domestic producers and to enable the more productive industries and enterprises to grow more rapidly.

Trade liberalization also has the effect of reducing the price bias against tradable goods caused by an overvalued exchange rate. Agricultural commodities in the Asian context are highly tradable, and their production has been penalized more heavily by distortionary trade and macroeconomic policies that lead to real exchange rate overvaluation than by sector-specific intervention policies (Krueger et al., 1992; Bautista and Valdes, 1993). Any movement towards a less restrictive trade regime will improve agricultural incentives and performance. It will also encourage export growth among internationally competitive domestic producers.

Would trade policy reform improve the effectiveness of agricultural growth in promoting overall income growth and equity? Based on counterfactual simulations using a multi-sector, computable general equilibrium (CGE) model of the Philippine economy, trade liberalization during the Green Revolution period (1965–80) would have favoured rural over urban households; small-farm households are the biggest gainers, while Metro Manila households are the biggest losers (Bautista and Robinson, 1996). Such positive redistributive impact is larger, the greater the extent of trade liberalization. The effect on gross domestic product (GDP) is also significantly positive, reflecting the strong macro-linkages of broadly-based rural income growth.

It must be emphasized that the chief beneficiaries of trade liberalization are the liberalizing countries themselves. The South Asian countries, China and Vietnam have undertaken important policy reforms in recent years towards greater openness in their trade regime with some visible signs of success.

Concluding Remarks

A large part of Asia is still heavily dependent on agriculture, including China and most countries in South and Southeast Asia. Although rapid economic growth has taken place recently, the rural population in many Asian developing countries continues to suffer from high levels of poverty, food insecurity and undernutrition.

This paper has argued that 'getting agriculture moving' can promote sustainable economic growth and poverty alleviation in these countries. Broadly based rural income growth is critically important, and can be facilitated by the removal of existing policy biases, particularly against small farmers and other rural producers, in terms of price incentives and access to infrastructure. Trade liberalization and expanded public investment in agriculture will help, whether they are done unilaterally by developing country governments or in concert with other countries.

Future perspectives on Asia's food and agriculture, discussed in this chapter in terms of various scenarios for the year 2020, indicate a relatively favourable global supply and demand situation if governments and the international development community continue to support the growth of agricultural production and trade at the recent levels. Two alternative scenarios show that greater multilateral benefits would result from a greater commitment to agricultural public investments and trade liberalization. However, the improvements in food security and nutrition status are also found in the IFPRI projections to be relatively small for the world's low-income regions, including South Asia. A clear implication is that the governments of individual countries in these regions need to exert even greater efforts in promoting agricultural growth and increasing the participation of the poor in that growth.

References

Agcaoili-Sombilla, M. and Rosegrant, M.W., *South Asia and the Global Food Situation: Challenges for strengthening food security*, unpublished manuscript (Washington, DC: International Food Policy Research Institute, 1995).

Bautista, R.M., 'Rapid Agricultural Growth is Not Enough: The Philippines, 1965–1980', in J.W. Mellor (ed.), *Agriculture on the Road to Industrialization* (Baltimore, MD: Johns Hopkins University Press, 1995), pp. 113–49.

Bautista, R.M. and Robinson S., 'Income and Equity Effects of Crop Productivity Growth under Alternative Foreign Trade Regimes: a CGE analysis for the Philippines', *Proceedings of the Second Conference of the Asian Society of Agricultural Economists*, Bali, 1996.

Bautista, R.M. and Valdes, A., *The Bias against Agriculture: Trade and macroeconomic policies in developing countries* (San Francisco, CA: International Center for Economic Growth and the International Food Policy Research Institute, ICS Press, 1993).

IFPRI, *A 2020 Vision For Food, Agriculture, and the Environment: The vision, challenge, and recommended action* (Washington, DC: International Food Policy Research Institute, 1995).

Krueger, A.O., Schiff, M. and Valdes, A., *The Political Economy of Agricultural Pricing Policy* (Baltimore, MD: The World Bank/Johns Hopkins University Press, 1992).

Kuo, S.W.Y. The Taiwan Economy in Transition (Boulder, CO: Westview Press, 1983).

Mellor, J.W., *Agriculture on the Road to Industrialization* (Baltimore, MD: International Food Policy Research Institute/Johns Hopkins University Press, 1995).

Ranis, G. and Stewart, F., 'Rural Linkages in the Philippines and Taiwan', in F. Stewart (ed.), *Macro-policies for Appropriate Technology in Developing Countries* (Boulder, CO: Westview Press, 1987).

Rosegrant, M.W., Agcaoili-Sombilla, M. and Perez, N.D., 'Global Food Projections to 2020: Implications for Investment', *2020 Vision for Food, Agriculture, and the Environment*, Discussion Paper No. 5 (Washington, DC: International Food Policy Research Institute, 1995).

Vyas, V.S., 'Asian Agriculture: Achievements and challenges', *Asian Development Review*, 1, 27–44, 1983.

Chapter 13

Comments on 'The Issue of Sustainability in Third World Food Production'*

The basic issue is how to increase significantly food and agricultural production in developing countries without degrading environment or depleting natural resources, in response to rising demand for food generated by increasing population and income. In recent decades increases in production were obtained through: (a) increases in yield per hectare as well as per capita, depending upon the relative scarcity of land and labour; and (b) to a limited extent, through the expansion of area under cultivation or increase in cropping intensity. In the future, in view of limited opportunities of expansion of land area or cropping intensity, much greater emphasis than in the past will need to be placed on increasing yield or productivity in order to obtain the required increase in production.

The relevant question, therefore, is whether it is possible to expand greatly food and agricultural production, let us say, at 3 per cent to 5 per cent per annum by increasing yield per hectare without at the same time depleting natural resources or degrading environment, i.e., without causing soil erosion, loss of soil nutrients, contamination and pollution of water, such as water logging or salinity and hazards to or deterioration in health through pollution caused by chemical inputs.

Advances in biological, chemical and mechanical technology in the past led to the expansion of the world food supply at a declining real price. In other words, technological advances more than offset the diminishing return to land. The question has now been raised as to whether it is possible to repeat the past experience in the future. For two reasons doubts have been expressed. One, is that the pace of technological progress slowed down since the 1960s. In fact, no significant breakthrough has occurred since the 1960s in the development of high yielding varieties of seeds for most foodcrops. Second, it is apprehended that the continuation of the high input technology of the

---

* First published in *The Pakistan Development Review*, Vol. 31, No. 4, 1992, pp. 530–34.

past will cause environmental degradation. Agricultural technology available currently or in the foreseeable future is based on highly intensive use of chemical inputs and mechanical equipment; this has seemed to have caused the varying degrees of environmental degradation. A breakthrough in biotechnology which will significantly raise yield without heavy reliance on chemical and mechanical inputs does not seem to be highly probable in the foreseeable future.

Looked at the point of view of technology, therefore, a pessimist may conclude that there is no way that the required increases in food production can be obtained in the foreseeable future without some amount of depletion of natural resources or degradation of environment. This, therefore, places the burden on the optimists as well as on the policy makers to expand considerably the support for agricultural research to develop technology which is both growth stimulating and environment friendly. To advise developing countries to reduce dependence in the short to medium run on high input and growth stimulating technology as some environmentalists tend to do would not meet the challenge facing the developing countries. Given the urgent need of increase in food supply, improving nutrition and alleviating poverty, there is a challenge which remains unmet. Apart from developing environmental friendly technology, what is needed is an improvement in the management of natural resources through the use of existing technology while avoiding its excessive environmental costs. Much depends on institutions and policies – sectoral, national and international. Policies affect the choice of technology, the cropping pattern and the intensity with which inputs are used. Secondly, since the main actors are farmers, herders and foresters, it is necessary to understand and examine the structure of incentives and constraints facing them; the latter determine the way in which and the intensity with which the natural resources are managed.

It is necessary to find appropriate policy and institutions which will modify the incentives governing behaviour of the households in such a way as to minimize the adverse impact on environment. The primary concern of the household, i.e., the attainment of food security and higher income need to be consistent with what is required for environmental safeguard.

Much has been said in the past about the relationship between poverty and environmental degradation. It is a complex relationship; not enough is known about it and more research should be done. There is a two-way relationship between poverty and environmental degradation each aggravating the other. With environmental degradation the productivity of land and water declines and, therefore, output and income is reduced and consequently poverty is accentuated. The poor have urgent needs for the present and discount the future more heavily than the rich. Given the pressing current needs, they tend to mine the soil or over graze the land in order to extract as high an output and income as possible from the use of resources at its disposal; they move to marginal and fragile areas as productivity on the better endowed lands declines and demand for food increases; they have no resources to replenish soil once it is mined or depleted, nor do they have ability to undertake soil and water conservation measures.

With an income increase in income, they require command over resources and technology to undertake conservation measures; the incentive to conserve resources for the future increases as the pressure to discount the future becomes less urgent. It is not clear that under all circumstances increase in income or resources at the disposal of the farmers will necessarily lead to investment in soil conservation.

It depends, for example, on alternative investment opportunities, including the investment opportunities in non-farm activities in rural or urban areas. However, as the productivity of and income from the existing land or better endowed areas increase, the pressure to move to marginal fragile areas will be reduced.

The poverty-environment nexus emphasizes the need for promoting poverty alleviating measures in developing countries; this is also the reason why a labour intensive and employment-generating programme which has beneficial effects on poverty reduction is also a very environment friendly strategy.

Closely linked to the issue of poverty-environment interrelationship is the issue of agricultural intensification in low potential areas vis-à-vis or in relation to high potential areas. Even though high potential areas, i.e., agri-ecologically more favoured regions have a high density of population and frequently have the very large absolute number of the poor, the low potential areas also have their own share of the developing world's poor. In some low potential areas the degree of poverty and the percentage of the poor is higher than in high potential areas.

Secondly, high input technology which is widely prevalent in high potential areas is not always appropriate in fragile and low potential areas. Heavy use of chemical inputs, i.e., fertilizers, pesticides, or a heavy reliance on mechanical equipment may not be suitable for the structure and composition of soil in the low potential areas. They, therefore, often need low input technology. Appropriate yield increasing technology suitable for low potential areas is urgently required. The alternative of out migration from low-potential areas is not without serious sociopolitical constraints; it involves large costs of settlement including development of infrastructure and related facilities. At the same time, an appropriate combination of income generating activities, both farm and non-farm, through a diversification of cropping pattern, livestock, forestry and non-farm activities, etc. needs to be developed.

On the issue of environmental degradation, one needs to distinguish between the depletion or degradation of resources which can be reversed or repaired in the long run and those which cannot be so reversed. If degradation is reversible, for the sake of growth and poverty alleviation degradation may be sustained in the short run with a view to recouping this as resources expand and technologies develop in the future. In this way, conflict between the short- and the long-run objectives can partially be mitigated

Among the policy reforms which are frequently mentioned as conducive to environmental preservation or enhancement are the appropriate pricing of natural resources. Market failures and government policies together or singly often contrive to undervalue natural resources such as forests or foodcrops as well as inputs such as chemical, mechanical, etc. which, if overused, lead to environmental degradation. There are two issues worth considering in this context. One is the question of divergence between market prices and socially optimum prices for natural resources owing to externalities and market failures. This can be corrected by taxes and subsidies. Second, private prices/costs are kept below market prices through subsidies of various kinds or through government intervention or regulation. This relates to the need for the removal of policy distortions used by the government action.

The policy induced distortions and subsidized prices which lead to overuse or excessive use of chemical inputs, for example, can be corrected through elimination or the reduction of subsidies on chemical inputs as has been done recently in many countries. This may lead to a temporary slowdown in the excessive use of such inputs, but would lead to their more efficient use if appropriate extension, training, education is devised and marketing and distribution systems are improved. In this category are policies relating to forestry, which, for example, include the concession rights, logging prices, tax concessions and credits governing logging of tropical forests, etc. These policies have important implications for the rate and pattern of deforestation. Appropriate management of agricultural inputs, including water and chemical inputs, provided at market prices, reflecting opportunity costs, assumes high priority in an environment friendly agricultural strategy. A high level of education and training of farmers is essential for an environmentally sound management of land and water.

To solve the second problem, i.e., the divergence between market prices and socially optimum prices, it is necessary to face the more difficult problem of quantifying the externalities imposed by intensive use of natural resources and internalizing such external costs for those who impose such costs on society. This will require pricing of fertilizer above the market price to include the contaminating effect it has on water and soil; it would include higher pesticide prices to take into account adverse effects on health and food contamination which may result from inappropriate use of pesticides. Similarly, it will include pricing of irrigation water to include the additional cost of designing projects which will prevent waterlogging and salinity.

Technology and policy changes can be fully effective only in an appropriate framework of institutions. These institutions vary between different geographical regions, sociocultural groups as well as agri-ecological zones. One institution which has been talked about a great deal in the context of environmental degradation is that of property rights. A system of secure and long-term property rights is essential for enabling internalization of externalities. It will also enable the resource user to take a long-run view so that they do not over-exploit them at present. The establishment of secure property rights may require changes in land tenure system and in the pattern of landownership.

In this context, the role of rate of discount used in evaluating future costs and returns of development projects is no less relevant. It has been mentioned that within the acceptable range of discount rates the present value of future net returns after 10 years becomes very small; but this does not, however, resolve the problem of dealing with intergeneration equity, i.e., how far the present generation is willing to sacrifice and preserve resources for the future. If costs are incurred today and returns accrue in the future over a period of time, the longer time investment needs are necessarily discouraged. The lending agencies require 10 per cent to 15 per cent rate of return on projects; this is much higher than the long-term real rate of interest of 4 per cent or so. This is done to take into account the unanticipated inflation and other costs associated with project development and implementation. This discourages projects with longer-term time horizon. Whether the rate of interest should be reduced in order to encourage investment in environmental preservation, raises the question as to whether rate of interest can be decided solely on environmental considerations. It leads us to a wider and more complex issue.

Chapter 14

Linkages between Agriculture and the Overall Economy*

The linkages between the agricultural and nonagricultural sectors of an economy are many and varied. They operate through intersectoral movement of production factors, such as labour and capital, as well as goods and services. With increased agricultural productivity and reduced labour requirements per unit of output, labour tends to move out of agriculture, especially when the population is increasing. In most developing countries, the percentage of the population engaged in agriculture, which is often 60 to 70 per cent, exceeds the percentage of gross domestic output that is contributed by agriculture, usually 30 to 40 per cent. Labour productivity is low in agriculture. However, in many developing countries in recent years, the proportion of the population engaged in agriculture has tended to decline, not so much in response to increased agricultural productivity, but rather as a result of the pressure from increased poverty and unemployment in agriculture.

Also, the growth in the agricultural sector generates surpluses which contribute to expanding investment in the nonagricultural sector. Either financial institutions mobilize and transfer savings for investment in the nonagricultural sector, or the public sector mobilizes agricultural surpluses by fiscal and commercial policies to invest in the nonagricultural sector.

Agriculture as a Source of Savings and Foreign Exchange

Because agriculture is important as a source of national income in many developing countries, attempts to mobilize substantial savings for national investment need to rely on the agricultural sector: the lower the level of per capita income, the higher the percentage of national income originating from agriculture. The relative importance of agriculture as a source of employment

---

* First published in Javier, E. and Renborg, U. (eds), *The Challenging Dynamics of Global Architecture* (The Hague: ISNAR, 1989); proceedings of a seminar sponsored by the International Service for National Agricultural Research, the German Foundation for International Development and the Technical Centre for Agricultural and Rural Cooperation.

and income is seen in Table 14.1. Even if agriculture is not an important source of overall national income or employment, it may still be a very significant source of income in a particular region within a country.

Table 14.1 Agriculture's share of GDP, employment and exports, selected years, 1964–84

	Share of agriculture (percent) in:					
	GDP		Employment		Exports[1]	
Country group[2]	**1964–66**	**1982–94**	**1965**	**1980**	**1964–66**	**1982–84**
Low-income countries	42.8	36.3	76.0	72.0	58.6	32.8
Africa	46.9	41.3	84.0	78.0	70.7	68.4
Asia	42.5	35.7	74.0	71.0	54.0	25.9
Middle-income oil exporters	21.8	14.8	62.0	50.0	40.8	13.6
Middle-income oil importers (excluding major exporters or manufacturers)	25.2	18.0	63.0	53.0	54.2	44.8
Middle-income oil exporters (major exporters or manufacturers)	19.3	12.1	50.0	36.0	56.9	20.2
Developing countries	30.2	19.91	66.9	63.2	52.3	22.0
Industrial countries	5.1	3.1	13.7	7.1	21.4	14.1

Notes

1 Includes re-exports.
2 Data for developing countries are based on a sample of 90 countries.

Source: World Bank (1986).

Frequently in a depressed or underdeveloped region of a middle-income or even a high-income country, the majority of the population derive their income from agriculture, and agricultural progress and its stimulating impact on the nonagricultural activities within the region are crucial for regional development. Substantial population migration is seldom a feasible way to solve poverty and unemployment, especially in the short- and medium-term, even though with improved infrastructure and education over time, expanding opportunities in the industrial sector would provide income and employment to the migrating population.

In most developing countries, investment resources are mobilized from the agricultural sector by means of commercial, exchange rate and fiscal policies. In many countries, taxes on agricultural exports are often the major source of government revenue, while an overvalued foreign exchange rate depresses the return on many agricultural exports in domestic currency below world market prices. At the same time, the domestic industries have access to agricultural inputs that are priced below the world market. Also, quantitative restrictions on industrial imports protect manufacturing industries. This generates the high profits in the import-substituting domestic industries, and provides resources for investment in the industrial sector. The heavily protected domestic industries squeeze agriculture in two ways. First, the prices of industrial inputs used by agriculture and of the manufactured goods consumed by farmers rise above world prices and shift the terms of trade against agriculture. Second, higher profits from investment in the industrial sector divert private savings from the agricultural sector to the industrial sector.

In general, the producers of export crops are paid less than world prices (Table 14.2). In some cases, domestic prices are as much as 30 per cent below world prices, as estimated at the official rate of exchange. In recent years, the price of food crops has been raised above the world price in several countries to encourage domestic food production. However, many cases, domestic prices of both food and export crops are lower than world prices, if the latter are estimated in terms of scarcity price or equilibrium rate of foreign exchange. With an overvalued exchange rate, the domestic price equivalent of world prices of internationally traded commodities is depressed below what farmers would have received at an appropriate exchange rate.

The mobilization of resources away from the agricultural sector is inevitable in the early stages of development. The critical factor is whether the resources mobilized in the agricultural sector are invested within that sector to an extent which is commensurate with its needs and the opportunities for profitable investment. Also relevant are the mechanisms through which resources are mobilized from agriculture for investment within and outside agriculture. To extract resources by turning trade terms against agriculture through overvalued exchange rates and quantitative import restrictions is not a very efficient measure. Furthermore, it acts as a disincentive and depresses agricultural growth – the very source of surplus in the early stages of development. Direct taxation of agricultural income and land is a more efficient way to mobilize resources. However, administrative and institutional

Table 14.2 Direct and total nominal protection rates, 1975–79 and 1980–84 (in %)

Country	Product[1]		1975–79		1980–84	
			Direct	Total	Direct	Total
Argentina	Wheat	(X)	-25.1	-41.4	-12.7	-49.4
Brazil	Wheat	(F)	35.2	3.4	- 6.5	-20.2
	Cotton	(X)	13.4	-18.5	2.6	-11.1
Chile	Wheat	(F)	10.8	33.2	9.3	2.0
	Grapes	(X)	1.0	23.4	0.0	-7.3
Colombia	Wheat	(F)	4.8	-19.7	8.9	-25.3
	Coffee	(X)	-7.0	-31.5	-4.9	-39.1
Dominican Republic	Rice	(F)	19.6	2.1	25.7	6.3
	Coffee	(X)	-14.9	-32.4	-32.3	-51.6
Egypt	Wheat	(F)	-18.6	-36.8	-21.0	-34.9
	Cotton	(X)	-36.3	-54.4	-21.8	-35.7
Ghana	Rice	(F)	79.2	13.2	118.4	29.4
	Cocoa	(X)	25.6	-40.4	34.0	-55.0
Ivory Coast	Rice	(F)	7.6	-24.9	15.5	-10.0
	Coffee	(X)	-31.5	-64.1	-25.2	-50.8
Korea	Rice	(F)	90.8	73.1	86.2	73.9
Malaysia	Rice	(F)	37.8	33.5	68.0	58.4
	Rubber	(X)	-25.2	-29.5	-18.3	-27.8
Morocco	Wheat	(F)	-7.4	-19.0	-0.1	-8.0
Pakistan	Wheat	(F)	-12.5	-60.8	-20.6	-55.2
	Cotton	(X)	-12.3	-60.6	-7.3	-41.8
Philippines	Rice	(F)	1.2	-26.0	0.1	-28.2
	Copra	(X)	-10.7	-37.9	-26.0	-54.3
Portugal	Wheat	(F)	14.5	9.2	25.9	13.1
	Tomatoes	(X)	17.1	11.8	17.1	4.2
Sri Lanka	Rice	(F)	17.8	-16.8	10.6	-20.8
	Rubber	(X)	-28.5	-63.1	-31.3	-62.7
Thailand	Rice	(X)	-27.7	-43.1	-14.9	-34.0
Turkey	Wheat	(F)	27.8	-12.5	-3.3	-38.6
	Tobacco	(X)	1.8	-38.4	-27.6	-62.9
Zambia	Corn	(F)	-12.8	-54.3	-8.8	-65.9
	Cotton	(X)	-13.4	-55.0	-4.6	-61.7

Note: X = export crops; F = food.

Source: Krueger et al. (1988).

constraints in many developing countries make them rely heavily on taxes from commodity exports and imports At the same time, this provides an opportunity for the government to channel resources back to agriculture through public expenditures.

Public expenditure devoted to agriculture in many countries is no more than 11 to 15 per cent, which falls far short of what is required in terms of agriculture's importance as a source of income and employment. The agricultural sector provides 60 to 70 per cent of total employment. Both private and public investment in agriculture must be raised beyond current levels in most developing countries if agricultural growth and, consequently, overall growth are to be accelerated.

An important link between the agricultural and the nonagricultural sectors (which is closely related to agriculture as a source of savings) is that in many developing countries agriculture provides the majority of foreign exchange for essential imports. The relative contribution of agriculture to export earnings in developing countries is shown in Table 14.3. Agriculture's contribution to foreign exchange earnings is high even in countries where the percentage of national income derived from agriculture or the proportion of employment provided by agriculture is small, for example in Latin America. As the rate of

Table 14.3 Agricultural exports from developing countries as a percentage of total exports, disaggregated by per capita income

	Per capita income[1]		
	Less than $400	$400–1,600	More than $1,600
Agricultural exports as a percentage of total exports	(Percentage of countries in income categories)		
More than 80	61.1	30.3	12.5
60–80	11.1	39.4	25.0
50–60	11.1	–	25.0
40–50	8.3	6.1	12.5
30–40	2.8	15.2	12.5
Less than 30	5.6	9.1	12.5

Note

1 Not all columns total 100% because of rounding.

Source: Based on UNCTAD (1981).

development accelerates, demand for imported investment goods and intermediate inputs goes up rapidly. Under these circumstances, increasing agricultural exports remains a key source of foreign exchange. Agricultural exports provide the needed foreign exchange component – capital equipment and intermediate inputs – to enable domestic savings to be fully utilized and thus help remove foreign exchange constraints on domestic investment.

Consumption and Production Linkages

In addition, there are intersectoral linkages of two different types: consumption and production. The part of the income generated in the agricultural sector which is spent on nonagricultural goods and services provides the consumption linkage: the higher the level of per capita income in agriculture, the higher the percentage of total expenditures spent on nonfarm goods and services.

At the same time, the nonagricultural sector provides markets for food and agricultural raw materials. The forward production linkages are processing, marketing, distribution and the further fabrication of agricultural goods for use in the nonagricultural sector. The trade and commerce sector in developing countries especially in rural areas, is predominantly engaged in marketing and distributing agricultural commodities.

Nonagricultural inputs provide a backward production linkage because the agricultural sector uses fertilizer, pesticides, irrigation equipment and other mechanical equipment to harvest, plough, weed and transport agricultural commodities.

In countries where agriculture either constitutes a large percentage of national income or provides a major source of employment, the growth linkages generated by agricultural development are likely to be strong because the impact on the total economy through production and consumption linkages is also likely to be strong technical progress, which contributes to increased production and higher incomes in the agricultural sector, stimulates overall growth. The higher the rate of technological progress, the more widespread is the impact on large and small farmers, traditional and cash crops, and arid or irrigated land. This affects not only agricultural growth but also the rest of the economy.

The impact of technological progress on employment in agriculture depends on growth rate and growth pattern in output; the latter is the composition of output and choice of techniques. Technological progress in the form of biological and chemical innovations is usually widely diffused with a time lag affecting both large and small farmers. Usually there are no economies of scale in such innovations, and therefore small farmers benefit from them, provided they have access to credit and to inputs that increase yields.

Mechanical innovations, however, tend to be labour saving. In many developing countries, mechanization has been encouraged by a public policy of keeping labour more expensive and capital cheap, resulting from either a high wage policy or a low interest rate policy, including an overvalued foreign exchange rate. Mechanization, however, does not necessarily have an overall negative impact on employment. The effects of direct negative employment through displacement of labour in some agricultural operations such as ploughing or harvesting may be compensated by the creation of additional employment through a higher cropping intensity, as well as in the production, maintenance and repair of equipment. A larger aggregate output and a higher cropping intensity, such as multiple cropping due to mechanization, increases agricultural labour requirements. Thus, the type and nature of technological progress in agriculture is relevant to the magnitude and intensity of intersectoral linkages. A recent study, for example, indicates that in many Asian countries during the 1970s, a 10 per cent increase in value added in agriculture led to a 3 to 4 per cent increase in employment (Ahmed, 1988).

Technological Progress, Food Price and Overall Economic Growth

One important way in which agricultural progress affects overall economic growth is by reducing food prices. When demand for food is fuelled by either increasing per capita income or population growth, cost-reducing technological progress reduces food prices and offsets any inflationary pressure in the economy.

In an open economy domestic prices equal border prices plus tariffs. In a closed economy, where domestic prices are higher than world prices due to quantitative restrictions on imports, an increase in food production leads to a fall in domestic prices. With price elasticity of food demand in developing countries less than unity, a fall in food prices also reduces the gross income of food producers. But since a fall in the value of output is matched by a fall in unit costs, there is neither net loss nor excess profit in a competitive market. Where there are differences in efficiency or costs among individual products, more efficient farmers earn extra profits. Frequently, however, technological progress does not take place in a vacuum; at the same time, population and income growth lead to an expanded demand, which keeps food prices from falling to the full extent of their reduced costs. In many instances, the government intervenes through a price-support programme to prevent a short-run decline in prices to the full extent of the reduced cost and also to prevent the corresponding fall in the income of food producers. However, to the extent that a country imports food and sells it domestically below the world market price, technological progress, which increases production at lower per unit costs, enables it to reduce food imports and distribute the benefits of cost reduction via lower prices and lower subsidies.

Technological progress does not uniformly extend to all food producers in all regions simultaneously. Those producers who do not enjoy the benefits of cost-reducing technological innovations will be confronted with a lower market price without having the simultaneous advantage of higher productivity and lower cost. They lose from technological progress in which they do not share. If the losers happen to be poor farmers with no secure access to new technology, and the gainers are a few large producers, then rural income inequality and poverty are aggravated. Therefore, the need for wide diffusion of technological progress to maximize it's beneficial impact on the producers can hardly be overemphasized.

On an international level, when many countries enjoy the benefits of cost-reducing technological innovations, the world price is likely to fall, thus benefiting food-importing countries. Historically, technological progress resulted in a downward trend in world cereal prices, which benefits both producers and consumers (Figure 14.1). As unit costs in the exporting countries fell, rising populations and per capita income in both exporting and importing countries partly offset the downward pressure on prices.

The impact of technological innovations on cost and price depends on the extent to which the marketable surpluses in a country increase in relation to demand. Only a portion of the increase in output is marketed. Medium-sized and large farmers have a higher ratio of marketable surplus in relation to output compared to smaller farmers. If an increase in output is concentrated among smaller farmers, they will consume most of it because of their low consumption levels and high demand elasticity. The downward pressure on market prices will be correspondingly less. On the other hand, if the largest part of the increase is concentrated on a few very large producers with large marketable surpluses, without any corresponding increase in income and demand from poorer farmers, then the downward pressure on prices is likely to be in large. If the fall in prices exceeds the fall in costs, the income of the surplus-producing farmers would be adversely affected. Their incentive to

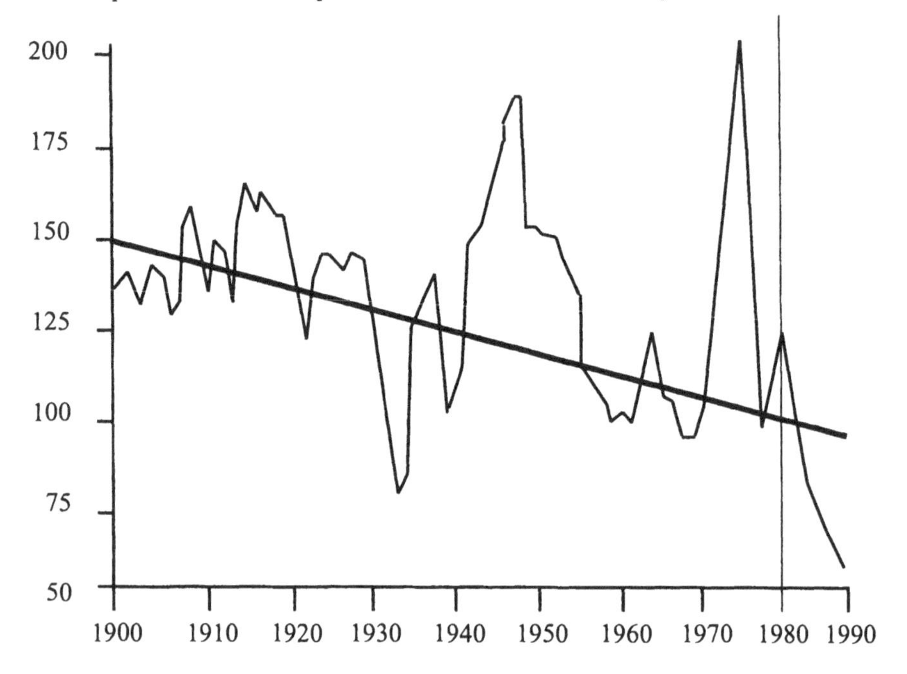

Figure 14.1 Long-run trend in food prices

Source: Tyers and Anderson (1992).

continue increasing food production will suffer unless there is an offsetting increase in demand by consumers.

Low food prices which follow cost-reducing innovations directly improve the real income of the poor, who are the net purchasers of food. Technological progress in food production, which enlarges the supply of the principal wage good – food – at a stable and low price, facilitates the adoption of an employment-based development strategy, especially in the nonfarm sector.

Furthermore, a fall in food prices improves the terms of trade of the industrial sector and lowers the real wage income in terms of the output of the nonfarm sector. This encourages labour-intensive industrialization, as well as a substitution of labour for capital, in the various processes and products in the nonfarm sector. However, the relative decline in the ratio of labour cost to capital cost in the nonfarm sector depends not only on the relative fall in the food prices, but also on a range of macroeconomic policies, which affect the relative prices of labour and capital.

The extent to which reduced food production costs and a relative fall in price improve the terms of trade in the industrial sector partly depends on the extent and the magnitude of marketing, distribution and other transaction costs in the movement of food from rural producers to urban consumers. An increase in these costs would offset the impact of the relative decline in food prices.

Low wages facilitated by cheap food help expand labour-intensive exports, both agricultural and industrial. This is because relatively cheap food strengthens the comparative advantage of labour-intensive activities in the world market.

Agricultural growth stimulates expansion in the nonagricultural sector through consumption linkages, as well as forward and backward linkages in the production process. The consumption linkage is stronger than the production linkages. This is partly because in developing countries the ratio of purchased inputs in agriculture is low, and therefore the expansionary impact on the demand for agricultural inputs is limited. The magnitude of backward linkages increases rapidly as industrial structure becomes more complex and agriculture becomes more modernized and uses more purchased inputs. The relative importance of purchased inputs in gross agricultural output in various developing regions is shown in Table 14.4.

Table 14.4 Ratio of purchased inputs over gross outputs in selected developing countries

Country	Ratio	Country	Ratio
Argentina	26.98	Pakistan	12.22
Benin	8.65	Peru	18.07
Colombia	18.22	Philippines	28.00
Ecuador	17.23	Sri Lanka	17.17
India	24.43	Sudan	27.42
Indonesia	9.66	Tanzania	13.13
Korea	22.06	Thailand	6.38
Mexico	29.41	Turkey	32.48

Source: FAO (1986).

In developing countries in general, the forward linkages from agriculture through processing and distributing agricultural output appear to be far larger than those from the originating side of inputs. The distribution of agricultural products undoubtedly generates the largest nonfarm production links for agriculture. If retailing agricultural produce is approximately proportional to its share in production and in total rural consumers' expenditures, then about

45 per cent of rural retailing can be assumed to be forward distribution links with agriculture (World Bank, 1987, p. 97).

Consumption Linkages and Overall Growth

Increased purchases of nonfarm goods made with increased income in the agricultural sector potentially serve as an important stimulant to the nonfarm sector. The stimulus to growth is not confined necessarily to the rural nonfarm economy, but is also extended to the urban sector, depending on the extent to which the latter is integrated with the rural sector through transportation, communication, marketing and distribution channels. Frequently the nonfarm goods and services on which increased income of rural agricultural households is spent are labour-intensive and produced in rural regions. The output of the nonfarm sector includes not only manufactured goods but also trade and other services. The role of trade and services in the rural areas as a source of employment and income has not been sufficiently emphasized. Much greater attention has been paid to manufacturing in areas such as handicrafts, cottage industries and other small-scale rural industries. The rural services on which increased rural income is spent include, among other things, housing, education, health, transportation and personal services (Hazell and Roell, 1983).

As increased farm productivity raises farmer incomes, not only is a higher proportion of income spent on nonfarm goods and services, but also farm household income begins to diversify. At the level of marginal or very small farmers with landholdings too small to provide a minimum income, nonfarm employment is not only a supplementary source of income, but is often a high proportion of total income. As farm size increases, self-employment or employment of hired labour on the farmer's own land increases. At the same time, agricultural income as a proportion of total income of the farming household increases. However, as farm size increases beyond a certain level, or as the farmer's income exceeds a certain threshold, the share of nonagricultural income in total income rises again. As the very large farmers increase the productivity of their land and labour, they invest the savings generated by their higher income in nonagricultural activities and derive higher nonagricultural income.

Nonfarm income constitutes an important proportion of the total income of rural households in both India and Bangladesh (Tables 14.5 and 14.6). In villages in Bangladesh with well-developed infrastructures, and in India, the

Table 14.5 Nonagricultural income as a percent of total income in rural households by technology and land ownership groups, Bangladesh, 1982

Land ownership group	Technology group[1]	
	Underdeveloped	Developed
Landless and marginal (0.5 ha)	61.9	43.9
Small (03–2.5 ha)	42.5	38.6
Medium (2.5–3.0 ha)	33.3	26.9
Large (5.0 h a)	24.1	30.0

Note

1 In developed villages high-yielding rice varieties have been sown on 80 per cent of the total planted area.

Source: Ahmed and Hossain (1988).

Table 14.6 Composition of income by income ranges in rural India, 1975–76

Income range (Indian rupees)	Total income		
	Agriculture %	Nonagriculture %	Share of wage income %
<3,600	40.1	59.9	45.0
3,601–7,500	58.5	41.5	16.2
7,501–15,000	64.5	35.5	2.1
15,001–30,000	74.5	25.5	0.2
<30,000	40.5	59.5	–

Source: National Council of Applied Economic Research (1980).

percentage of nonfarm income is highest among the smallest farmers or farmers in the lowest income groups – 44 per cent in Bangladesh and 60 per cent in India. With the increase in income among Indian farming families, the share of nonagricultural income in their total income first goes down and then goes up again for those in the highest income bracket (National Council of Applied Economic Research, 1980). In Bangladesh, the change takes place at a slower pace. Although the share of nonagricultural income is higher for the large farmers than for the medium-sized farmers, it is only by a small margin. This may be because large farmers in Bangladesh are not as large nor do they earn as high an income as those in India because the average level of poverty is higher in Bangladesh.

A study in Pakistan found that 40–50 per cent of all marginal farmers had a nonagricultural occupation which substantially added to their incomes or significantly reduced their poverty. Furthermore, the productivity of marginal farmers increased because some members of farming households had nonagricultural income sources. Farmers invested their nonagricultural income in their farms. Thus, nonagricultural income was a source of capital for investment in the agricultural sector to purchase agricultural inputs, as well as for livestock development (Klennert, 1986).

Intersectoral consumption linkages depend partly upon income distribution in rural areas, especially the distribution of the incremental income that accrues in the agricultural sector to different sizes of farming families. It is the medium-sized farmers who tend to have the expenditure patterns that have a greater potential to stimulate demand for mainly labour-intensive, nonfood goods and services. They spend a higher proportion of their incremental income on nonfarm goods and services than do small and very small farmers. Because of their low income, the small farmers spend a much higher proportion of incremental income on food rather than on nonfood items. The direct stimulation effect of their consumption expenditures on the rest of the economy is therefore limited. However, the increased demand for food by poor farmers and by landless labourers stimulates expanded food production, including production by medium-sized and large farmers. The increased production and income of these two medium groups in turn stimulates the demand for nonfarm goods.

In many instances, the medium-sized farmers are numerous and their absolute aggregate expenditure on nonfood items is often as great as, if not greater than, the aggregate expenditures of a larger number of small farmers. Expenditure of a higher percentage of an incremental income on nonfood items by the medium-sized farmers, starting with a large base, strongly stimulates nonfarm sector employment and income

In a number of studies in India, Malaysia, and Nigeria (Tables 14.7– 14.10), it was found that if increased production is concentrated on very rich households, and not on the medium-sized and small farmers, expenditure patterns are most likely to be skewed towards goods and services that are imported or are frequently capital intensive. In Malaysia, 63–66 per cent of the incremental income of the medium-sized farmers, between the fourth and seventh decile in farm size, was spent on nonfarm goods and services in 1972–

Table 14.7 Food expenditure elasticities for low-income families

Country or region	Urban	Rural
Bangladesh	1.06	1.06
Brazil	0.83	0.83
Egypt	0.71	0.68
Indonesia	0.88	0.98
Malaysia		
Muda	–	0.88
Nigeria		
Funtau	–	0.89
Gusau	–	1.04
Sri Lanka	0.72	0.86
Sudan	0.74	0.84
Thailand	0.62	0.65

Source: Alderman and Braun (1984).

73. This went up to 74 per cent for the highest income deciles. The percentage of locally produced nonfarm goods in incremental aggregate expenditure was 38–40 per cent, whereas the expenditure on imports from outside the region was about 26 per cent. It is noteworthy that the average size of middle-income farmers in this context was quite small because the largest farm size in this study was no more than 15 ha. to 20 ha.

In Africa, on the other hand, marginal budget shares – the expenditures out of incremental income that are spent on nonfarm goods and services – were lower than in Asia. Twenty-four per cent of the incremental income was spent on nonfarm goods and services in 1976–77 and 11 per cent on locally produced goods and services. Moreover, the variations between different-sized groups in Africa were not high. This is because of the low absolute level of income in the African example compared with that in Asia. A much higher percentage of incremental income was spent on cereals, and marginal budget shares of incremental income spent on noncereal foods such as livestock and horticultural production were as high as 30 per cent. They were also highly labour intensive or locally produced. The expenditure patterns of different-sized farming households in a sample of developing countries are shown in Table 14.8.

The stimulating effect on the rest of the economy by the consumption expenditures of medium-sized farmers on labour-intensive nonfarm goods and services and, therefore, on the income and employment of the poor, is

Table 14.8 Rural household exper

		A
	Gusau N. Nigeria	*Rura Sierr Leon*
Commodity group		
Food, alcohol and tobacco	80.7	73.7
Clothing and footwear	7.2	7.0
Consumer expendables	4.3	–
Housing	0.3	–
Transport	1.9	2.2
Durables	1.1	–
Education and health	1.1	1.4
Services and social and religious obligations	3.3	4.3
Locational group		
Locally produced		
Foods	75.3	69.0
Nonfoods	8.4	8.8
Regional imports		
Foods	5.4	–
	22.2	21.6
Nonfoods	10.9	–
Nontradables	24.7	–

Table 14.8 (cont'd)

	Gusau N. Nigeria
Commodity group	
Food, alcohol and tobacco	–
Clothing and footwear	–
Consumer expendables	–
Housing	–
Transport	–
Durables	–
Education and health	–
Services and social and religious obligations	–
Locational group	
Locally produced	
Foods	0.93
Nonfoods	–
Regional imports	
Foods	1.07
Nonfoods	–
Nontradables	–

Table 14.9 Marginal budget share

Group	1
Commodity group	
Food, alcohol and tobacco	67.
Cereals and cereal products	21.
Fruits, vegetables, and nuts	9.
Meat and fish	12.
Eggs and dairy products	2.
Clothing and footwear	7.
Consumer expendables	4.
Housing	2.
Transport	2.
Durables	-1.
Education and health	2
Personal services and entertainment	1.
Social obligations	12.

Table 14.9 (cont'd)

Locational group	
Food	
Home-produced	22.47
Locally produced	21.40
Imported	23.53
Nonfood	
Locally produced	17.87
Imported	14.74
Nontradables	23.72
Average farm size (acres)[2]	2.14
Average family size	7.07
Per capita expenditure (M$)	150.00

Notes

1 All household characteristic variables are e

2 Farm area is the operated paddy area.

Source: Hazell and Roell (1983).

Table 14.10 Marginal budget share

Group	1s
Commodity group	
Food, alcohol and tobacco	77.8
Cereals and cereal products	50.9
Fruits, vegetables, and nuts	4.5
Meat and fish	7.2
Eggs and dairy products	7.3
Clothing and footwear	8.8
Consumer expendables	4.2
Housing	0.4
Transport	1.4
Durables	0.7
Education and health	1.4
Personal services and entertainment	1.3
Social obligations	3.5

Table 14.10 (cont'd)

Locational group	
Food	
Home-produced	59.15
Locally produced	14.49
Imported	4.23
Nonfood	
Locally produced	10.95
Imported	11.17
Nontradables	26.96
Average farm size (acres)[2]	8.52
Average family size	12.52
Per capita expenditure (M$)	42.00

Notes

1 All household characteristic variables are ev

2 Farm area is the operated paddy area.

Source: Hazell and Roell (1983).

subject to three sets of leakages. First, larger farmers may have a higher propensity to save. The savings by large farmers, even though they constitute in the first instance a leakage from the consumption linkage, can serve an essential role as a source of investment to expand the productive capacity in the nonfarm sector and thus to increase the output of nonfarm goods and services in response to increased demand. Second, the increased output may lead to a fall in prices and a fall in income due to demand inelasticity. Third, the medium-sized farmers may follow capital-intensive techniques, and therefore may not provide much employment either for the small farmers seeking employment or for the landless labourers.

The demand pull provided by agricultural growth needs to be matched by an elastic supply response from the nonfarm sector. In order for the rural nonfarm sector to respond strongly and positively to the stimulus provided by the increased expenditure of the farm sector, a few preconditions need to be met. Among them are infrastructures such as roads, transportation, communication systems and electricity; rural credit to finance both current and investment costs of nonfarm activities, and education. The role of government policy in providing infrastructure, credit facilities, education, extension and training to those engaged in the nonfarm sector cannot be overemphasized. Infrastructure reduces marketing costs, creates competition in the marketing structure by facilitating easy access and encourages specialization within regions of a country in accordance with cost advantages (Ahmed and Hossain, 1988). Infrastructure development maximizes intersectoral linkages by enabling a decentralization of industrial and other nonfarm activities through a country. The development of market towns and small industrial cities is a very cost-effective way not only to prevent the growth of large industrial concentrations with their high social and economic costs, but also to decentralize industrial activities and bring them nearer to the source of demand and supply of raw materials and food.

An infrastructure that is highly dispersed throughout the country is needed. But this does not imply building roads or providing electricity in areas where either population densities or agroecological circumstances do not warrant profitable investment in either agricultural or nonagricultural activities. Agricultural research and research to design appropriate technology for labour-intensive, nonagricultural activities deserve high priority. The development of entrepreneurship is closely linked with the growth of institutions that are able to mobilize rural saving as well as provide credit to finance a variety of nonfarm activities. Education, both primary and secondary also stimulates nonfarm entrepreneurial activities.

The critical role of infrastructure and rural institutions in strengthening intersectoral linkages is emphasized by the example of Africa, where the incremental share of nonfarm goods and services in country expenditures is low. The poor transportation and communication links between villages and towns have an important impact on the intersectoral linkages. These impede access to nonfarm goods and services and increase their cost relative to food prices. The role of infrastructure in the development of nonfarm sources of income and employment is illustrated with an example from Bangladesh (Table 14.11).

Table 14.11 Percentage increase in average income per household of developed villages over underdeveloped villages by income source

Income source	Increase (%)
Business and industries	20
Business	10
Industries	53
Wage income per capita	88
From agriculture	55
Not from agriculture	108

Source: Ahmed and Hossain (1988).

Attempts have been made to quantify the impact on overall growth of a certain percentage increase in agricultural output. For example, it is estimated that an increase of $1 in agricultural value added would lead to an overall increase in GNP of $18 in Asia and $1.5 in Africa.

In a recent study of 34 developing countries in which agriculture was 20 per cent or more of the GDP for the period 1961–84, it was found that a 10 per cent increase in agricultural value added led to a 13 per cent increase in nonagricultural value added. For the period 1973–84, the increase in nonagricultural value added was 14 per cent. In a number of Asian countries the increase in nonagricultural value added in response to a 10 per cent increase in value added in agriculture varied from 2 per cent in South Korea to 16 per cent in Malaysia (Bautista, 1988).

The variations are due to differences in the relative importance of agriculture in national economies and in the state of infrastructure and other factors that facilitate investment in the nonfarm sector. In a small country that depends predominantly on the export market for its nonfarm sector, in the way South Korea depends on its industrial exports, agricultural growth would be less of a stimulus to the development of the nonfarm sector than would be the case in other countries.

Similarly, in countries that depend on enclaves or exclusive export zones where agricultural exports are concentrated, and where at the same time technical progress is encouraged, the linkages of the export sector with the domestic, industrial or nonfarm sector would be limited. This happened in colonial times, especially when production of a large-scale export crop was capital intensive and often owned by foreign investors, whose income was spent on imported goods and whose savings were transferred abroad.

References

Ahmed, R., *Employment and Income Growth in Asia: Some Strategic Issues* (Washington, DC: IFPRI, 1988).

Ahmed, R. and Hossain, M., *Infrastructure and Development: Agriculture and Rural Economy of Bangladesh* (Washington, DC: IFPRI, 1988).

Alderman, H. and Braun, J.V., *The Effects of Food Price and Income Change on the Acquisition of Food by Low-income Households* (Washington, DC: IFPRI, 1984).

Bautista, R.M., *Agricultural-based Development Strategies: Issues and Policy Framework* (Washington, DC: IFPRI, 1988).

FAO, *Economic Accounts for Agriculture, 1976–1985* (Rome: FAO, 1986).

Grilli, E. and Yang, M.C., 'Primary Commodity Prices, Manufactured Prices, and the Terms of Trade of Developing Countries: What the Long Run Shows', *World Bank Economic Review*, 2(1), p. 20, 1988.

Haggblade, S., Hazell, P. and Brown, J., *Farm/Nonfarm Linkages in Rural Sub-Saharan Africa: Empirical Evidence and Policy Implication*, unpublished discussion paper (Washington, DC: World Bank, 1987).

Hazell, P.R. and Roell, A., *Rural Growth Linkages: Household Expenditure Patterns in Malaysia and Nigeria* (Washington, DC: IFPRI, 1983).

Klennert, K., 'Off-Farm Employment in Marginal Farm Households', *Quarterly Journal of International Agriculture*, 25(1), pp. 37–48, 1986.

Krueger, A.O., Schiff, M and Valdes, A., 'Measuring the Impact of Sector-specific and Economy-wide Policies on Agricultural Incentives in LDCs', *World Bank Economic Review*, 2(3), pp. 255–72.

National Council of Applied Economic Research, *Household Income and its Disposition* (New Delhi: National Council of Applied Economic Research, 1980).

Tyers, R. and Anderson, K., *Disarray in World Food Markets* (Cambridge: Cambridge University Press, 1992).

UNCTAD, *Handbook of International Trade and Development Statistics* (Supplement) (Geneva: UNCTAD, 1981).
World Bank, *World Development* (Washington, DC: World Bank, 1986).

Additional Reading

Cavallo, D. and Muntlak, Y., *Agricultural and Economic Growth in an Open Economy: the Case of Argentina*, Research Report No. 34 (Washington, DC: IFPRI, 1982).
George, P.S., *Public Distribution of Foodgrains in Kerala*, Research Report No. 7 (Washington, DC: IFPRI, 1979).
Johnston, B.F. and Mellor, L.W., 'The Role of Agriculture in Economic Development', *American Economic Review*, 51(4), pp. 566–93, 1961.
Johnston, B.F. and Mellor, J.W., 'The World Food Equation: Inter-relations among Development, Employment, and Food Consumption', *Journal of Economic Literature*, 22, pp. 531–74, 1984.
Lee, T.H., *Intersectoral Capital Flows in the Economic Development of Taiwan, 1895–1960* (Ithaca, New York: Cornell University Press, 1971).
Mellor, J.W., *The Economics of Agricultural Development* (Ithaca, New York: Cornell University Press, 1966).
Mellor, J.W. and Lele, U., 'Growth Linkages of the New Foodgrain Technologies', *Indian Journal of Agricultural Economics*, 28(1), pp. 35–55, 1973.
Mundlak, Y., *Intersectoral Factor Mobility and Agricultural Growth*, Research Report No. 6 (Washington, DC: IFPRI, 1979).
Rangarajan, C., *Agricultural Growth and Industrial Performance in India*, Research Report No. 33 (Washington, DC: IFPRI, 1982).
Wanmali, S., *Service Provision and Rural Development in India: A Study of Miryalguda Taluka*, Research Report No. 37 (Washington, DC: IFPRI, 1983).
Wanmali, S., *Rural Household Use of Services: A Study of Miryalguda Taluka*, India. Research Report No. 48 (Washington, DC: IFPRI, 1985).

PART V

FOOD SECURITY: POLICY REFORMS IN AGRICULTURE

Chapter 15

Government: Issues in Policy Reform*

My comments touch on several of the issues discussed by Odin Knudsen and John Nash: the influence of price and nonprice factors; the relationship between prices, aggregate supply response and growth; input subsidies and government expenditures on agriculture; price stabilization; protecting the poor during adjustment; agricultural protectionism; and the sustainability of reform.

Reforms of agricultural pricing policies, including those affecting the level and stability of prices, liberalization of agricultural trade and reduction in government involvement in the marketing and distribution of agricultural outputs and inputs, have been among the most widely discussed and controversial aspects of agricultural structural adjustment programmes. The analysis of Knudsen and Nash deals more with policy recommendations in the World Bank's agricultural adjustment lending than with the impact of such reforms on performance. Still urgently needed is an empirical analysis of how the reforms were actually implemented and what their impact was on the agricultural sector.

Extensive research has quantified the extent to which economywide policies such as those dealing with trade and the exchange rate affect relative prices not only within agriculture but in other sectors as well. Research has also shown that the effects of these policies dominate those of sector-specific pricing policies implemented through subsidies and taxes or public intervention. For reforms of price policy to be effective, they must be combined with or even preceded by macroeconomic policy reforms, especially in the areas of trade and the exchange rate. High inflation or volatile exchange rates can soon erode or offset reforms of government support prices or agricultural taxes and subsidies.

Although some years ago experts considered 'getting the prices right' to be the most effective way of generating efficient growth, opinion today seems to be shifting toward the belief that both *price and nonprice factors* are important. Experience with structural adjustment programmes may have

* First published in Thomas, V. and A. Chhibber (eds), *Agricultural Policy in Restructuring Economies in Distress: Policy Reforms and the World Bank* (Washington, DC: The World Bank, 1991).

contributed to this pragmatic conclusion. Controversy persists, however, over two issues: 1) the relative importance of price and nonprice factors; and 2) the sequencing of corrective measures on these two fronts.

The relative priority that should be attached to price and nonprice factors depends on the particular circumstances of each country. If price distortions are severe and long-standing, pricing policy reforms may greatly improve efficiency and growth even if nonprice factors are inadequate. It would be a one-shot improvement, however, unless nonprice factors were significantly strengthened. In other cases, the strengthening of nonprice factors may be the dominant consideration, partly because, as Knudsen and Nash point out, the aggregate supply response to price changes is itself a function of *nonprice factors*.

The interrelations between price and nonprice factors can be illustrated in a number of cases. For example, despite favourable coffee and tea prices in Kenya, productivity per hectare increased little among smallholders, primarily because of inadequate access to research and extension services, credit and inputs; in contrast, productivity increased on large estates because of more ample access to these nonprice factors. In Cameroon, despite lower cotton prices than in Kenya, Malawi and Nigeria (both in nominal terms and in purchasing power parity) and despite a fall in the price of cotton relative to the price of maize, yields per hectare were larger than in other countries because Cameroonian farmers had access to improved technology and extension services. Similarly, in Nigeria new technology and organizational efficiency increased cocoa production despite low returns from exports due to an overvalued exchange rate (Lele and Mellor, 1988).

In the view of many developing countries, the recent heavy emphasis of international development agencies on pricing policy reforms, without commensurate attention to nonprice factors, is an excuse for inadequate assistance and inaction on nonprice issues. Everyone recognizes that improving nonprice factors requires substantial resources in both the short and long runs. The supply response is not only higher in the long run than in the short run, but it is also higher in the presence of improved nonprice factors such as infrastructure. Achieving this response requires a long-term commitment from both lenders and borrowers.

Another issue discussed by Knudsen and Nash is the *relationship among prices, aggregate supply response and growth*. As they mention, aggregate supply elasticities in agriculture are rather modest, frequently 0.6 or less. The higher estimates of price responses that are obtained in a general equilibrium framework come only after a period of 10 to 15 years and are based on models of technological change and capital accumulation that include intersectoral capital flows and labour migration. These models seem to suggest that favourable price changes in agriculture are the principal engine of growth. Such price changes set in motion technological progress, which is incorporated in physical and human capital on the basis of farmers' selection of the most appropriate technologies from among a wide set of those readily available. This model, however, tends to ignore or underplay the role of public action in the field of technology, institutions and infrastructure. Empirical verification of these models, including their implicit assumptions, is still limited and requires further research (Mundlak, 1989; Cavallo and Mundlak, 1982; IFPRI, 1988).

In this connection, the controversy regarding the relationship between relative price changes in agriculture and the generation of technology needs some clarification. First, it is generally agreed that price incentives accelerate the diffusion of technological innovations once they are available; what is being debated is the generation of technology in response to price incentives. Second, the relevant policies and organizations differ for basic or fundamental research and applied and adaptive research. Third, no distinction is made between changes in relative prices within agriculture and changes in agricultural prices relative to nonagricultural prices, or between the effects of short- and long-term price changes on the generation of technology. Systematic analysis and clarification of the various aspects of the process of technological innovation and their interrelations are needed. For example, although the emphasis of private research may be affected by relative price changes within agriculture, it is unlikely that aggregate government research expenditures on agriculture or basic science are affected to any significant extent.

In the context of adjustment lending, government expenditures on agriculture, particularly *input subsidies*, have received much attention in recent years. Loans for agricultural adjustment programmes have usually included the condition that input subsidies be reduced or eliminated.

Among input subsidies, the most controversial is the one for fertilizer – the most important ingredient in the Green Revolution. Several arguments have been advanced in favour of a fertilizer subsidy: it compensates for imperfections in the capital market and for the risks associated with adopting a new technology with high cash costs; it is required to stimulate the process of trial and error needed initially to determine the optimum uses of fertilizer; and it is needed temporarily to offset high costs of distribution and transportation in the early stages of fertilizer use, when the volume of sales is low and the unit cost of distribution is high.

The importance of these considerations depends on the circumstances of each country and the effect of fertilizer use on output relative to the prevailing ratio of fertilizer price to output price. Although it is often argued that, however justified, temporary subsidies are inadvisable because they tend to become permanent, several countries have shown that subsidies can be reduced or even eliminated over time. It has also been argued that the resources spent on subsidies would be better spent on agricultural investments with higher returns such as research, extension, infrastructure and so on, and that such expenditures would in turn help increase returns from the use of fertilizer. Experience with structural adjustment programmes in many countries demonstrates, however, that the reduction of subsidies is often accompanied by a reduction rather than an increase in public investment in agriculture, with adverse effects on growth and employment in agriculture.

Economists continue to debate whether the *stabilization of agricultural prices* has a desirable impact on consumer and producer welfare. Some argue that substantial benefits may accrue from price stabilization if macroeconomic and dynamic considerations are taken into account. Price instability adversely affects investment not only by farmers but also by those engaged in marketing, distribution and processing. Moreover, consumers, especially poor consumers, incur high transaction costs every time prices change because they need to reallocate their expenditures in order to fulfil their minimum nutritional requirements (Timmer, 1988; Ahmed, 1989). Furthermore, the very poor – who in many countries constitute a large part of the population and have almost no access to credit – cannot hold back sufficiently on consumption when prices are low to save enough to avoid hunger when prices are high. Even farmers who do not produce enough food for their own needs must sell some of their output immediately after harvest when prices are low to meet urgent cash requirements and must buy more during other times when prices are higher.

Most governments, in both developing and developed countries and whatever their position on the role of markets, intervene to achieve some price stability. Governments are unwilling to stand idle and let people go hungry when world prices have risen, even if their action may worsen the problem at some time in the future; nor do they allow their farmers to go bankrupt and join the urban unemployed if prices collapse in the short run, even if the commodities they produce have limited markets in the long run. Policy makers therefore need a mechanism for addressing price fluctuations. Analytical studies and practical measures are needed so that a cost-effective method for stabilizing prices can be designed.

Prices are affected not only by fluctuations in world prices and exchange rates but also by variations in domestic production due to the weather or the availability of inputs and technological innovations. Knudsen and Nash concentrate on price instability arising from external causes rather than from fluctuations in domestic production. This emphasis naturally leads to a discussion of taxes and subsidies on foreign trade as the principal mechanism for stabilizing domestic prices. Experience, however, is very limited in this respect, and the subject requires further study. In addition, countries differ widely in their capacity to formulate and implement such taxes and subsidies.

The fluctuations in domestic output and prices can be reduced by varying exports and imports, by varying domestic stocks, or by doing both. For countries that are not consistent exporters or importers, an exclusive reliance on trade adjustment mechanisms for stabilizing prices is not necessarily the effective approach. Occasional exporters cannot sell their surplus output in the world market except at a considerable loss, while occasional importers cannot import a small amount at short notice to offset shortfalls in domestic production without incurring delays and high costs. Furthermore, a country's staple food products may not commonly be traded in international markets. These considerations support the case for using domestic stocks, in appropriate combination with variations in exports and imports, wherever feasible as a method for stabilizing domestic prices.

What is needed is to analyze for a given country the trade-offs among: 1) the degree of price stability being sought; 2) the size and cost of stocks; and 3) the degree of reliance on imports. The narrower the price band (the margin between ceiling and floor prices) within which stability is sought, the larger the size of the necessary stocks and the larger the cost of stabilization. For example, a study in Kenya found that holding the producer price of maize between US$129 and US$151 per ton would cost between US$1.0 million and US$1.5 million more annually than would holding prices between US$125 and US$155 per ton (Pinckney, 1989). Even a moderate degree of flexibility in response to changes in world prices or domestic production can considerably reduce the costs of price stabilization. In the Kenyan example, a policy that translates a US$10.00 change in the world price to a US$1.00 change in the domestic price would save about US$350,000 annually on average, while a policy that allows the official price to increase by 6 per cent when production falls by 10 per cent would save about US$600,000 annually (Pinckney, 1989).

A narrow price band that does not meet the costs of private traders for marketing, distribution and processing will discourage private stockholding and so increase the amount of public stocks required. In view of the high cost of stabilizing prices through the management of domestic stocks, such efforts should be confined to a very few vital commodities. It is unlikely, however, that a public stock policy for price stabilization can be carried out without some control over imports and exports. If exports and imports are unregulated while domestic prices are held within a band, the cost of holding stocks will be enormous. The actual importing and exporting can be undertaken by private traders under a system of licensing, and domestic trading in food grains can also be left to private traders, with occasional intervention by the public stockholding agency.

At least in the early stages of the adjustment process the need to *protect* and compensate the poor is increasingly being recognized – for example, when food prices rise after the elimination of subsidies or employment and income fall at least in part because of a reduction in government expenditures. Targeting subsidies to the poor is easier said than done, however. It confronts not only the political opposition of those excluded from the subsidy but also the problems of logistics and inadequate information about the target group. If the very poor constitute a high proportion of households, marketwide subsidies may be more cost-effective than narrowly targeted schemes that are subject to large leakages. A recent study by the International Food Policy Research Institute found, for example, that the total cost (the amount of subsidy plus administrative costs and leakages) of transferring US$1.00 to the target groups in an economically depressed rural area in the Philippines was US$1.19 if all households received subsidies, US$1.60 if only households consuming less than 80 per cent of their calorie requirements were targeted, and US$3.61 if only households with malnourished preschool children were targeted. The major reason for the differences in costs was leakages to nontargeted groups in the latter two cases (IFPRI, 1988).

The discussion by Knudsen and Nash of *agricultural protectionism* in developing countries makes it appear stronger and more pervasive than it is. It is true that agricultural protectionism is on the rise in the middle income developing countries, particularly for import-competing food products. In other developing countries, however, usually only one or two commodities have domestic prices that are higher than world prices at the official exchange rate. The current depression in world prices has contributed in several instances to an improvement in the relative price of import-competing agricultural commodities in domestic markets. In some cases, exchange rate adjustments have led to a rise in domestic prices. This emphasizes once again that trade reforms cannot be divorced from exchange rate reforms.

The reform of protectionist policies in agriculture or of trade policies generally should take place within a framework of overall trade liberalization so that intersectoral discrimination does not persist. One forum for reciprocal trade liberalization is the Uruguay Round of Multilateral Trade Negotiations. In the Uruguay Round, developing countries, especially high-income developing countries, will be requested to grant reciprocity in trade concessions. In this context, they should receive credit for unilateral trade liberalization undertaken under structural adjustment programmes. Furthermore, as Knudsen and Nash emphasize, the incorporation within the GATT framework of concessions already given by developing countries is likely to improve their durability and reduce the likelihood of policy reversal.

To contend that the only way to deepen and broaden recent policy reforms in agriculture is to dismantle all government undertakings or public enterprises in the agricultural sector seems an extreme position. Recent experience confirms two major risks or difficulties of rapid and wholesale privatization: 1) the public monopoly may be replaced by a private monopoly or oligopoly, with government failures replaced by market failures of a greater magnitude; and 2) the public marketing and distribution system may collapse before a viable private one can be developed. Examples of such outcomes abound. International development agencies need to maintain a nonideological approach on this matter so that efforts can be made both to improve the efficiency of public enterprises and to encourage competition between public and private enterprises. The *durability of agricultural policy reforms* depends less on the dismantling of all government institutions capable of interfering with the market mechanism and more on each country's conviction concerning the desirability and extent of policy changes. Such reforms should be based on a reasoned analysis undertaken by the recipient countries in an atmosphere of open debate in which dialogue with the lending agencies is only one input. This process requires a national capability for analyzing and formulating policy. Assistance by international organizations in building up such a capability will make an important contribution to the sustainability of policy reforms.

References

Ahmed, R., *Rice Price Fluctuations and an Approach to Price Stabilization in Bangladesh*, Research Report No. 72 (Washington, DC: IFPRI, 1989).

Cavallo, D. and Y. Mundlak, *Agricultural and Economic Growth in an Open Economy: The Case of Argentina*, Research Report No. 36 (Washington, DC: IFPRI, 1982).

IFPRI, *Commentary: Achieving Growth with Equity Through Appropriate Agricultural Pricing Policy*, Research Report No. 2 (Washington, DC: IFPRI, 1988).

Lele, U. and J.W. Mellor, 'Agricultural Growth, Its Determinants and Their Relationship to World Development: An Overview', paper prepared for the Twentieth Conference of the International Agricultural Economics Association, Buenos Aires, Argentina, 24–31 August 1988.

Mundlak, Y., *Intersectoral Factor Mobility and Agricultural Growth*, Research Report (Washington, DC: IFPRI, 1989).

Pinckney, T.C., *Storage, Trade, and Price Policy under Production Instability: Maize in Kenya*, Research Report No. 71 (Washington, DC: IFPRI, 1989).

Timmer, P.C., *Agricultural Pricing and Stabilization Policy* (Cambridge, MA: Harvard Institute for International Development, 1988).

Chapter 16

Tensions between Economics and Politics in Dealing with Agriculture*

Theodore W. Schultz has provided a clear statement of his major contributions to the theory of economic development. Since the majority of the world's population is poor and the world's poor are largely concentrated in the agricultural sector, he has devoted much attention to the economics of peasant agriculture. He distances himself from what he calls the traditional development economics, with its emphasis on the import substitution strategy of industrialization as the principal source of technical change and economic growth. Most will agree with much of what he says on the limitations of this strategy.

However, what he calls the 'bad start' of development economics has been followed in the past 20 years or so by considerable theoretical and empirical analysis of the economics of being poor. This has been enriched – at least some of us would like to believe – by learning from experience, or learning by doing. To this large body of literature he makes no reference, nor does he put his original contributions in the context of subsequent developments in the analysis of peasant economy.

Today, we perhaps have a more balanced view of the role of agriculture in overall economic growth. The growth in per capita income is associated, on the one hand, with a decline in the share of national expenditure on agricultural output and, on the other, with an increase in the productivity of labour in agriculture. As a result, there is a decline in the relative importance of agriculture in the process of economic development.

Since agriculture employs most of the labour force in a poor country and constitutes the most predominant source of income, it is necessarily the principal source of savings or capital accumulation and labour supply for the rest of the economy. Since over time the nonagricultural sector has to expand faster than the agricultural sector in order to absorb an increase in the labour force and to meet an increase in demand for its output, investment in

---

* First published in Braun, J.V. and E. Kennedy (eds), *Agricultural Commercialization, Economic Development, and Nutrition* (Baltimore: Johns Hopkins University, 1994).

nonagriculture has to accelerate at a rate that would frequently exceed the rate of saving in this sector. It therefore has to be financed partly by savings generated in the agricultural sector. The question is not whether the nonagricultural sector should draw on agriculture in the process of economic development; rather the relevant questions are at what pace, by what method, and in what time sequence or at what stage in the process of growth such intersectoral transfer of resources one way or another should take place. However, a stagnant agricultural sector cannot provide resources for investment elsewhere; only a dynamic agriculture enjoying sustained productivity growth can contribute to overall economic growth. Although this is part of the 'received doctrine', it is not necessarily observed or practised by many developing countries.

Of relevance to these questions is the tension between politics and economics, to which Schultz refers, but he does not elaborate on it. What he calls the political market is indeed dominated by special interest groups, who seek to further their own welfare – not only their income, but also their power over resource allocation and capital accumulation. The urban interest groups – industrialists, workers and urban consumers who desire access to cheap food and raw materials – seem to wield greater political power than the peasants. That adverse terms of trade against agriculture can depress agricultural growth is not perceived by the dominant interest groups, whose time horizon is inevitably short run. How to reconcile the conflicting pulls of divergent interest groups and of short- and long-run considerations is a problem in political economy that is common to both developed and developing countries. Economists have not been of great help in finding a solution. They can, however, highlight the conflict between sectoral interests and social welfare as well as between short- and long-term interests of various groups.

Schultz believes that economists often either accept the value judgment of politicians or fail to convince them of the irrationality of their decisions. He exaggerates the potential role of economists in the political decision-making process. His own experience contradicts his position in this regard. Perhaps the economists' failure may be accounted for by their inability or unwillingness to acknowledge the politicians' goals and constraints. It is not enough to understand why the politicians do what they do; it is necessary to identify the circumstances under which they are likely to act in ways considered economically rational. It is frequently not a lack of knowledge on the part of the policy makers, although in a few cases they may be the slaves of the 'defunct theorists' of yesterday; the truth, however, is that what appears politically optimal is often not economically rational or optimal.[1]

Schultz's most significant early contribution relates to the efficiency of the peasant economy. Since farmers are rational and efficient in tile use and allocation of resources, neither a reorganization of institutions nor a reallocation of resources is likely to improve efficiency or expand output. The marginal product of labour in a peasant economy is not zero; consequently any withdrawal of labour from agriculture reduces output. Growth occurs with the introduction of new technology, new inputs and new markets. Farmers' response to prices is positive and substantial. Whatever inefficiencies exist are due to government interventions; they create price distortions which inhibit growth, among other things, by constraining both technological innovations and their diffusion.

The seminal value of Schultz's contribution – at a time when the hypotheses of conservative, tradition-bound farmer behaviour and a backward-sloping supply curve were widely accepted – can hardly be exaggerated. His challenge probably stimulated an unprecedented amount of both theoretical and empirical research on the economics of peasant agriculture. In subsequent years there was a great spurt of writings on the nature of rural product and factor markets, the role of uncertainty and risk and the impact of land tenure, land distribution and rural institutions on the efficiency of resource use and on technological innovations.

The unequal control over or ownership of land leads to inequitable access to credit, services and inputs, and thus may inhibit the optimal utilization of land. Redistribution of land increases aggregate production whenever output per hectare on small farms is higher than that on big farms. A consolidation of fragmented and scattered holdings may improve productivity because of increased efficiency in the use of inputs such as water and draft animals and release labour without an adverse effect on output.

In the growing literature on agricultural growth and agrarian change, new hypotheses or theoretical constructs have been and are being devised. A small farmer's behaviour is to be distinguished from that of a commercial farmer or a landlord. The small farmers are highly vulnerable to production and market risks; they often consume a large fraction of what they produce and supply most of their own labour requirements.

On a small farm, the family as a decision-making unit combines consumption and production decisions. In many countries the family is not a homogeneous decision-making unit; frequently women and men constitute separate decision-making units. The integration of the decision-making process leads to conclusions different from those reached separately, even if the family acts as a profit maximizer. The family decisions on production and consumption may be the outcome of some sort of cooperation-cum-bargaining game.

In many instances, markets themselves may be inadequately formed. Although the traditional system of obligatory payments and custom-determined rewards is being eroded by the steady penetration of markets, the extent of this penetration is a matter of empirical judgment. Even when markets function, they are limited by inadequate information and high transaction costs. For example, in the rural labour market, the employer often places a high premium on the prompt availability of labour at the right time of the season and therefore searches for labour only within his own village, counting on the goodwill or loyalty of the villagers.

The theory that agricultural wages are competitively determined by forces of supply and demand cannot explain the persistence of unemployment among poor farm families, whose 'appetite' for voluntary unemployment is not likely to be great. The transactions taking place in the labour market at less than the market-clearing wage can be ascribed to market failure caused by the play of noneconomic factors such as 'social conventions or principles of appropriate behaviour whose source is not entirely individualistic' (Solow, 1980, p. 3).

Both analytical and empirical research over the past several years has attempted to put the price responsiveness of farmers in a proper perspective. First, the price elasticity of supply of aggregate agricultural output is to be distinguished from, and is usually lower than, the elasticity of supply of a particular crop in response to relative price changes. The empirical estimates of aggregate supply elasticity put them between 0.2 and 0.6 per cent; they are small but not totally negligible. Second, expanding farmers' output requires not only price incentives but also adequate inputs, appropriate technology, credit and marketing and distribution infrastructure, including the availability of consumer goods. Even in the absence of these complementary supply-augmenting factors, setting prices right expands output but by much less. Third, although there is general agreement that prices are not a sufficient condition and that other nonprice measures are needed for increasing production, there is disagreement as to the relationship between other measures and correct prices. Some argue that putting prices right by itself will call forth appropriate or requisite improvement in nonprice measures such as public investment in research and irrigation; this line of reasoning suggests that public investment is optimally allocated at current prices. On the contrary, it is reasonable to argue that the circumstances which lead to an undervaluation of agriculture and keep prices low are also responsible for inadequate agricultural research or insufficient public investment.

Schultz seems to ascribe a prominent role to adequate price incentives in explaining the success of the Green Revolution in the Indian Punjab. Obviously, this is a simplification. A well-developed irrigation system, a long-standing system of landowning cultivators, adequate rural infrastructure, including roads, markets, and rural electrification, a seed and fertilizer technology supported by national agricultural research – all contributed to the highly favourable environment in which price incentives played a role. Similarly, although a lack of price incentives in certain African countries has discouraged agricultural production, it is misleading to ignore a large number of factors such as droughts, the unrelenting erosion of Africa's relatively fragile soil, wars and civil disturbances, the deterioration of the external terms of trade, inadequate physical infrastructure and lack of technological packages appropriate to diverse agroecological circumstances. Techniques of empirical research have not yet advanced far enough to disentangle conclusively the separate contributions of these factors to the African agricultural crisis.

The depressing effects of rising food prices on income and food consumption of the poor and on social and political stability are too well known to be repeated here. To devise cost-effective supplementary measures to cushion the adverse effects on the poor is not always easy. It may be argued that price distortions, by offering low-cost food to poor consumers, seek to correct other distortions caused by the unequal distribution of income and absolute poverty. To quote A.C. Pigou: 'A man ordered to walk a tightrope carrying a bag in one hand would be better off if he were allowed to carry a second bag in the other hand, though, of course, if he started bagless to add a bag would handicap him.'[2]

In an ideal world one instrument, price policy, should not be used for two objectives: efficient resource allocation and equitable income distribution, or alleviation of poverty. In the real world, however, such neat or elegant choices between several policy instruments are not readily available for lack of political feasibility or implementation capacity. To devise an appropriate policy package in such a situation, one is constrained to seek a second-best policy, while striving to minimize its adverse effects on allocational efficiency. To quote Robert Solow (1980, p. 1):

> Most of us are conscious of a conflict that arises in our minds and consciences because, while we think it is usually a mistake to fiddle with the price system to achieve distributional goals, we realise that the public and the political processes are perversely more willing to do that than to make the direct transfers we prefer. If we oppose all distorting transfers, we end up opposing transfers altogether.

> Some of us seem to welcome the excuse; but most of us feel uncomfortable. I do not think that there is a very good way of resolving that conflict in practice.

Schultz's emphasis on human capital as an important input in economic growth and agricultural development is widely acclaimed. His reminder of the economic value of education is eminently relevant at the present time. Faced with the scarcity of resources and with the need for reductions in government expenditures, many developing countries are more willing to cut education, health and social expenditures than to cut directly productive physical investment, a trend that jeopardizes long-run growth prospects.

Schultz's analysis seems to yield two propositions. First, private education expenditures should be treated as investment rather than consumption. Second, knowledge and skill (acquired from education and training) are productive factors whose physical return can be measured exactly like that of any other productive factor. Furthermore, the growth of knowledge helps explain much of the unexplained residual in econometric growth equations.

Schultz seems to ignore that education is both a consumption and an investment good – and is considered as such by individuals as well as by societies. Human capital is as heterogeneous as physical capital and raises similar problems of measurement. In addition, it is not easy to isolate the contribution of new knowledge from that of new inputs (Green Revolution techniques are a clear case); both often form a single indivisible technological package.

An entire body of theoretical and empirical literature has been built up around the costs and benefits of investment in education. For example, farmers' education needs to be distinguished from the training provided by the extension service. The latter disseminates information relating to the methods and techniques of farming, whereas education imparts the knowledge and skills needed to process and use this information. In a situation of technological change when new inputs and new methods rather than routine farming operations have to be mastered, education assumes added importance. Extension and training stress practices and 'packages' rather than principles. Education provides command over principles, which enable farmers to screen and evaluate the components of packages. When education is combined with new technology and complementary inputs, farmers' productivity is estimated to increase almost twice as much as with education alone. It is not clear to what extent the success of the Green Revolution – new technology embodied in high-yielding inputs – is due to formal education of the farmers rather than to extension and training through nonformal ineans such as demonstrations on government or on neighbours' farms and the dissemination of information by mass media.

The contribution of education to economic growth is not confined to its microeconomic effects but extends to its macroeconomic consequences. There is some evidence that literacy, for example, contributes to increased investment and affects life expectancy and fertility. Developing countries with higher literacy rates tend to grow faster, other things remaining the same; they also have higher investment rates. Analytical research and empirical work on these macroeconomic aspects are still in an early stage.

Do developing countries underinvest in education? Public education expenditures rose from 2.4 per cent of their GNP in 1960 to 4 per cent by the late 1970s. The adequacy of educational expenditures cannot be fully evaluated, however, without reference to the costs of educational services and to the levels and types of education. The pendulum has swung from an emphasis on secondary and higher levels of education in the 1950s and 1960s to an emphasis on primary education in the 1970s. Investment in primary education is estimated to yield higher returns than that in the higher levels of education, especially in promoting agricultural development.

The increase in expenditure on agricultural research in developing countries in recent years – supported by national and international efforts – is a tribute to the success of pioneers like Schultz who contributed significantly to the growing recognition of the role of research in agricultural development. The suggestion that the allocation of research efforts among different crops has been guided or misguided by wrong price signals – that crops with depressed prices received less research efforts than crops that enjoyed favourable prices – does not fully conform to the pattern of public expenditure on agricultural research. Schultz admits that recently research expenditures have shifted toward food crops, even though in many instances their prices were not favourable or even at the world level.

The development experience of the four East Asian economies has been the subject of much discussion in recent years. Different observers have emphasized different aspects of their development policies. Indeed, in the early years countries such as Korea did pursue an import-substitution policy by restricting trade and by biasing their incentive structure more toward domestic than toward export markets. In the next stage they shifted toward an export promotion strategy, which in some cases was carried, often with the use of subsidies and other financial incentives, beyond what was warranted by comparative cost considerations. In both cases they followed, on the whole, a policy of labour-intensive industrialization in both import and export sectors.

They did not climb the ladder of comparative advantage solely in response to market signals and by a policy of laissez-faire; government intervention and the public sector did play a role – in some cases a very important role – in their development strategies.

The relevant lesson of their successful development strategy seems to be not that government intervention or participation in economic activity should be minimized or excluded but that it should be focused on priority areas and be consistent with the implementation capacity. The relative roles and patterns of the public and private sectors change over time in response to changing circumstances. The chief difficulties with the import substitution strategy in the past seemed to be in conforming to comparative cost considerations and eliminating discrimination against exports.

The success of a trade-oriented outward-looking strategy depends no less on the policies of developed countries. Schultz does not refer to the domestic agricultural support policies of developed countries – their effect on the level and stability of world prices and their adverse consequences for growth in developing countries – or to the growing trade barriers against the labour-intensive manufactured exports of developing countries. If all developing countries were to repeat the intensity and the scale of the export-promotion strategy of the East Asian economies, their manufactured exports would increase sevenfold and make up approximately three-fourths of the imports of industrialized countries (Cline, 1982, p. 85). This degree of penetration in the latter's domestic markets would be likely to provoke a large-scale retaliatory action, if current experience is any guide.

Not all countries, it is argued, will succeed at the same time and in the export of the same commodities. There would be 'pioneers' and 'followers' in a well-regulated progression. In the real world a fine tuning to obviate any competition among developing-country exporters is not likely, as is evident from recent examples of textiles and leather goods. What is likely – as in the case of Korea and Taiwan – is that the export expansion phase will follow the import substitution phase in each country; while some are still in the first stage, others will enter the second stage. In the future, the rate of export expansion is unlikely to be of the same order of magnitude as that of the four East Asian economies. Their experience indicates that export promotion through subsidies and incentives could be as inefficient as an overly protectionist import-substitution strategy.

It is in this context that one has to view Schultz's comparative analysis of the economic growth of Jamaica and Singapore. Admittedly, Singapore has followed a more outward-looking developing strategy than Jamaica and has encouraged a policy of labour-intensive industrialization to a much greater extent than Jamaica. This difference has undoubtedly contributed to the difference in growth. But this is only a part of the story. Singapore has no agricultural sector worth its name; almost all its population is engaged in industry, trade and services, unlike Jamaica where agriculture and mining are the predominant sources of income and employment. More than 24 per cent of the exports of Singapore were entrepot exports. Moreover, 87 per cent of Singapore's exports in 1980 were manufactured goods, whereas about 37 per cent of Jamaica's exports were bauxite; bauxite, sugar, and bananas together contributed 41 per cent of its exports. Services constitute 35 per cent of total exports in Singapore and 43 per cent in Jamaica. Jamaica did specialize in exports (such as coffee, sugar, bananas, bauxite and tourism), in which it had a comparative advantage. The slow growth in its exports was due in part to exogenous factors, including world market conditions, and in part to inappropriate domestic policies. The world trade in commodities exported predominantly by Jamaica did not achieve anywhere near the rate of growth enjoyed by the commodities which were Singapore's principal exports. In addition, the political instability of Jamaica – which had no parallel in Singapore – discouraged private investment in all sectors and adversely affected both exports and overall growth. The relative openness of an economy is too simplistic an explanation for its performance, however. There is no single policy change that can account meaningfully for differences in economic performance among different countries.

To conclude, one cannot but join many others in admiring the vision and insight of Schultz as one of the pioneers in development who highlighted some of the crucial but neglected factors in economic development and opened up new vistas of research and analysis for many generations of economists.

Notes

1 The need to pursue second-best solutions is not much evident in Schultz's analysis. He recognizes that his strength lies in the realm of ideas and in thinking about the ' long pull'. I only hope that he does not disparage or denigrate too strongly the economists (among whom can be counted many of his followers and admirers) who work in international organizations or in national governments. One likes to believe that those who 'stand and wait' in these institutions may also 'serve' by bringing to bear the complexities of real life on the analysis of development problems.

2 A.C. Pigou on the theme of second best, as quoted by Streeten, 1984, p. 346.

References

Cline, W.R., 'Can the East-Asian Model of Development be Generalised?', *World Development*, February, 1982.

Solow, R.M., 'Theories of Unemployment', *American Economic Review*, Vol. 70, No. 1, March, 1980.

Streeten, P., 'Development Dichotomies', in G.M. Meier and D. Seers (eds), *Pioneers in Development* (New York: Oxford University Press, 1984).

Chapter 17

Commercialization of Agriculture and Food Security: Development Strategy and Trade Policy Issues*

Introduction

The choice between subsistence food crops, on the one hand, and cash crops, especially nonfood cash crops predominantly meant for exports, on the other hand, is a subject of considerable debate among policy makers as well as development specialists. The debate raises issues not only at the level of farming households but also at the level of national and international policies, including macroeconomic policies such as trade and exchange rate policies. This chapter reviews and focuses on those aspects of principal relevance in the context of an overall agricultural development strategy and food security.

The controversy regarding cash or export crops versus food crops is part of a bigger debate relating to the commercialization of agriculture in developing countries. Commercialization can broadly be defined as a rise in the share of marketed output or of purchased inputs per unit of output. A shift from basic food crops, which are produced and predominantly consumed on the farm, to cash crops, which are produced mainly for sale in the market, is therefore viewed as part of the commercialization of agriculture process. It is associated with an extension of the market or of the exchange economy, leading to increased specialization and division of labour. Cash crops can be both food and nonfood crops. Even though nonfood crops have been the principal component of cash crops, their relative importance has been on the decline recently.

Choice Between Food Crops and Nonfood Cash Crops

To put the debate at the national policy-making level in its proper perspective, it is important to stress a few salient factors influencing the between food

* First published in Braun, J.V. and E. Kennedy (eds), *Agricultural Commercialization, Economic Development, and Nutrition* (Baltimore: Johns Hopkins University, 1994).

crops and nonfood cash crops. First, there are constraints on allocation of land and other resources between various crops; not all lands are agro-ecologically suitable for all crops. Second, different crops are frequently grown in different seasons and are, therefore, grown in combination over the year. Mixed farming, hence, is often common. Third, mixed farming, including a combination of different crops, is often preferred because monoculture leads to soil degradation or, at least, does not contribute to renewal or maintenance of soil fertility without a supply of additional nutrients. Therefore, cash or export crops and food crops can both, in certain agro-ecological environments, be necessary components of a crop rotation.

Fourth, food crops, including basic staples (cereals, roots, tubers and pulses) and cash food crops (sugar, fruits, vegetables, oils and oil seeds), constitute the largest share of the value of total crop output in developing countries. In 93 developing countries, excluding China, food crops of both types constituted 89 per cent of the total crop output in 1983–85, whereas the share of nonfood cash crops was 11 per cent (Table 17.1). Basic food crops constituted 50 per cent of the total crop output in the same period; Asia and Africa had the highest share, 59 and 57 per cent, respectively. Africa and Latin America had the highest share of nonfood cash crops, but this was no more than 15 per cent. Over the last 25 years, the proportion of basic food crops in the total crop output has hardly changed. However, within the group of cash crops, the share of food cash crops has expanded from 35 to 39 per cent, whereas that of nonfood cash crops has declined from 14 to 11 per cent.

Fifth, over the last 25 years – that is, during the 1961–85 period – basic food crop output increased at a rate of 2.7 per cent per year, a much higher rate than the 1.4 per cent rate of growth of output of nonfood cash crops (FAO production data tapes). However, food cash crops grew at 3.4 per cent per year during this period, and the annual rate of growth of all cash crops was about 2.9 per cent. During the 1960s, the output of basic food crops increased at a faster rate than cash crops (both food and nonfood), but this situation was reversed during 1976–85, mainly because of an acceleration in the growth rate of food cash crops from 3.0 per cent to 3.8 per cent per year.

Sixth, is there an association between growth in food production and growth in cash crop production? Do countries that do well in cash crop production also do well in food production? A cross-section study of 78 countries over the 1968–82 period found that growth in acreage under cash crops was positively associated with growth in staple food production, and that, moreover, growth in the proportion of total crop area devoted to cash crops was associated with growth in per capita staple food production (von

Table 17.1 Distribution of total crop output between basic food crops and cash crops, developing country regions, 1961–63 and 1983–85

Region/year	Basic food[a]	Cash crops Food[b] (per cent)	Nonfood[c]
Sub-Saharan Africa			
1961–63	54	30	16
1983–85	57	27	15
Near East and North Africa			
1961–63	42	49	9
1983–85	38	54	8
Latin America			
1961–63	36	40	24
1983–85	34	51	15
Asia (excluding China)			
1961–63	59	31	9
1983–85	59	33	8
Developing countries (excluding China)			
1961–63	51	35	14
1983–85	50	39	11

Source: FAO production data tapes.

Notes

a Cereals, roots and tubers, and pulses.
b Sugar, vegetables, bananas, fruits, vegetable oils and oilseeds.
c Tea, tobacco, coffee, cocoa, jute, cotton, fiber, rubber etc.

Braun and Kennedy, 1986).[1] Many countries that had positive growth rates in basic food production also had positive growth rates in nonfood cash crop production and vice versa. Fifty-seven of the 60 countries that achieved positive growth rates in total nonfood output between 1961–85 also attained positive rates of growth in basic food production. Similarly, of 82 countries that recorded positive growth rates in food production over the same period, 57 countries (70 per cent) also attained positive rates of growth in nonfood production. Therefore, on the whole even though high growth rates in basic food production were not necessarily associated with high growth rates in nonfood production, the direction of change was correlated.

Seventh, what is the relationship between growth in production of nonfood crops and growth in overall food supply? Is an increase in nonfood crop production associated with a decrease in total or per capita food supply? In 89 of 90 developing countries for which data are available (FAO data tapes), aggregate food supply (domestic food production plus food imports) increased during the 1961–85 period regardless of change in nonfood production. However, the situation was different as far as per capita food supply was concerned. In the majority of 60 countries that recorded an increase in nonfood production, per capita basic food supply increased; only in 23 countries did per capita food supply decrease.

The foregoing analysis indicates that food crops, including both basic and cash crops, dominate crop production. In the last several years, food cash crops have increased in importance, but total cash crop production has remained unchanged in relative importance due to a decline in the importance of nonfood cash crops. The decline in the relative importance of nonfood cash crops was most significant in Latin America, where the share declined from 24 to 15 per cent of total crop output between 1961 and 1985, although it was observed in almost all regions of the developing world. The analysis also indicates that the relative rates of growth in food and nonfood production were not the dominant factors determining overall food supply. Food supply in countries with inadequate domestic food production was a function of access to food imports. Access to food imports was, in turn, a function of food aid, foreign exchange availability and food import policies of various countries.

At the national development strategy level, a number of arguments have been advanced to explain why specialization in export crops may be undesirable. First, export crops face a long-run adverse trend in the terms of trade. Second, export prices of cash crops are highly unstable and, therefore, have an adverse effect on development, as well as on the stability and availability of food supplies to be obtained in exchange of export crops. Third, long-run dynamic comparative advantage does not seem to favour export crops. Fourth, export crop production habitually tends to be large scale and capital intensive and, therefore, does not promote employment or equitable distribution of income. Finally, an emphasis on export crops as opposed to food crops may detract from household food security in terms of food consumption and nutritional status of poor families. The following sections investigate the validity of these arguments.

Long-Run Trends in Relative Prices in World Markets

It is argued by 'food first' advocates that production of basic food crops deserves to be accorded higher priority even if nonfood export or cash crops enjoy a comparative advantage and yield higher returns. The reason usually advanced for such a proposition is that export crops, especially agricultural raw materials and tropical beverages (the principal exports of developing countries), tend to suffer from a long-term decline in the terms of trade and do not have bright market prospects in the future. Although much has been written on long-run movements or trends in the international terms of trade of agricultural exports in general, the evidence is not conclusive about whether a decline, if it occurs, is due to adverse demand conditions or reductions in cost.

A recent comprehensive study (Grilli and Yang, 1988) of long-run trends in terms of trade of different groups of agricultural commodities found that from the beginning of this century up to 1986, the real prices or terms of trade of agricultural raw materials, nonbeverage food commodities and cereals have declined by about -0.82 to -0.77 per cent, to -0.51 per cent, and 0.68 to -0.62 per cent, respectively, per year (trend rate of change) (Grilli and Yang, 1988). The largest decline was in the real price of agricultural raw materials and, hence, there was a downward trend in the real price of agricultural raw materials vis-à-vis that of food and cereals. However, the annual decline was very small. Furthermore, the real price of beverages rose over the 1900–86 period at about 0.6–0.7 per cent per year, with the result that the terms of trade of beverages with respect to agricultural raw materials, nonbeverage food and cereals moved in its favour over the period. Thus, the long-run movements in the terms of trade of the two groups of export crops vis-à-vis food and cereals crops were very different.

In more recent years, the real prices of nonbeverage foods and cereals have fallen faster than in earlier years (Table 17.2). During the 1962–86 period, the rate of decline in the price of cereals was faster than that of agricultural raw materials and other food. Therefore, the terms of trade of agricultural raw materials, tropical beverages and other food all improved vis-à-vis cereals during that period.

Neither the long-run decline in the relative price of agricultural raw materials vis-à-vis food and cereals nor the faster decline in cereal price in recent years (1962–86) necessarily provides a signal or guide for long-run investment in any one crop. First, the relative decline in the price of an export crop may have been due to falling costs rather than to declining demand. Second, even if net returns from export crops – for example, agricultural raw materials – declined over time, they may still be higher than those obtained from other crops, such as cereals. A decline in food (cereal) prices in recent years relative to other crops does not suggest reliance upon food imports in

Table 17.2 Recent trends in terms of trade of different commodity groups, 1948–86 and 1962–86

Commodity group	Trend rate of change per year (%) 1948–86	1962–86
Agricultural raw materials	-2.17	-1.83
Tropical beverages	–[a]	–[a]
Other food	-0.64	1.12
Cereals	-1.85	-2.46

Source: Computed from data provided in World Bank (1988a).

Note: a = no significant trend.

exchange for agricultural export crops. The depressed prices of cereals in recent years are mainly due to heavy export subsidies by developed, exporting countries such as the United States and the European Community (EC) that are vying with each other to dispose of the large surpluses generated by domestic price and other support programs. Therefore, the low world food prices of the 1980s, which may not endure with cessation of competitive export subsidies and reduction of surpluses in the developed, exporting countries, do not provide an appropriate signal for long-run allocation of resources between food and export commodities in developing countries.

Also, agricultural exports, which face long-run stagnation or growth in world demand, do not provide a viable avenue for specialization on developing countries. Many commodities are faced with the threat of synthetic substitutes; others, especially agriculture raw materials, face a decline in demand due to increased economy in their use per unit of finished output. Trade restrictions in importing countries, mostly developed countries, further constrain the expansion of agricultural exports.

Countries that face inelastic export demand in response to price changes can improve their combined real income by acting in unison to manage or restrict export supply. However, such cooperation among agricultural exporting countries has been rare. In fact, they frequently compete aggressively to secure a larger share of an inelastic and slowly growing export market; the resulting fall in prices tends to reduce aggregate export earnings for all of them. In such situations, the less producers are confronted with the need to diversify their exports to focus on commodities that face an expanding future world demand, while the same time to strengthen their competitive advantage through cost-reducing innovations and technological progress. While traditional agricultural exports such as sugar and hard fibres do not hold out prospects for rapidly expanding world demand, commodities such as horticultural products, oilseeds and livestock products show brighter market prospects (Islam, 1990).

Relative Prices and Food and Export Crops in Domestic Markets

Even though relative prices of different agricultural commodities in world markets, in relation to their relative domestic costs of production, are expected to provide guidance for selection of an appropriate combination of crops, most developing countries, in fact, intervene through border measures so that domestic relative prices differ, sometimes significantly, from relative prices in world markets. The relative prices of food and export or cash crops confronting producers in their domestic market are of relevance to their decision-making process.

Most developing countries traditionally discriminate against agriculture, including both food and export crops, through a combination of economy-wide and sector-specific policies. Frequently, an overvalued exchange rate, combined with a high level of industrial protection, reduced relative returns from agriculture vis-à-vis nonagriculture, which kept the real prices of both food and export crops low.

The economy-wide policies that tended to discriminate against agriculture were reinforced in the case of export crops by export taxes, the incidence of which often fell on producers, and by export marketing boards, which paid farmers less than export prices or world prices. In several countries, export taxes and export marketing boards were a major source of revenue for the government. At the same time, food price policies pursued through state interventions in food marketing kept prices to farmers low in the interest of urban consumers.

Discrimination against agriculture has been most acute in Africa and, to a lesser extent, in Asia and Latin America. The relative domestic prices of food versus export crops have, however, varied widely among countries, depending on the net result of a multiplicity of policies that impinge on the agricultural sector. In recent years, there has been some decline in the discrimination against agriculture, especially against food crops, in many countries, which can be seen from changes in producer export crops and in nominal rates of protection (Table 17.3). Discrimination against food crops has declined faster than against export crops. Increases in the nominal rate of protection have been slower for

Table 17.3 Changes in farmgate prices in selected developing countries

	Index of farmgate price (1969–71 = 100)			Percentage change[a]
Crop group	*1973–75*	*1978–80*	*1981–83*	*1969–83*
Cereals	121	114	120	+18
Export crops[b]	110	117	100	not significant

Source: FAO (1987).

Notes

Thirty-eight developing countries, including 13 in Africa, nine in Asia, 10 in Latin America and six in the Near East, are included in this analysis.

a The percentage change is calculated on the basis of a trend equation for 1969–83.

b Coffee, cocoa, tea, cotton, rubber, jute, soybean, palm oil and tobacco are included in this group.

Table 17.4 Indices of nominal rates of protection 1969–71 = 100

	1973–75		1978–80		1981–83	
Region	**Cereals**	**Export crops**	**Cereals**	**Export crops**	**Cereals**	**Export crops**
Africa[a]	91	89	161	91	170	109
Africa	80	95	134	94	143	101
Asia	69	90	92	100	117	103
Latin America	70	100	117	97	137	105
Near East	74	88	116	91	128	102
Total[c]	74	94	116	96	133	103

Source: FAO (1987), pp. 193–5.

Notes

Nominal rate of protection is defined as the ratio of farmgate price to border price. Border price, in the case of an export competing crop, is f.o.b. price minus transport, handling and marketing costs between the border and the farmgate. In the case of an import competing crop, the border price is c.i.f. price plus transport, handling, and marketing costs from the border to the farmgate.

a Includes countries that favoured higher food prices.

b Includes countries that relatively favoured higher export crop prices.

c Thirty-eight developing countries.

export crops than for food crops (Table 17.4). The increase in the nominal rate of protection for cereals has been fastest in Africa, followed by Latin America and the Near East.

A different measure of nominal protection for different crops is expressed in terms of the relative domestic price of each crop vis-à-vis that of nonagricultural commodities, where the world price is into converted domestic price at an equilibrium rate of exchange, rather than at a nominal rate of exchange (Krueger, Schiff and Valdés, 1988). Data from 16 countries indicate that food crops enjoyed a positive rate of direct nominal protection[2] in 1975–79 and that this continued roughly at the same level, 20 per cent, in 1981–83 (Krueger, Schiff and Valdés, 1988). Export crops suffered from a negative rate of direct nominal protection of roughly the same magnitude – 10 per cent, in both periods, 1975–79 and 1981–83. The positive direct nominal protection for food crops was more than offset by the exchange rate overvaluation, resulting in a negative, though small, total protection of about -5 per cent. In the case of export crops, the exchange rate overvaluation aggravated the direct negative protection, so that the total protection was not only negative but also more than five or six times higher than in the case of food crops. Furthermore, the total protection against export crops increased over the years, from -36 per cent to -40 per cent.

Why did export crops suffer greater discrimination in terms of direct as well as total negative protection than staple foods? There were three main reasons. First, as mentioned earlier, developing countries faced with limited sources of revenue found taxes on export crops to be an attractive source of revenue and one that could be relatively easily collected, either directly through export taxes or indirectly through profits of the export marketing boards. In several instances, surpluses siphoned off by parastatals or marketing boards ended up largely paying for the inefficient management and high overhead costs of these bodies. Second, in several countries it was presumed that export demand for such crops was price inelastic, and, hence, taxation of exports, by limiting supply, would contribute towards increased export earnings. For individual countries, it was seldom that price elasticity of export demand was low. Third, in recent years there has been an increasing emphasis on food self-sufficiency and the need to increase domestic food production in the context of rising food imports, as a consequence of increasing population and rising per capita income, especially in middle-income countries. This has led many countries to raise domestic prices either through higher procurement prices of the public marketing agencies or through taxes on imports of food or cereals.

Price and Income Instability: Food Versus Export Crops

The instability of prices of both export and food crops in the world market has been advanced as an additional argument against dependence on the world market for a basic need such as food. Prices of, and earnings from, export crops, which determine the capacity to import food, have been highly unstable. This has been aggravated by fluctuations in world prices (Table 17.5) and supplies of food. These two tendencies do not necessarily offset or compensate each other.

Table 17.5 Instability of world market prices, 1962–87

Commodity group	Percentage variation from trend
Food	29
Rice	31
Wheat	23
Maize	20
Tropical beverages	25
Agricultural raw materials	18

Source: UNCTAD (1987).

Differences in price instability are observed among the commodity groups: agricultural raw materials, in general, seem to have a lower degree of price instability than do cereals and tropical beverages. However, there are variations within each commodity group. For example, price variability around the trend for maize is considerably lower than that for rice. Price variability for sugar is as high as 74 per cent around the trend, whereas for soybeans it is 21 per cent and for bananas it is 16 per cent. Among tropical beverages, the index of instability of cocoa prices is as high as 33 per cent. Among raw materials, the index of instability of prices for rubber is as high as 25 per cent and for sisal it is 37 per cent (UNCTAD, 1987).

There is no *prima facie* reason why the choice between alternative crops at the margin should depend on the relative degree of price instability, even though returns from less stable products are often higher than from more stable products. Risk aversion may lead farmers to discount the higher returns from more unstable crops by a margin determined by their subjective evaluation of risk. Evaluating the risk of instability depends upon the sources of instability. Price instabilities originating from supply fluctuation offset each other and, as supply and income fluctuations are inversely related, stabilize income. This is usually the case with agricultural commodities. Uncertainty or instability of relative prices would make farmers adopt a certain degree of diversification of farm products but would not result in farmers' specializing to the extent warranted by average comparative cost considerations.

However, more pertinent questions may be raised at the national policy level. Should world prices, in view of their volatile and fluctuating nature, be used as a guide to long-term allocation of resources among different crops? How should the long-term trend in prices be determined? The pragmatic answer is to rely on three- or five-year moving averages of prices. There is scope for developing-country governments to intervene to stabilize prices around such moving averages to provide appropriate price signals to farmers. Intervention can be in the form of import taxes, import subsidies or stocks, the latter being more costly. Most countries have taken domestic measures to achieve a degree of stability of domestic prices that is greater than that of corresponding world prices. Such a policy of 'exporting' domestic instability contributes to world market destabilization. The policy measures adopted for such domestic price stabilization vary widely in terms of resource costs. The greater the desired degree of price stability, the higher cost, for example, of the stocks that have to be held, or the higher the loss in efficiency in diverting from relative prices as a guide to resource location (Pinckney, 1989; Ahmed and Bernard, 1989).

Subject to the provision of using long- or medium-term prices, either comparative costs or domestic resource costs of earning or saving foreign exchange should continue to be used as a guide for resource allocation. Long-term projections of relative prices are needed for making long-term investment decisions such as investment in irrigation or in mechanization projects.

To the extent that price instability leads to instability in export earnings and, hence, in the ability to procure food supplies from abroad, it creates instability in the access to food supplies. Stable food supplies are essential for food security. Food insecurity may result from a rise in world prices of imported food, thus reducing food imports and, hence, food supplies, given unchanged foreign exchange earnings to import food and given typically low short-term supply response.

Even if prices of nonfood exports and food imports move in opposite directions, they do not necessarily compensate each other. Countries confronted with the need to import food or cereals from a highly volatile world market, on the basis of fluctuating agricultural export earnings, tend to emphasize the production of food beyond the point that is warranted by comparative cost considerations. However, reliance on domestic production does not eliminate or necessarily reduce fluctuations, in food supply that originate from weather-induced or policy-related fluctuations (Sahn and von Bratin, 1989). Countries, however, feel more secure dealing with variability in domestic food production than with uncertain supplies and prices in the world market.

The risk of reliance on the world market arising from volatility of export prices or earnings or from food import prices can be reduced or mitigated by an assurance of access to foreign exchange resources to meet rises in food import prices or shortfalls in domestic food production. The Cereal Financing Facility of the International Monetary Fund (IMF) was originally conceived to meet such a need. However, it has been greatly reduced in scope in recent years, partly by its integration with the IMF's Compensatory Financing Facilities for Export Shortfalls and partly by the introduction of 'conditionalities' for economic policy reforms. Alternatively, international commodity agreements may be used to stabilize prices in world markets, either by internationally coordinated, nationally held stocks or by supply management by exporting countries through allocation of export quotas among themselves with a view to adjusting supply to expected demand, along the long-run price trend.

The risks of dependence on the world market arise not only from variability of prices but also from disruption of supplies due to war or civil unrest, breakdown of international shipping and transportation arrangements, or export embargoes initiated to overcome domestic shortages in exporting countries or to achieve political and strategic objectives.

The need for assured and uninterrupted access to food supplies has been a powerful motivation behind the search for greater food self-sufficiency and the build-up of large domestic food stocks in both developed and developing countries. In order to encourage developing countries to depend on world trade to an optimal extent, it is necessary to ensure an open and liberal world trading regime so that import restrictions and export subsidies or embargoes do not reduce the level and stability of agricultural export earnings or of world agricultural prices.

Long-run Dynamic Comparative Costs

The reliance on comparative costs needs to be viewed dynamically by individual countries: relative costs change across countries and commodities in response to technological change fuelled by research and development efforts, both national and international. These dynamics raise a difficult question about investment strategy in commodities that are currently in excess supply in the world market – and expected to be in the future – resulting in a long-run downward trend in prices. Countries whose domestic costs of production exceed long-run export or world prices should gradually move out of production of these commodities unless they expect their costs to go down in the future due to cost-reducing innovations and increased productivity. There would be short-term costs of unemployment and income loss on the part of those employed in the unprofitable sector. There is, thus, scope for structural adjustment assistance for the sector in transition.

New producers with comparative advantages may emerge on the world market, either because of their ability to exploit hitherto untapped agro-ecological advantages or because of their success through research and development efforts in achieving cost-reducing innovations. Increasingly, comparative advantage is less a matter of a given endowment of agro-ecological resources and more a matter of technological innovations that improve or modify cost advantages derived from resource endowments. Also, new crops may be introduced in countries that did not grow them earlier due to lack of domestic demand or unfamiliarity with the world market. The rising comparative advantage of East Africa in tea production to the disadvantage of the traditional tea producers in South Asia, the emergence of Malaysia as a low-cost producer of cocoa, and the shift of palm oil production from Africa to Malaysia and then to Indonesia, are all illustrations of shifts in comparative advantage of different commodities among countries. To freeze the pattern of global production of a particular commodity among its traditional producers would be tantamount to sacrificing the advantages of lower costs and cheaper supplies of agricultural commodities in the future.

In recent years, the focus of international and national agricultural research efforts has tended to shift away from export or cash crops toward food crops in response to pressing food needs in many developing countries. This shift in focus is more true of international than of national research expenditures. During colonial times, research on export crops received high priority in most developing countries. Consequently, while a reallocation or shift of emphasis in research efforts was needed in the post-independence period, given the past neglect of research on food crops, the balance may have shifted too far, thus sacrificing opportunities for efficient export crop production. Research on export crops needs to be strengthened to meet competition from synthetic substitutes, expand new uses or markets, and maintain the market share of developing countries in world trade. This is especially important for those export crops that face low or declining demand or price.

The relative importance of research efforts on various crop groups can be observed from the distribution of crop scientists engaged on crops in different regions (Table 17.6). The dominance of food scientists is obvious in all regions during the 1980s. Even among scientists working on cash crops, those working on food cash crops dominate, except in Asia. A smaller share of scientists work on nonfood cash crops in Latin America and the Near East, whereas, in Africa, the share of scientists working on food and nonfood cash crops does not differ significantly.

Table 17.6 Distribution of crop scientists working on different crops in different regions, and importance of respective crop groups, 1980–85

		Crop	
Region	**Basic food**	**Food cash (%)**	**Nonfood cash**
Asia (excluding China)	54	18	28
	(59)	(33)	8)
Sub-Saharan Africa	50	27	23
	(57)	(27)	(15)
Near East and North Africa	35	47	18
	(38)	(54)	8)
Latin America	52	37	11
	(34)	(51)	(15)

Source: Oram (1988).

Note: The figures in parentheses represent the percentages of total crop output among diffierent groups of crops.

The distribution of scientists does not suggest a relative diversion of scientists from basic food crops to cash crops. Given the inadequate scientific research efforts in agriculture in many countries, what is needed is an increase in total efforts and, at the same time, an increase in emphasis on cash crops in cases of significant past neglect.

If the relative share of different crops in total output is any guide for the allocation of research efforts as indicated by the distribution of crop scientists, there is no serious incongruence, except in the case of Latin America. But, then, it is highly questionable how far the distribution of scientists is a rational criterion; research expenditures, rather than proportion of scientists, weighted by the efficiency or quality of research efforts, would be a better indicator of relative research efforts on different crops. The relative rates of growth in productivity – that is, yield per hectare – could be a measure of the impact of research and development efforts, on the one hand, and of investment, on the other, on different crops. During the 1960s, the rate of growth in productivity for developing countries as a whole was higher for cash crops than for basic food crops (Table 17.7). However, in the 1970–84 period, the rate of growth in productivity increased considerably for basic food crops in all regions and was much higher than that for cash crops. In fact, growth in productivity of cash crops declined for the developing world as a whole during 1970–84, compared with 1962–70. Among the regions, it increased only in the Near East and Asia; in Africa, there was a decline and in Latin America growth was less than one-third of what it was in the 1960s. Moreover, there was a considerable divergence in performance between food cash crops and nonfood cash crops in different regions.

Table 17.7 Percentage change in yield value per hectare of different crop groups in developing-country regions, 1962–70 and 1970–84

Region	Basic food	Cash crops		
		Food	Nonfood	Total
Sub-Saharan Africa				
1962–70	5.4	0.7	15.9	6. 1
1970–84	26.2	-35.7	0.2	-24.8
Near East and North Africa				
1962–70	12.2	-13.1	41.1	0.1
1970–84	27.2	11.6	35.5	20.1
Latin America				
1962–70	-2.5	58.9	-4.9	31.2
1970–84	18.8	-10.4	29.1	7.3
Asia (excluding China)				
1962–70	15.3	11.8	11.9	12.4
1970–84	38.3	21.9	18.7	22.2
Developing countries (excluding China)				
1962–70	10.7	17.5	7.2	14.8
1970–84	32.3	2.2	17.6	9.6

Source: FAO data tapes.

Note: Values in dollars per hectare in constant prices of 1979–81 are used for the computations.

Implications of Choice Between Food Crops and Export Crops for Employment, Poverty and Nutrition

Questions have been raised about whether efficiency considerations or comparative costs are the only criteria for determining choice between crops and export crops. Are the effects of different choices on income distribution, employment, poverty and nutrition not relevant and should they not influence the choice? It is generally agreed that multiple objectives, such as growth, equity, alleviation of poverty and undernutrition, cannot all be achieved by one single policy such as choice of cropping pattern, and a variety of policy would be necessary.

Whether cash or export crops produce more employment than food crops depends on the choice of technique. An export crop may be more labour intensive than a particular food crop – for example, jute versus rice in Bangladesh. Furthermore, it is often argued that food production is predominantly undertaken by small farmers, whereas export crops are frequently produced on large-scale, plantation-type farms. This is not necessarily so; both crops can be produced on large or small farms, depending on the institutional framework of agricultural production in a country. In many countries, export crops are produced by small farmers. Hence, the choice of crops between food and export crops does not necessarily imply a choice of the scale of production. Rarely are economies of scale specific to a particular crop; often they relate to the precise operations, irrespective of whether the crop is an export crop or a food crop. Similarly, intersectoral linkages – that is, the 'spread' effects the overall economy of the production of a particular crop – depend on the choice of technology that determines the nature and amount of purchased inputs, and that in turn determines the backward production linkages through input-output relationships. Consumption linkages, on the other hand, depend on the extent of increase in income and on who receives the increase in income. If the increase in income accrues to the very poor, it would stimulate demand for food, since the poor spend a higher percentage of their income on food. If the increase in income accrues predominantly to the medium-scale farmers, it is more likely to stimulate demand for nonfood items, such as manufactured goods, as well as for hired labour, leading to a higher level of employment for the landless poor.

The overall impact of commercialization, including the expansion of cash or export crops, on food security depends basically on the nature of technology, institutions, infrastructure and policies, including macroeconomic policies. Commercialization of agriculture based on large-scale, capital-intensive, plantation-type export crop production, as was the case in the past when it was centred on an urban export enclave without intersectoral linkages to the rest of the economy, does usually enhance food security. This is especially so if enhanced export production is neither associated with a rise in food imports nor with an increased productivity in food production. Commercialization of agriculture that is based on smallholders and on technological progress in food and export crop production leads to an expansion of income employment directly as well as indirectly in the rest of the economy through intersectoral linkages and can strengthen food security.

Notes

1 It should also be noted that in 1982 cash crops – both food and nonfood – occupied more than 3 per cent of the cropland in 28 out of 78 developing countries, whereas nonfood cash crops occupied more than 30 per cent of cropland in only three countries.

2 The *total nominal rate of protection* is defined as the percentage difference between the relative domestic price (ratio of prices of agricultural commodities to those of nonagricultural commodities) and the relative border price at the equilibrium nominal rate of exchange. The *indirect nominal rate of protection* is the percentage difference between the relative border price at the official nominal rate of exchange and the relative border price at the equilibrium nominal rate of exchange. The *direct nominal rate of protection* is the difference between the total and the indirect nominal rate of protection.

References

Ahmed, R. and Bernard, A., *Rice Price Fluctuations and an Approach to Price Stabilization Bangladesh*, Research Report No. 72 (Washington, DC: IFPRI, 1989).

Grilli, E.R. and Yang, M.C., 'Primary Commodity Prices, Manufactured Goods Prices, and the Terms of Trade of Developing Countries: What the long run shows', *The World Bank Economic Review*, 2(1), pp. 1–47.

Islam, N., *Horticulture Exports of Developing Countries: Past performances, future prospects, and policy issues*, Research Report No. 80 (Washington, DC: IFPRI, 1990).

Krueger, A.O., Schiff, M. and Valdés, A., 'Agriculture Incentives in Developing Countries: Measuring the effect of sectoral and economy wide policies', *World Bank Economic Review*, 2(3), pp. 255–71.

Pinckney, T.C., *The Demand for Public Storage of Wheat in Pakistan*, Research Report No. 77 (Washington, DC: IFPRI, 1989).

Sahn, D.E. and von Braun, J., 'The Relationship between Food Productionand Consumption Variability: Policy implementations for developing countries', *Journal of Agricultural Economics*, 38(2), pp. 315–27.

von Braun, J. Kennedy, E., *Commercialization of Subsistence Agriculture: Income and nutrition effects in developing countries*, Working Paper on Commercialization Agriculture and Nutrition 1 (Washington, DC: IFPRI, 1986).

United Nations Conference on Trade and Development, *Commodity Yearbook* (Geneva: UNCTAD, 1987).

PART VI

FOOD SECURITY: POVERTY AND UNDERNUTRITION

Chapter 18

Agricultural Growth, Technological Progress and Rural Poverty*

How do the pace and the pattern of agricultural growth affect poverty? Are there ways in which an agricultural growth strategy can be designed to have a positive impact on poverty?

Theoretical and empirical analysis has led to divergent answers to these questions. Much depends on the nature of technological progress and its diffusion among different groups of farmers, on its impact on the growth of income and productivity, as well as on the initial conditions under which the growth process is initiated – particularly the distribution of assets, especially land. Frequently, access to land as well as to fixed and working capital for financing modern inputs is unequal, as is access to education, training and research. For growth to reduce poverty and promote equity, the access to all of these factors must be widely diffused. In many countries, experience with the direct redistribution of assets, particularly land, and with public measures directing inputs or resources to poor farmers, has not been very promising. The political and administrative obstacles in the way of direct assaults on poverty are very substantial. This enhances the importance of the poverty-alleviating effects of the growth process itself. To the extent that the latter are positive, the need for a more direct redistribution of assets for poverty-alleviating measures may be less urgent.

In fact, it is advisable to pursue efforts on both fronts – that is to a) maximize the so-called 'trickle-down' effect of growth on poverty while b) undertaking direct poverty-alleviating measures. An appropriate balance between the two is needed so that, in the long run, direct poverty-alleviating measures to not adversely affect the savings, investment, and technological progress that contributes to growth.

* First published in Lewis, John P. (ed.), *Strengthening the Poor: What have we Learned?* (New Brunswick, NJ: Transaction Books in cooperation with the Overseas Development Council, 1988), pp. 121–31.

Direct Effects of Growth on Poverty

Technology

The biological and chemical innovations that have generated most of the rapid growth in Third World agriculture have been scale-neutral in terms of cost per unit. Technologically speaking, therefore, big farmers do not enjoy a differential advantage because of economies of scale. Mechanical innovations, however, have tended to be capital-using and labour-displacing, especially when incentive structures and macroeconomic policies lower the relative cost of capital below its opportunity costs – either because of a subsidized interest rate policy, or because of overvalued exchange rates (where capital equipment comes from abroad). This affects the poor adversely, since most of the poor are either wage-labour or sharecroppers.

Yet not *all* mechanization has a net adverse effect on employment. In some instances, mechanization reduces the time required for land preparation and harvesting, and it increases cropping intensity and hence labour requirements and employment. Even if labour input per unit of output declines, larger output increases the net volume of employment. This is apart from the indirect employment generation in the manufacture, maintenance and repair of equipment.

The direct poverty-alleviating impact of technological progress depends on: a) how far it spreads to the small farmers, sharecroppers, or tenants; and b) how much it increases the employment of the rural poor. Empirical evidence to date confirms that technology *has* spread, though with a time lag, to the small farmers and sharecroppers, who are usually more risk-averse and hence wait until after the income-augmenting effects of new technology are well demonstrated (Hossain, 1987). However, for the spread to occur, two obvious requirements are that high-yielding modern inputs be plentifully available and that infrastructure (i.e., a road and transportation system for the marketing and distribution of improved inputs as well as training, education and extension facilities) likewise be widely available. Another matter of high priority is development – through adaptive and region-specific research – of new technologies suitable for diverse agro-ecological conditions (e.g., differing water, soil, and climatic conditions). In the past, technological progress has often been confined to principal cereals – for example, wheat, rice, maize – and to regions well endowed with irrigation potential and good quality soil, bypassing poor or rain-fed soils and so-called 'inferior' crops such as sorghum, millet and roots and tubers, etc. Yet the agro-ecologically disadvantaged regions are often very poor; poor farmers are also often to be found producing crops that have scarcely benefited from yield-increasing technological innovations. In addition, in contrast to the high yielding varieties of wheat and rice, those improved crop varieties that reduce dependence on purchased inputs generally facilitate easier access to new technology, and its faster adoption, by resource-poor small farmers with limited access to credit. Examples are drought-tolerant or pest- and disease-resistant crop-varieties, short-maturing varieties that increase cropping intensity and the biological fixation of nitrogen. These and similar technological innovations enable a rapid increase in the productivity and income of small farmers and thus tend to reduce the conflict between growth and equity and poverty alleviation.

Credit

How effectively do the poor participate in the market mechanism so that they can gain fully from the benefits of technological progress? Especially relevant in this context is the access of the poor farmers to the markets for credit, labour, inputs and outputs.

Rural credit markets are inadequately developed, fragmented and fraught with imperfections. Nor are there adequate institutions for the investment opportunities mobilization of rural savings, or profitable investment opportunities for such savings – due partly to lack of information or knowledge as well as of transport and communications infrastructure, and partly to the discriminatory pricing and exchange rate policies that depress returns to agricultural investment.

The risks and uncertainties of providing small farmers access to the credit market are frequently overestimated by private lenders; tenants, sharecroppers and small farmers either lack secure title to or do not own sufficient land to serve as collateral for loans. Furthermore, the transaction costs of dispensing and supervising small loans are very high, thus limiting the access of small cultivators to institutional sources of credit.

Delinking credit from the ownership of land as collateral has been tried in a very limited way; the uncertainties of crop failure on the one hand, and of price fluctuations on the other, increase the risk of default and constrain the supply of crop-based loans. Public measures designed to reduce, insure against, or compensate for the price fluctuations can lower the risks for credit institutions – but they involve budgetary costs. Similarly, public subsidy of credit agencies can offset the high cost of small loans – but likewise cost budget money. On the other hand, innovative credit schemes based on group lending or group guarantee by borrowers have contributed to the reduction of both the risks administrative costs.

Experience to date indicates, however, that what is important is not a low or a subsidized rate of interest but access to an elastic supply of loanable funds. Small farmers have demonstrated their ability not only to pay high rates of interest but also to repay on time. What is needed is vigorous competition in the provision of credit through a multiplicity of lenders, both public and private. The credit market is often interlinked with the markets for land and for output; landowners and traders are frequently moneylenders. Different credit institutions developed through public initiative, growth of marketing intermediaries and possible institutional change in the land structure – all of these will help increase competition in the credit market, delink the different factor markets and widen opportunities for small farmers. No less important are legal procedures and institutions for the enforcement of contracts; their weaknesses have adversely affected the viability of credit institutions.

Markets for Outputs and Inputs

In the markets in which they sell and buy commodities, small farmers suffer from a relative disadvantage vis-à-vis big farmers. Often, small farmers with limited resources to hold stocks must sell immediately after harvest at a price lower than the annual average price or the price that prevails later in the year; they buy later in the year, when prices are higher, to meet their own consumption needs. In remote or distant regions, they either receive lower output prices, or pay higher prices for inputs because of high costs of transportation, marketing and distribution. Furthermore, when inputs are in short supply and there are few suppliers – often public distribution agencies – small farmers are likely to lose out, since big farmers with higher status in the power structure have greater access to the public distribution system. Hence what is necessary in a crucial sense is an abundant supply of inputs and credit, as well as a strengthened marketing system to expand and stabilize supplies over time, especially to meet the seasonal variations in requirements of inputs and credit.

Innovation, Employment and Poverty,

The impact of accelerated agricultural growth on rural labour through increased employment and/or an upward pressure on real wages depends on the pattern of growth, including the composition of output and the choice of techniques, as briefly mentioned earlier. In many developing countries, a high rate of growth in population, and therefore in the rural labour force, tends to exceed growth in employment opportunities. The labour market is frequently fragmented or imperfect, with the result that few employers, sometimes acting as both traders and/or creditors, can choose from among a large number of workers seeking employment. Frequently, wages are determined by convention, old habits, or family links. Big landowners often prefer to employ workers known personally to them, and about whose availability, especially at times of peak seasonal demand, they feel sure. Personal contacts and family links influence employment and wages, promote preferential treatment for those within rather than outside a village, and constrain inter-village mobility and employment opportunities, both short- and long-run. Two factors can help strengthen workers' participation in the market. First, education can improve the perception of alternative opportunities and encourage mobility of labour; it can also enhance awareness of workers' rights and help farmers to organize and build bargaining strength. Second, employment opportunities in the nonfarm sector, including urban employment, improve the relative wages of rural labour.

Technological progress requires a careful husbanding and application of modern inputs and increases the use of labour in such operations as land preparation, transplanting, weeding and harvesting, etc. In many instances, small farmers, sharecroppers, or tenants make more intensive use of modern inputs, i.e., higher inputs of fertilizer, modern seeds and water per acre, than do large farmers, at least in the long run, and produce a greater output per acre – even though they frequently pay higher prices for inputs and high interest rates for credit. They maximize net returns (net of purchased inputs) per unit of land by driving down the marginal product of labour to a level at which the average income of the family (total product divided by numbers of family members) is adequate to provide an acceptable subsistence wage. Moreover, with tenants or sharecroppers competing for very scarce land, tenancy arrangements with landlords are likely to involve very intensive use of inputs, especially when the owners share a part of costs and sometimes also provide credit. Higher output per acre can compensate for higher per acre application of purchased inputs and still leave net income per acre at a level higher than that obtained on large farms (Lipton and Longhurst, 1985a and 1985b).

Technological progress in agriculture can have different employment effects, depending on the size of the farms involved. From the point of view of very poor or landless labourers, employment opportunities tend to be positively related to the share of middle or large farmers in the expansion of output; with increased income, these better-off farmers tend to reduce their own rate of participation in the labour force and to hire additional labour. Conversely, small farmers meet requirements for labour by working more hours themselves. Therefore the interests of small farmers and those of landless labourers without access to nonfarm employment opportunities do not coincide.

The concentration of technological progress on small farms alone may not expand the wage-employment opportunities of the landless as much as if landowners, especially the middle farmers, also participate in technological progress. This may imply, however, that income distribution in fact worsens as the rich gain proportionally more than the poor, while the absolute condition of the poor improves.

Indirect Effects of Growth on Poverty

Technological change in agriculture has indirect effects of two kinds on the alleviation of poverty. The first is the impact of technological change on the supply of food and on its relative price and, consequently, on real wages and employment in both the agricultural and non-agricultural sectors. The second is the multiplier effect on income and employment in the non-agricultural sectors set in motion by the increase in agricultural production and income. Both of these indirect effects are greater if the gainers in the first instance include many not-so-poor middle farmers. But the direct effect on poverty is greater if the beneficiaries in the first instance are exclusively the poor, especially the smallest farmers and landless labourers.

Growth Patterns, Food Supplies, and Prices

The effect of increased agricultural production on domestic food prices depends on how closed or open the economy is to international trade. In an open economy, domestic prices will be equal to world prices, whereas in a closed economy, variations in domestic production will affect the domestic price. In reality, the economies in developing countries are in an intermediate position; they are neither completely closed nor totally open to international trade. Domestic prices diverge from world prices due to tariffs and quantitative restrictions on exports and imports. For a commodity that is domestically produced and partly imported under tariffs, the domestic price is higher than the world price by the amount of the tariff. The domestic price falls when the shift in supply function or increase in productivity is large enough to cause domestic supply at the prevailing price to exceed domestic demand. The new domestic price, however, may remain at a level higher than the world price. There will still be no imports at that price; tariffs remain prohibitive. But with a further shift in the supply function, the domestic price may fall to the level of the world price, and the country may change from importer to exporter status.

The domestic price can be lower than the world price either because there is an export tax, the incidence of which falls fully or partially on the exporter, or because consumers pay subsidized prices that are implemented either through public procurement of domestic output at less than world prices or through subsidies on imports. With an export tax, an increase in productivity and a shift in the supply function would not affect the domestic price. In the second case, what happens to the domestic price depends on the government price policy: it can let prices fall below the previously fixed prices to match the reduction in cost while subsidies are maintained, or it can abolish subsidies, keeping prices unchanged. If the cost reduction is large enough, it can reduce both price and subsidies.

The higher the rate of growth in marketable surplus in relation to demand, the greater the decline in the relative price of food – insulated partially or fully from the world market. The magnitude of the marketable surplus depends not only on the rate of growth in food production, but also on how the increase in production is distributed among different categories of farmers. Medium and large farms have a higher ratio of marketable surplus to output than do small farms. If the increase in food output is generated largely by small farms, this will result in a small marketable surplus. Small farmers have high income elasticity of food demand; they may frequently have a larger family size and a greater dependency ratio. Accordingly, an increase in output on the small farms tends to be consumed by the farmers themselves. If, on the other hand, a substantial increase in food production is provided by a very few large producers, then the fall in price is likely to be so large as to discourage production and detract from the favourable effects on employment and income flowing from increased production by large farmers. This is because the market demand for food will be constrained in the absence of an adequate increase in income of the small fanners, who are far more numerous. There has, therefore, to be an appropriate balance in the distribution of increased output among the different categories of farmers so that a fall in price resulting from an increase in output is not so severe as to discourage production.

The indirect favourable impact on food prices is most likely to be realized if the increase in production takes place mainly among the middle and small farmers, but *not only* among middle and small farmers. In many countries, governments intervene to stabilize farmers' prices or to support minimum prices through buffer stock operations. The politically vocal or organized large and middle farmers may exercise pressure to support prices at levels that allow farmers – especially big farmers – to reap excess profits from cost-reducing innovations rather than pass them on to the consumers. The optimum policy therefore calls for allowing prices to fall to an extent that maintains producers' incentives while conferring benefits on the net purchasers of food. Given the differential impact of technological progress on different categories of farmers, or different farm sizes, a steep decline in price may reduce the net income of the small farmers who, at least in the early stages of the diffusion of new technology, do not benefit, or do not benefit adequately, from cost-reducing innovations. Thus the price of progress may be that, initially, the pioneers or the early innovators earn excess profits.

Low food prices resulting from cost-reducing innovations directly improve the real income of the poor, who are the net purchasers of food (Mellor and Desai, 1985; Mellor, 1978). By making available a large supply of the principal wage good – food – at a stable and low price, technological progress in food production facilitates the adoption of an employment-based development strategy, especially in the nonfarm sector. Otherwise, as employment and income of the poor expand, given their high income elasticity of demand for food, a rapid increase in food demand is likely to put an upward pressure on food prices.

Furthermore, a fall in food prices improves the terms of trade of the nonfarm, industrial sector and tends to lower wages in the nonfarm sector and to encourage labour-intensive industrialization and the substitution of labour for capital in various processes and products. Thus food price declines also enhance the competitiveness of labour-intensive products in the world market and thus promote exports, which in turn adds further stimulus to the expansion of output and employment in the nonfarm sector.

The role of marketing, distribution and transaction costs in the movement of food from rural producers to urban consumers is crucial. An increase in such costs would offset the cheapening effects of technological progress on the relative price of food.

Intersectoral Linkages

Increased agricultural output and income have a multiplier effect on income and employment. There are certain linkages within the agricultural sector itself: for example, an expansion of income in staple food or cereal production leads to a higher percentage being spent on horticulture and livestock products. When carried out on small farms – as they are in many developing countries – these activities are labour-intensive and help expand employment and income of the poor.

An increase in agricultural production affects the rest of the economy via intersectoral linkages. On the one hand, it leads to increased demand for inputs such as fertilizers, pesticides, irrigation equipment and other tools and implements; on the other hand, it stimulates demand for processing, marketing and distribution of agricultural output. Empirical evidence suggests that the forward linkages (i.e., marketing, distribution and processing, etc.) of agricultural products provide a much stronger linkage than the backward linkages (i.e., expanded use of production inputs). More important than the production linkages are the consumption linkages – stemming from the expenditure of incremental agricultural income on non-agricultural goods and services – produced in both rural and urban/semi-urban areas (Haggblade, Hazell and Brown, 1987, pp. 113–25; Hazell and Roell, 1983, p. 12). These linkages include the stimulation of not only a wide variety of manufactured consumer goods, but also trade and other services; they are frequently highly labour-intensive. The employment-creating role of small and cottage industries has received much attention. But the role of rural trade and services (which include housing, education, health, transportation and personal services, etc.) as a source of employment and income for the poor has not been sufficiently recognized.

The growing demand for non-agricultural goods and services (upon which, of course, employment expansion depends) must originate primarily from agricultural growth, especially the expenditures of middle and large farmers who devote larger shares of their income gains to nonfarm, labour-intensive goods and services. The expenditure patterns of very small farmers, who spend a higher proportion of incremental income on food than on non-food items, provide a weak intersectoral linkage. Wherever the middle farmers are numerous, their absolute aggregate expenditure on non-food items is often as great as, if not greater than, the aggregate expenditures of a larger number of small farmers. Starting from this large base, the fact that middle farmers channel higher percentages of whatever increments on income they get into expenditures on non-food items provides a strong stimulus for the nonfarm sector's output and employment. In Malaysia, a study in which the largest farm size covered was 15–20 hectares showed that middle farmers (defined as those from the fourth through the seventh deciles in terms of farm size) spent 63–66 per cent of their incremental income on nonfarm goods and services in 1972–73. Of the nonfarm goods, about three-fifths were produced locally and two-fifths came from outside the region. In Africa, the sharees of incremental income that medium-sized farmers spent on nonfarm goods and services were lower, i.e., 24 per cent in 1976–77; those on locally produced goods and services were about 11 per cent. But the incremental income shares spent on livestock and horticultural products, which were also highly labour-intensive and locally produced, were higher – as high as 30 per cent. In Bangladesh, the shares spent by middle-sized farmers on nonfarm goods and services were about 25–30 per cent.

The expansion of the nonfarm sector in the rural areas provides employment and income for the rural poor; this, in turn, creates additional demand for food as well as non-food commodities, thus setting in train successive rounds of income and employment expansion. In Malaysia, an increase of $10 in agricultural income was associated with an increase of $8 in rural nonfarm income; in India, a 10 per cent increase in agricultural employment was estimated to lead to a 10 to 13 per cent increase in nonfarm employment (Hazell and Roell, 1983, p. 12; Haggblade, Hazell and Brown, 1987, pp. 113–25).

However, the consumption expenditure patterns of the richest or largest farmers do not have strong positive interlinkages for income and employment expansion in the nonfarm sector. First, the propensity of these farmers to spend on imported goods is high – constituting a leakage from the domestic multiplier effect. Second, the saving propensity of these farmers is also high and constitutes a leakage from the consumption expenditure stream. It is true, of course, that if these savings are invested in labour-intensive activities, whether rural, or urban, they would have positive employment effects. Nevertheless, the net leakages tend to be greater than in the case of the medium farmers, whose consumption patterns are more domestically oriented.

The demand pull provided by agricultural growth needs to be matched by an elastic supply response from the nonfarm sector. A few preconditions need to be met if the rural nonfarm sector is to respond strongly and positively to the stimulus provided by increased expenditures by the farm sector. Most important among these is the availability of physical infrastructure – a road and transport system, a communications system, rural credit to finance both current and investment costs of nonfarm activities, as well as education, extension and training relating to nonfarm activities. The role of government policy in either directly providing – or stimulating the private sector to provide – these prerequisites is crucial.

Conclusion

Targeting rural technologies and public investments on small farms leads to immediate gains in equity and production – although not necessarily to gains for the landless or for marginal farmers, who cannot respond by expanding output. If the focus is exclusively on *small* farms, the indirect impact on growth in income and employment through the expenditure multiplier is not likely to be as high as it would be if the increase in production largely took place in the first instance on middle-sized farms. Focus on *middle-sized* farms, on the other hand, may have a high multiplier effect on growth, with a favourable impact on absolute poverty – but at the cost of worsening the relative distribution of rural incomes.

Therefore, there is a trade-off between growth and equity or poverty alleviation; a balance needs to be struck by targeting technology and investment on a broader range of farm-size groups, including *both* small and medium farms. The degree of absolute poverty initially, the relative distribution of farm sizes and the urgency of immediate action to relieve poverty will all affect the appropriate policy choices.

References

Haggblade, S., Hazell, P. and Brown, J., *Farm/Nonfarm Linkages in Rural Sub-Saharan Africa: Empirical Evidence and Policy Implication*, unpublished discussion paper (Washington, DC: World Bank, 1987).

Hazell, P.R. and Roell, A., *Rural Growth Linkages: Household Expenditure Patterns in Malaysia and Nigeria* (Washington, DC: IFPRI, 1983).

Hossain, M., 'Fertilizer Consumption, Pricing, and Foodgrain Production in Bangladesh', in B. Stone (ed.), *Fertilizer Pricing Policy in Bangladesh* (Washington, DC: IFPRI and Bangladesh Institute of Development Studies, 1987).

Lipton, M. and Longhurst, R., 'Labor and the MVs', in *Modern Varieties, International Agricultural Research, and the Poor*, Consultative Group in International Agricultural Research, Study Paper No. 2 (Washington, DC: The World Bank, 1985a), pp. 49–64.

Lipton, M. and Longhurst, R., 'Putting Together the MV-Poverty Mystery', in *Modern Varieties, International Agricultural Research, and the Poor*, Consultative Group in International Agricultural Research, Study Paper No. 2 (Washington, DC: The World Bank, 1985b), pp. 87–100.

Mellor, J.W., 'Food Price Policy and Income Distribution in Low-Income Countries', *Economic Development and Cultural Change*, Vol. 27, No. 1, October 1978, pp. 25–6.

Mellor, J.W. and Desai, G.M., 'Agricultural Change and Rural Poverty: A Synthesis', in J.W. Mellor and G.M. Desai (eds), *Agricultural Change and Rural Poverty: Variations on a Theme by Dharm Narain* (Baltimore, MD: The Johns Hopkins University Press, 1985), pp. 192–9.

Chapter 19

Undernutrition and Poverty: Magnitude, Pattern and Measures for Alleviation*

Magnitude and Pattern of Undernutrition

What is the magnitude and pattern of undernourishment in the developing world? The nutritionists are not all unanimous on the methodology of estimating undernutrition. Controversies cover a number of aspects. First, there is the controversy relating to the relative importance of calorie deficiency, on the one hand, and deficiency of protein, as well as of vitamins, on the other. It is now agreed among the majority of nutritionists, however, that in most countries the composition of diets is such that whenever they provide enough calories, they also meet the minimum protein requirements. This is true so long as the diet does not consist exclusively of one or two very bulky carbohydrates, so that the amount of carbohydrates which provides adequate amounts of both calories and protein is too large to be consumed and absorbed per day by an average individual. The lack of adequate vitamins and minerals, even if adequate calories are available, causes specific diseases but does not cause an endemic lack of energy and general ill health, which is at the core of undernutrition. The dominant problem in most developing countries is a lack of adequate calories.

Second, even if deficiency in calorie intake is recognized as the most important aspect of undernutrition, dietary energy requirements vary depending upon age, sex, and level of physical activity of individuals (Osmani, 1985; FAO, 1985). There is no consensus among experts as to whether the actual or some optimum level of physical activity should be used in determining dietary energy requirements. The estimates of undernutrition by the Food and Agriculture Organization of the United States (FAO), for example, have been based on the assumption of a minimum level of physical activity, that is, energy expenditure involved in eating, washing, dressing and sleeping. The basic metabolic rate (BMR) of an individual is a rate of energy expenditure in

---

* First published in Helmuth, J.W. and Johnson, S.R. (eds), *World Food Conference Proceedings, Vol. 2: Issue Papers* (Ames: Iowa University Press, 1989).

a fasting rate to complete rest in a warm environment, which varies between individuals depending upon age and sex (Osmani 1985; FAO 1985). Satisfactory data on energy expenditure rates for different types of activities in different circumstances and countries or on time-allocation of individuals/ groups of individuals among different types of activities are hard to come by.

There are two additional problems in the estimation of dietary energy requirements. First, individuals can adapt to certain minimum levels of caloric intake without danger to their health or without 'physiological dysfunction' by means either of metabolic regulation – that is, varying the efficiency of dietary energy utilization by changing body composition – or by reducing weight. Among these alternative possibilities of adaptation, the latter is more likely than the others (Osmani, 1985). This phenomenon results in intra-individual variations in energy requirements and may thus cast doubt on the concept of average calorie requirements of individuals. Since BMR is calculated from body weight and since body weight adjusts within limits to changes in calorie intake without danger to health, energy requirements are estimated by FAO on the basis of the lower limit of the range of desirable weight for a given height. Second, the use of an average requirement does not allow for variations in requirements arising from genetic differences between individuals. This creates the possibility of misclassification but the direction of bias is indeterminate. To use an average requirement to estimate the undernourished population implies that there will be some among those classified as undernourished who will not be undernourished and some who are undernourished will be excluded.

The possibility of metabolic adaptation to food intake tends to suggest that the calorie requirements should be set not at the average level but at the lower end of the calorie distribution, i.e., standard deviations away from the average caloric (1.2 BMR). On the other hand, the presence of inter-individual differences in energy requirements supports the use of average energy requirements as the cut-off point (1.5 BMR) on the assumption that statistical errors of overestimation or of underestimation cancel each other out (FAO, 1985; Osmani, 1985).

Direct observations on the distribution of food intake by individuals are difficult to come by in all countries. In many cases, therefore, indirect estimates of calorie intakes are made on the basis of a country's distribution of income or of aggregate food expenditures. This is on the assumption that the distribution of income or aggregate food expenditure are both monotonic correlated to the distribution of calorie intake. This correlation does not always hold in practice. Higher income does not always lead to a higher calorie intake.

With increasing incomes, higher-valued food commodities are consumed which do not necessarily have higher nutrient content; that is, the same nutrient value is obtained with a higher expenditure. Additionally, when a certain amount of minimum requirements as perceived by individuals is met, even at a relatively low level of income, an increase in income may lead to a temporary decline in the percentage of income spent on food with a view to attaining a minimum level of non-food basic items, after which high income elasticity of demand for food is restored.

Estimating the Number of Undernourished

On the basis of above assumptions, FAO estimates the number of the undernourished population in the developing countries during 1979–81 was between 348 and 512 million for 89 developing countries, excluding China. For 1983–85, between 15 per cent and 22 per cent of the population of these countries was undernourished. The percentage of their undernourished population declined over the years, that is, from 19 per cent and 27 per cent during 1969–71. While differences exist among different regions in terms of percentages of the population that are undernourished, ranging from 35 to 26 per cent in Africa and 14–22 per cent in Asia to 10–14 per cent in Latin America and 9–6 per cent in Near East/North Africa, the large majority of the world's undernourished population lives in Asia (58 per cent) and in Africa (28 per cent) (FAO, 1987a).

There are wide divergences among countries and regions in the developing world in terms of per capita caloric availability. Asia (excluding China) and Africa, with per capita daily calorie supplies at 2,250 and 2,050 respectively during 1983–85, are below the average for the whole developing world; the difference in per capita calorie supplies between Africa and Asia is not as striking as the difference in the percentage of the population that is undernourished (FAO, 1987a).

The various estimates of the magnitude of the undernourished population are based on the assumption of ideal environmental conditions, that is, ideal sanitary and health conditions including pure drinking water. Ill health and infectious disease both reduce energy intake and increase energy requirements, and thus may contribute to undernutrition. Malabsorption of food, especially in diarrheal diseases, reduces the effective utilization of food. Parasitic elements residing in gastrointestinal tracts claim a share of the food and thus reduce the amount available to the body. Energy is also diverted to the production of protective factors or elements in the body for resistance against infection. All of these factors increase an individual's energy requirement. Loss of appetite through illness and infection reduce the intake of food and, in some cultures or societies, food allocation is reduced during illness.

In summary, the quantitative measurement of undernutrition thus depends upon assumptions regarding the degree of adaptability of individuals to variations in food intake, inter-individual differences in calorie requirements, levels of physical activity of individuals and the individual's health and sanitary environment. None of the prevailing estimates of the undernourished population include environmental factors. The FAO estimates of undernutrition in 1983–85 include a high and a low estimate; the low minimum estimate ($350 million) allows for intra-individual variability and for adaptation to low food intake through the adjustment of body weight; the high estimate ($512 million) is based on the concept of average requirements, which allow inter-individual differences, and hence may include some who are not undernourished and exclude some whose requirements are higher than the average. Neither the low nor high estimates of FAO allow for any economically or socially necessary level of physical activity. Once allowances for an average level of physical activity are made, as in the World Bank estimates, the estimated number of the undernourished population may go up to 700 million or more. (What is an average level of physical activity, and what are the calorie requirements for different degrees or types of physical activity are all matters of controversy, as has been mentioned earlier.)

Whatever the exact number, the number of the undernourished population in the developing world is significantly high. But it is not so staggeringly high as to discourage efforts to alleviate it. In this context, one can in some sense consider the low estimate as representing those who are the most severely undernourished.

The Seasonality of Undernutrition

A number of other issues relating to undernutrition have important policy relevance. There is a seasonality in the pattern and magnitude of undernutrition. This is derived from the seasonality in: a) the pattern of physical activity or energy expenditure, which determines energy requirements; and b) in food intake, arising from seasonal variations in the price and availability of food supplies as well as in employment and income. During the lean season, when food stocks are depleted and new supplies have yet to come to the market, employment opportunities may be limited, and for certain farming operations, there may be an increased energy expenditure.

In terms of nutritional status, the vulnerable groups frequently include women, especially lactating and pregnant women, and children. In these cases the gap between requirements and supplies of calories is often larger than the average gap for all households. This is partly related to the intra-household distribution of food and other basic non-food requirements. There is some evidence in a few Asian countries, such as the Philippines and Bangladesh, that selective deprivation of women takes place, whereas in Africa, for example, Kenya and Gambia, and in parts of Latin American, where women do the most work, they are often fed better. In some African countries women get paid a 'bride price' because they are considered to be of great economic value. Also girls, more than boys, accompany their mothers during the day and benefit more from traditional nurturing and feeding.

Access to food and access to hygiene, including personal and environmental hygiene, complement each other. In particular, an increase in per capita income or household income does not lead to a fall in the incidence of illness, especially in children. Diarrhoea has a significant negative effect on a child's growth. Although increases in income have a positive impact on children's calorie intake, the impact is small. Diarrhoea, energy intake, and mother's height and/or her nutritional status all affect children's nutrition.

During the 1980s, as a result of the impact of structural adjustment policies undertaken in many developing countries, including heavily indebted countries, poverty and undernutrition have tended to worsen. Reductions in public expenditures, including those on health and nutrition measures as well as on development policies, frequently combined with a cut in food subsidies and a corresponding rise in food prices, have contributed to the fall in employment as well as the rise in poverty in several countries (Helleiner, 1987; Demery and Addison, 1987).

The major thrust of the remainder of this chapter is on the general macroeconomic and the agricultural sector issues bearing on poverty, hunger and undernutrition; other issues such as health and environmental aspects, though important, are not dealt with except in passing.

Measures for Alleviating Hunger and Undernutrition

In spite of the high hopes and solemn resolutions of the UN World Food conference in 1974 held in the aftermath of the World Food Crisis of 1974 to ensure that no child should go to bed hungry and that no family should fear for tomorrow's bread, widespread hunger and poverty persist. Adequate and stable food supplies and assured access on the part of the poor and hungry to the basic foods that they need are the three basic foundations/cornerstones on which poverty-alleviating measures need to be based.

At all levels, whether global or regional, and in most developing countries, per capita calorie supplies have increased over the past two decades. The rate of growth of food production in developing countries as a whole since the 1970s has exceeded the rate of growth in population, except in Africa in the past decade or so. However, at the same time, per capita demand exceeded per capita food production, resulting in increasing food imports by developing countries; the largest share in the food imports of developing countries (60 per cent or more) belongs to middle-income countries. The food imports of the poorer developing countries are constrained by their low income and limited export earnings to pay for imports. Thus worldwide adequacy does not ensure sufficient supplies for poor countries with a foreign exchange constraint; food aid finances more than 20 per cent of cereal imports of low-income food deficit countries (FAO, 1987b).

National self-sufficiency in food production does not by itself promote greater access to food on the part of the poor. In the Near East and in Latin America, the percentage of the undernourished population fell significantly between 1969/71 and 1983/85 (from 15.7 per cent to 5.6 per cent in the Near East, and from 12.7 per cent to 9.5 per cent in Latin America) while at the same time their food (cereal) self-sufficiency declined, more drastically in the Near East than in Latin America, which turned from a net exporting region (105 per cent self-sufficiency) to a net importing region (96 per cent self-sufficiency). Africa, for example, is a region where both food self-sufficiency declined and undernutrition expanded during this period. India has achieved self-sufficiency in recent years but contains vast numbers that are undernourished (FAO, 1987a).

The Failure of Entitlement

Adequacy of food supplies at the aggregate level derived from domestic production and/or imports does not by itself contribute to the alleviation of poverty nor does it prevent the occurrence of the extreme form of food deprivation typified by famines. This is because the poor do not have adequate purchasing power to secure access to food. The failure of 'entitlement' or effective demand can occur in a variety of ways which are linked to the system of production and distribution of income in an economy (Sen, 1981). Those who are food producers need adequate land and other inputs to produce food;

those who purchase food require access to other assets through the use and exploitation of which they can acquire command over enough income to purchase food. Alternatively, they require employment at an adequate wage. Among the purchasers of food, those who produce non-food or non-agricultural commodities must sell enough of other commodities at prices which secure an income high enough to ensure access to adequate food. A shortfall in food production can cause deprivation or hunger, either because food production is the main source of income and purchasing power, or because a fall in output causes a rise in food prices which, without a corresponding increase in the income of food purchasers, results in a decline in food consumption. A fall in entitlement may be caused by a fall in the price of other assets, such as cattle or commodities to be exchanged against food (a fall in the terms of trade), a decline in employment or the wage rate, or a rise in the price of food caused by a general expansion of monetary demand.

The multiple source of either inadequacy or a decline in entitlement and of access to food outlined above, both in the short- and long-run, emphasize that first, alleviation of undernutrition requires action not only on the supply side but also on the correction of various causes of demand failure. Second, the failure of entitlement and effective demand arises not only within the agricultural sector but also outside it, as evidenced by an increasingly large number of poor people residing in the urban areas.

Urbanization has important implications for: a) the pattern or composition of food demand; and b) the marketing and distribution of food which is produced in rural areas but consumed in urban areas. Urbanization, independently of changes in the level of income, causes a shift of the consumption pattern away from traditional crops such as sorghum, millet, roots and tubers, toward cereals such as maize, wheat, and rice. Not only the rich but also the poor urban consumers are increasingly dependent on cereals, which are mostly imported and which, in many cases, cannot be substituted by domestic production due to agro-ecological constraints, especially in Africa. National food pricing policy, combined with an overvalued exchange rate, often cheapens imported cereals vis-à-vis traditional crops, which in the past have not benefited from cost-reducing technological innovations either in production or in storage and processing. The latter suffers, in addition, from the high costs of marketing and distribution. Traditional crops, unlike cereals, cannot be processed for the preparation of convenience foods, a necessity in the context of an increasing participation of women in the labour force, which leaves little time for elaborate methods of preparation required by traditional foods (Reardon and Delgado, 1986).

Changes in food consumption patterns are inevitable in response to increasing incomes and changes in consumers' preferences. What is needed is to ensure that the speed of such changes does not outstrip the capacity for growth and adjustment of the domestic economy, including agricultural potential, in line with comparative cost considerations and patterns of food demand. The agricultural sector needs to adjust changing demand structures through technological change, in both production and processing. This has to be encouraged by appropriate incentives, including incentive prices, to exploit its comparative cost advantage.

Neither self-sufficiency in food production nor the pursuit of a policy of 'Food First' emphasizing food crops to the exclusion of cash crops, is crucial for the alleviation of hunger and undernutrition. The degree of self-sufficiency sought by a country should be governed by its comparative costs of production, the equilibrium rate of exchange and degree of security of access to world supplies. The greater the uncertainty surrounding access to world supplies, as perceived by importing countries (due maybe to export embargoes or restrictions, threat of war, or interruptions in transportation, etc.), the greater is the incentive for reliance on domestic production. The controversy regarding food crops versus cash crops is greatly exaggerated; they are often jointly produced in a system of mixed farming, dictated by considerations of efficient crop husbandry. Furthermore, farmers who are predominantly producers of cash or export crops diversify their cropping patterns to include food crops as well in order to reduce risks associated with crop failures, or with interruptions and fluctuations in the supplies and market prices of food. Frequently, specialization in export/non-food crops is warranted by comparative cost considerations, and/or by agro-ecological characteristics.

Past empirical evidence indicates a positive association between the growth in overall or per capita staple food production and growth in the area devoted to cash crop production (Braun and Kennedy, 1986). Food crops are often sold as cash crops, and there are developing countries with food exports as major sources of foreign exchange. The adverse impact of non-food cash or export crop production on poverty/hunger is often traced to a number of associated circumstances, which are not by themselves the inherent or immutable characteristics of export crop production. It is argued, for example, that export/cash crops are often produced in large-scale plantations or with capital-intensive techniques which have limited or no favourable impact on incomes and employment of the rural poor. Also, earnings from export crops are spent not on food purchases for the producers of export crops, but on imports of consumers' goods (including luxury goods) for the urban rich. In addition, an expansion in any one region in the production of the export cash crop, which diverts land and other inputs away from food production, is not always offset by an increase in the supply of food from other regions. This is often due to the underdeveloped state of the transportation, marketing and distribution systems.

The above are examples of either policy failures or institutional bottlenecks which can be resolved or removed given an appropriate policy framework. This should promote the participation of the growers of export crops in the overall national gains which accrue from an efficient expansion of export/cash crops. A deterioration in income distribution or an increase in poverty arising from policies which discriminate against the poor, who happen to be producers of export crops, requires additional or offsetting measures for the alleviation of hunger. A similar case is one in which food prices for those who are engaged in expanding the production of export crops rise due to shortcomings in the organization of food production and marketing. Evidence based on a number of case studies in developing countries on the introduction of new export crops or of a shift from food to cash/export crops indicates that the income of the farmers as well as their calorie intake increased as a result of increased commercialization. The impact on calorie intake and on nutritional status was positive but small; incremental income was spent on non-food items as well. There was in no case a decline in the nutritional status of the average farmer. However, changes in cropping pattern or technology adversely affected the income of particular groups in some instances. For example, in the Philippines large landowners acquired land for sugarcane cultivation away from corn-growing tenants, who had no secure rights, turning the tenants into landless wage labour in sugar plantations (Braun and Kennedy, 1986).

Instability in food supplies and prices, as well as in the entitlements/incomes of the poor, aggravates hunger and malnutrition. This instability hits the poorest countries and the poorest people the hardest since they have very little reserve either of food or financial resources to fall back upon. The loss or sale of assets to meet short-term food crises leads to long-run deprivation and poverty. Infectious diseases or ill health suffered during periods of food shortages often impair long-term health and ability to work. A major source of instability is in food and agricultural production, which is a source both of food and income for the poor. To this is added, at the national level: a) instability in export earnings which causes instability in income and employment in the export sector as well as in import capacity; and b) instability in macroeconomic factors affecting both rural and urban income. Frequently, technological progress increases not only the average level of crop yields but also the variability of yields due to a variety of reasons, that is, narrowing of the genetic base of crops, dependence on a few high-yielding varieties, variability in the availability and price of purchased inputs which assume growing importance, greater vulnerability to the impact of the weather fluctuations or pests and diseases, etc. Measures are needed both to reduce the variability of output as well as to offset its adverse impact on the poor.

Patterns of Growth, Technological Progress and Poverty

The challenge facing developing countries is how to design an agricultural development strategy which would augment food supply and, at the same time, increase the income and employment of the poor so that they can secure access to food at a price which they can afford. The poor are not only the small farmers, often with land of a size which does not provide adequate income, but also the rural landless labourers and the urban jobless. Frequently, the poor farmers are located in arid or infertile regions with uncertain or inadequate rainfall. In particular, the questions besieging policy makers are the following: how should one increase the productivity of the small farmers, often subsistence farmers, with an inadequate endowment of resources or agro-ecologically inferior lands? What are the policies to provide incentives to surplus producers to expand output and the marketable surplus as well as to provide an increased employment for the landless? How can employment be expanded for the landless and jobless in both agricultural and non-agricultural activities in the rural areas?

Distribution of land to those without land, or with very little land, provides them with a source of employment and income as well as of food supply. Whenever land-holdings remain unutilized or are inadequately utilized and small farmers carry on an excessively intensive cultivation, leading to soil degradation, a redistribution of land contributes to growth, sustainability and equity. While the political difficulties in the way of land redistribution are often enormous, it is also evident that without a redistribution of land under circumstances where: a) land distribution is highly unequal; and b) the feasible rate of expansion of nonfarm employment to absorb the landless falls short of what is needed to make a dent on the income of the poor, a major assault on rural poverty is unlikely and the chances of the rural poor shifting to the cities to add to the urban unemployment and poverty are very high. (The Gini coefficients of land distribution in Latin America and Near East are much higher (0.7–0.8) than in Asia and Africa (0.4–0.6) (FAO, 1986, 1987).)

Where the majority of the rural poor are small farmers, a strategy designed to expand their access to high-yielding inputs, credit and water would help reduce poverty. Experience to date confirms that the small farmers not only adopt but also make an efficient use of modern technology, provided supplies of inputs are available and extension and educational efforts reach them. Inequality will probably increase but poverty may be reduced: a) if sufficient supplies of services and inputs are available; and b) if institutions are developed to organize small farmers to enable them to take advantage of the opportunities available to them. That a part of the services and inputs intended for the small farmers will be diverted to the rich landowners is a necessary consequence of an unequal, rural sociopolitical structure. Furthermore, new technology needs to spread to widely diversified agro-ecological regions, especially in rain fed, low-productivity regions as well as to a wide variety of crops, including traditional crops which are often the poor's main source of employment and food (Lipton and Longhurst, 1988). When the rural poor are predominantly the landless, expansion of employment in both agricultural and non-agricultural activities remains the major policy instrument.

The Critical Role of Women

Both analysis and experience tend to confirm that in any poverty-oriented development strategy, women occupy a critical role. In many countries women play an important role in agricultural production in both food and cash/export crop production; in some cases, they play a greater role in food production than in the production of cash crops. Insofar as the nutritional impact of an increase in household income and expenditure is concerned, what is important, however, is not which particular crops or activities generate family income but who controls income and household expenditures, especially food expenditures.

Modernization of agriculture affects women's role as a producer and provider of food; it affects the allocation of their time between various activities within and outside the household. The greater their participation in agricultural operations outside the household, the less time is available for other household chores including the provision and preparation of food and the care of children, both of which affect the nutritional status of the family. Traditionally, women's access to inputs and services for agricultural production has been constrained by the sex-bias in the delivery system and institutions; women's lack of command over land restricts their access to credit obtained on the security of land. Measures to expand their access to requisites for increased agricultural production are critical in promoting both equity and growth, especially when women-headed households are often the poorest families.

Evidence in a number of developing countries indicates that increasing women's control over household income and expenditure is likely to have a favourable impact on nutrition. It is the absolute increase in income rather than their relative share which is important. For example, with the introduction of irrigation in rice production in Gambia, there was a shift in relative control over production and income from women to men, as the communal land was assigned to men away from women in order to promote an efficient management of the new irrigation technology. But, at the same time, new technology greatly increased women's labour inputs and their wage earnings; it also improved total calorie consumption and the nutritional status of the family. Furthermore, a greater availability of food, especially during the 'wet or hungry season', prevented a drop in food consumption and a loss in weight during that season. This resulted in an overall improvement in the nutritional status of the family.

The Role of Modern Technology

Modern technology, especially biological and chemical innovations in agriculture, increases employment; intensive application of modern inputs requires increased amount of labour per hectare and secondly, it makes possible greater cropping intensity. Macroeconomic policies regarding the price of capital and labour, that is, the rate of interest and the wage rate, have an impact on the choice of techniques, that is, a lower relative price of labour will favour a more labour-intensive technique, and activities which can contribute towards employment expansion. Price incentives play an important role in the spread of modern technology; the responsiveness of supply to price incentives is greater, the higher the level of technology and infrastructure. Cost-reducing technological innovations reduce the price of food and thus increase the real income and food consumption of the poor. This process is facilitated if landowning middle farmers enjoy a rapid technological progress, expand their marketable surplus and are willing to share part of their gains in productivity with the consumers through a reduction in price, rather than if they themselves acquire the entire benefits through higher profits (Mellor and Ahmed, 1987; Mellor and Desai, 1985). A low price of food facilitates labour-intensive development, including labour-intensive industrialization and export expansion.

Furthermore, a higher percentage of the increased income of the middle landowners, unlike in the case of the richest landowners, is spent on labour-intensive services and consumer goods often produced in the rural areas. The net result is an increase in nonfarm employment and hence in the income and food demand of the poor which in turn stimulates food and agricultural production. This is in addition to the forward and backward linkages of agricultural growth stimulating the production of inputs for agriculture, on the one hand, and the processing of agricultural output, on the other. In some instances, an increase of $10 in the income of the agricultural sector has resulted in an increase of $8 in non-agricultural income as a result of linkages between agricultural and non-agricultural sectors (Hazell et al., 1983; Wanmali, 1985).

An employment-oriented agricultural development strategy needs to identify and encourage an expansion of production among those landowning-farming classes who, given the distribution of land and rural income in a country, share the twin characteristics of: a) highly labour-intensive food and agricultural production techniques and activities; and b) a pattern of consumer demand which emphasizes labour-intensive consumer goods and services. Infrastructure, credit and technical education are essential components of a policy designed to stimulate nonfarm agricultural employment in response to agricultural growth. Also important is a pricing policy for inputs and capital which stimulates labour-intensive techniques/activities.

The uneven spread of technological progress, especially among the small farmers, while the large landowners reap the initial benefits of new technology, may cause a loss in real income for the farmers as they confront a decline in output price without a corresponding reduction in cost. Furthermore, the high profitability of new technology may lead the large landowners to acquire land from tenants for self-cultivation by converting the latter into landless wage labourers.

The adverse side-effects of technological progress on the poor such as these need to be counteracted among others by: a) a fast diffusion of technology to the small farmers; b) an output price support policy which prevents precipitous price declines following the introduction of new technology; c) provision of security of tenure for tenant farmers; and d) organizations of landless labourers to counter the imperfections of the labour market adversely affecting the conditions of their employment.

The foregoing illustrates that in the absence of institutional change and asset redistribution, under conditions of high concentration of land and economic power in rural areas, the promotion of growth concurrently with equity or, in some cases, with a reduction in the incidence of poverty, is a very challenging and complex task.

Direct Measures for the Alleviation of Poverty

The impact of overall growth on poverty is seldom fast and large enough to make a dent on poverty and undernutrition in the near or medium-term future. Furthermore, growth itself may be slowed down by an undernourished population which lacks energy and is ill-equipped to take advantage of opportunities for work or employment. Among the measures which have a direct impact on poverty and undernutrition are food subsidies and/or direct transfers of income to the poor. Food subsidies can be used to provide food to the poor at less than the market price. Also, direct income transfers can be made to the poor, but such transfers need to be combined with an increased supply of food or else rising food prices tend to offset the transfer of income.

Since low-income groups spend 40–50 per cent of their income on food, a reduction in the prices of food staples through subsidies by 20 per cent raises the average income of the poor by 8–10 per cent. Since expenditure on cereals frequently amounts to about 15 per cent of the gross domestic product (GDP) in low-income and 8 per cent in middle-income countries, a 20 per cent subsidy on the price of cereal constitutes 3–1.6 per cent of their GDP (Alderman and von Braun, 1984). The financial transfers involved in any sizable food subsidy scheme are considerable and can scarcely be afforded by many developing countries.

How to design and implement a food subsidy programme targeted to the poor without a very high administrative cost is a daunting challenge. The narrower and the more sharply focused is the subsidy, the higher is the administrative cost. For example, in one case, the administrative cost of securing a given increase in caloric consumption of preschool, malnourished children through the targeting of food subsidies to the household with malnourished preschool children was estimated to be half of what it would have been if the subsidy were to be targeted exclusively to the malnourished children themselves (Garcia and Pinstrup-Andersen, 1987). The cost of a subsidy can be kept down if the beneficiaries happen to be concentrated in a well-defined geographical area, if commodities subsidized are cheap and are mainly consumed by the poor and if the implementation of the scheme can be integrated within an existing administrative machinery – that is, agricultural extension workers or child or maternity welfare centres – or if delivery of food is undertaken through existing marketing systems such as local stores. It is unlikely that food subsidy increases the food consumption to an equivalent extent of the subsidy. Necessarily, a part would substitute for existing food consumption. However, the net increase in food consumption obtained through subsidized food is usually higher than the increase obtained through an equivalent increase in monetary income (Reutlinger, 1986).

Among the measures directly augmenting the income of the poor are public work projects. They provide self-targeting for the poor and are effective if wages are below market wages. The subsidy involved in public work projects is reduced if the gap between the value of assets created by such projects and the cost of execution of projects can be kept down (Reutlinger, 1986). In view of the need in many developing countries for large variety of rural and urban infrastructure projects, that is, roads, drainage, irrigation, markets, storage, schools, hospitals, etc., which are highly labour-intensive, they potentially constitute a significant source of employment, especially if repairs and maintenance are also included. The basic challenge confronting many developing countries in expanding the effective use of such projects is how to develop institutional and technical capacity to design and implement such projects, and to provide enough resources not only for their execution but also for their repair maintenance.

International Assistance for Poverty Alleviation

Undernutrition in developing countries coexists with large food surplus and, in some cases, overnutrition in developed countries. Is there a straightforward and simple way of transferring surplus food from developed countries to feed the poor and hungry in the developing countries?

There are two ways in which excess food in developed countries can be used in developing countries for relief of poverty. Food aid in the form of grant or concessional sales can substitute for commercial imports and release foreign exchange resources which can be directed towards poverty-oriented projects and programmes. Many poverty-oriented programmes suffer from a lack of external capital inputs.

Furthermore, an employment-oriented development strategy which expands the income of the poor leads to increase in the demand for food. This will be true even if the main thrust of development strategy is towards increasing domestic food production. In many developing countries a large majority of farmers are involved in food production so that their income and employment can be expanded by focusing mainly on increasing food production. An increase in food production, when it is associated with an increase in output and income in the non-food and non-agricultural sectors, frequently leads to a rate of increase in aggregate income and food demand which, given the high income elasticity of food in low-income countries, is in excess of a feasible rate of increase in domestic food production in the near- or medium-term future. This increases requirements of food imports. For example, 10 per cent growth in per capita agricultural productivity is associated with 9–11 per cent growth in per capita GDP; 10 per cent in per capita GDP is associated with 10–11 per cent increase in food imports. A recent study found a 10 per cent growth in total GDP to be associated with 19–25 per cent growth in cereal (not food) imports (Bautista, 1987). Food aid by expanding aggregate food supply helps meet the increase in food demand, which results from an accelerated growth in income and employment of the poor; this moderates or prevents a rise in food prices and obviates the adverse effects on the poor. For low-income food deficit countries, the share of food aid in food imports was 20 per cent during 1983–86; for some countries the ratio was as high as 50 per cent or more (FAO, 1987c).

Food aid can finance employment-oriented development projects designed to promote growth in income and employment. It can finance the wage component or food expenditures of those engaged in development projects. The disincentive effect of food aid on domestic prices is offset if an additional demand is created either through food aid-financed projects or by an employment-oriented strategy. The additional income flow from food aid projects must exceed the amount of food aid by a multiple, so that extra demand for food matches increases in food supply. To the extent that only 40–60 per cent of additional income is spent on food in low-income countries, the creation of additional income of $1.60 to $2.50 is needed to absorb, without a price decline, $1 worth of food aid. Food aid projects require supplementary inputs other than labour, that is, both imported inputs as well as local non-labour inputs; the former in turn necessitates additional non-food financial aid, and local inputs would require financing through domestic savings or programme food aid which replaces commercial imports and generates resources for use in food aid projects. If food-for-work projects are to generate additional income and demand, they must be in addition to those that would otherwise have been undertaken. An effective use of food aid, as is obvious from the foregoing, would require close coordination with external financial assistance, in the context of an overall development strategy oriented towards expansion of income and employment of the poor.

Public expenditures in many developing countries on agriculture, on current developmental inputs like fertilizer, seeds, pesticides, as well as on capital expenditures such as irrigation, mechanization, land development and research, etc., are not commensurate with either agriculture's importance in overall economy or with requirements for an accelerated growth in agriculture.

Private investment tends to be discouraged by inadequate price incentives or by the discriminatory impact of macroeconomic policies like exchange rate overvaluation or industrial protectionism which reduce relative rates of return on investment in agriculture. Above all, investment, public and private, is constrained by the inadequacy of overall domestic resources in many developing countries.

External assistance to agriculture in 1983–85, estimated to be about $8.3 billion in nominal prices and $5.9 billion in 1975 prices, fell considerably short of what the World Food Conference in 1974 set as a target, i.e., $8.3 billion per year in 1975 prices, which was later on revised upwards to $11–12.5 billion (FAO, 1987c). Furthermore, on a per capita basis, low income countries received a lower amount, i.e., $3.16 in 1983/85 compared to $4.74 for all developing countries. The predominant source of external assistance has been the international agencies like the World Bank, Regional Development Bank and International Fund for Agricultural Development (IFAD), etc., devoting 25–30 per cent of their assistance to agriculture.

In the last two decades, external development assistance tended to emphasize poverty-oriented development projects in the agricultural sector. While this was in general desirable, there were two problems. One related to the design of such projects which focused on what was called 'integrated rural development projects'; they were more complex, comprising multiple components and requiring sophisticated administrative and management capacity. Second, in many cases poverty-oriented projects concentrated on the needs of the small farmers only and did not pay adequate attention to: a) rural infrastructure like roads, communications, transport and markets, electricity, etc., especially in the 1970s and 1980s; b) landless rural poor; and c) rural nonfarm employment. That public work projects relating to irrigation, drainage and rural roads often tended to favour the non-poor rather than the poor was unavoidable in the context of a highly unequal distribution of land and economic and social power. Projects or programmes which provide greater access for the poor to non-land assets, and which either expand their self-employment in agriculture or increase nonfarm employment, deserve a greater emphasis, however.

External assistance to strengthen local institutions, including local administrative machinery and rural people's associations or organizations which are able to design and implement agricultural and non-agricultural projects, has a high priority. Availability of credit holds a key to this approach. How to provide credit not only for productive purposes but also for temporary consumption needs for small farmers or landless labourers who can offer security, not of land but of their own labour, or of crop only; how to minimize the administrative costs of small loans provided at nonsubsidized rates of interest; and how to ensure high rates of repayment through various innovative approaches involving farmers' groups or associations all constitute a challenge confronting both the donors and recipients in developing countries. There are successful examples like Grameen Bank in Bangladesh, the lessons of which need to be learned and spread far and wide (Hossain, 1988).

Can external assistance be helpful in implementing land redistribution where land-holdings are highly skewed and constitute a serious bottleneck in the way of a successful assault on poverty? In many countries an important constraint is the lack of domestic financial resources to compensate the landowners; external assistance can be integrated with a programme of land redistribution to compensate the large landowners, as well as to finance programmes designed to increase agricultural productivity. Also, food aid may be useful in the transitional period to meet any shortfall in food production caused by temporary dislocation.

External financial and food assistance can also be used to reduce the instability of food supplies that arise from rises in food import price, declines in domestic food production and falls in export earnings, which reduce the availability of foreign exchange for the import of food. Food aid has often responded in the past to varying import requirements of food aid-dependent countries, but there is no systematic mechanism by which food aid, especially bilateral food, responds to changes in recipients' food supply. Globally, however, the flows of food aid were positively associated in the past with the variations in supplies and stocks in donor countries, which did not necessarily coincide with the opposite variations in the recipient countries. Efforts to enhance the role of food aid in stabilizing food supplies of recipient countries on a more systematic basis, possibly by expanding the multilateral food aid for this purpose, need to be intensified.

IMF Cereal Import Facility

The International Monetary Fund (IMF) Cereal Import Facility (CIF), launched in 1981 as a component of the IMF Compensatory Financing Facility for Export Shortfalls (CFF), was intended to compensate for the exceptional, but temporary and reversible, rise in cereal import bill, arising either from a shortfall in domestic production or a rise in import price. It was expected that availability of CIF would encourage developing countries to pursue food security, not through food self-sufficiency, but through reliance on world markets as dictated by comparative cost considerations. The reliance on and confidence in world markets can be encouraged only if they are assured of food supplies at times of high prices or domestic shortfalls, irrespective of the state of their external balance of payments. By linking it with the CFF, and furthermore, by relating drawings on CFF increasingly in recent years to stricter IMF conditionality about domestic policy reforms, significant departures have been made from the original purpose.

First, to argue that a rise in export earnings at a time of rise in import price of food, or a shortfall in domestic food production reduces or eliminates the need for cereal import financing, is to argue that first claim under all circumstances on a rise in export earnings is import of cereals, even though such an action may reduce imports of fertilizers and consequently food production the next year. Furthermore, a rise in import price of cereals may be associated with a rise in other import prices, both agricultural and non-agricultural, and may more than offset the rise in export earnings. Linking cereal imports with CFF is tantamount to making access to cereal facility a part of the IMF overall financial assistance for meeting short-term balance of payments difficulties, irrespective of their sources.

Second, the temporary shortfalls in export earnings are not primarily due to policy failures of developing countries. The policy failures in developed importing countries, including a recession or raw material inventory policies, or monetary or exchange rate policies, also cause short-run declines in export earnings. Furthermore, assistance for meeting short-term export shortfalls need not be linked to policy reforms to deal with long-run adverse trends caused by structural changes in demand and supply conditions, or by inappropriate policies, even though they may not strictly be distinguished. In any case, it is not necessary to use, for at least two reasons, CFF to seek to correct long-term factors, if any, which may cause short-term export shortfalls. Short-term shortfalls are never fully compensated by CFF and there are other IMF facilities which are intended to deal with long- and medium-term structural issues. At least in view of the special nature of food, the political sensitivity to and implications of an absence of its assured supply, as well as its humanitarian aspects, the Cereal Import Facility should be available almost semiautomatically.

The access to the IMF Compensatory Financing Facility for export shortfalls, once considered as an important source of multilateral short-term finance, has been restricted in recent years. Following a recession in developed countries in 1981/82, the drawing on this facility during 1982/83 went up to as high as $6 billion. But a subsequent change in rules in 1983 limiting the access to this facility and imposing the conditions for domestic policy reforms, resulted in a fall in drawings to less than $1 billion a year between 1984 and 1986 (Kaibini, 1988).

The principal focus of international development assistance in very recent years is on debt and structural adjustment problems in the developing countries. This focus, combined with a general climate of 'aid fatigue' in donor countries, has caused a major shift on emphasis in domestic policy failures, rather than the lack of resources, including external resources, as reasons for inadequate growth or poverty.

While the important role of international development assistance in support of domestic policy reforms is recognized, its amount and pattern have not lived up to expectations. Several countries, including some in Africa, have undertaken domestic policy reforms in the process of structural adjustment, but have been unable to receive external assistance adequate enough to generate a sufficient growth momentum needed to sustain the national acceptability of sometimes painful policy reforms. On the other hand, appreciable amounts of bilateral assistance continue to flow to countries without significant policy reforms, casting doubt on the credibility of aid for policy reforms.

Among the external constraints on development are restrictions on access to world markets faced by developing markets for their exports. The progress in the reduction of trade barriers to agricultural exports of developing countries is slow and expectations are hinged on the outcome of the outgoing General Agreement on Tariffs and Trade (GATT) negotiations. A significant reduction in restrictions on their exports, both raw and processed agricultural commodities, would substantially add to the exchange earnings of developing countries in amounts which can compare very favourably with the flow in recent years of external assistance to the agricultural sector. Unlike in the past, trade liberalization needs to be applied across the board and to commodities of special interest to developing countries, including not only cereals, oilseeds and oils and meet, but also sugar and tropical products, etc., to be of significant benefit to developing countries. Trade liberalization by itself does not contribute to the alleviation of hunger. It is true, however, that some alleviation is effected when small farmers or farm labourers become employed in the export sector as a result of trade liberalization.

If trade liberalization in certain products causes a rise in world price through a reduction of excess supplies in world markets, food deficit low-income countries may suffer a rise in import bill. However, such a rise in world prices is more likely in respect of dairy products and meat, the most highly protected sectors in developed countries' agricultural sector, than in cereals. The low-income countries scarcely import such commodities in any significant amounts. Moreover, if trade liberalization takes place for a wide range of agricultural commodities, gains from increased income from other commodities will, in many cases, more than offset the losses from a possible rise in cereal prices.

Conclusions

The number of undernourished populations in developing countries, excluding China, is estimated to range between 450 and 700 million. The numbers vary depending upon assumptions regarding: a) the extent to which individuals adapt to a decline in calorie intake without adverse effects on health; and b) the levels of physical activity. The higher the level of physical activity undertaken, the higher is the magnitude of calorie requirements. The percentage of the total population undernourished is the highest in Africa (about 30 per cent), with Asia a close second (22 per cent). However, about 60 per cent of the world's total undernourished population resides in Asia.

The low estimate of the undernourished population in developing countries given above may be said to represent those who are most severely undernourished. Furthermore, the above estimates do not take into account the impact of inadequate health and sanitary environment on the state of undernutrition; hence, in some regions and countries the above numbers may underestimate the magnitude of undernutrition. Access to food and access to hygiene are the twin pillars on which progress in the elimination or reduction of undernutrition depends.

Assurance of access to food on the part of the hungry and undernourished requires measures, on the one hand, to expand their food entitlement through the expansion of employment and income and, on the other hand, to ensure adequate and stable food supplies. Self-sufficiency in food production either at the national or the household level is not necessary; what is needed is self-reliance in food, that is, ability either to produce or procure food through an optimum cost-effective combination of domestic production and imports.

How to design an agricultural development strategy which simultaneously expands food production, on the one hand, and raises the income of the poor, on the other, is not an easy undertaking. Inequality may increase in the medium term, yet poverty may decline provided enough care is taken to ensure that inputs, services and credit (that is, preconditions of modern technological progress) are available to small farmers; that landowning-middle farmers are encouraged to expand output and provide increasing employment through labour-intensive activities and techniques, supported by appropriate macroeconomic policies; and that nonfarm employment in the production of labour-intensive manufactured goods, trade, and services is stimulated in response to increasing demand resulting from agricultural growth.

Hunger and undernutrition at the household level are also dependent on such factors as who in the family earns income, who controls the household food expenditure and how the additional income is spent. This is why women's role in food production and household food consumption is so crucial, because in many societies they play an important role. Measures to increase their employment and income-earning opportunities as well as to increase their access to education play an important role in alleviating undernutrition.

The hard core of the poor and the undernourished in many countries consists of the rural landless labour, the small farmers, farmers in arid or infertile regions and the urban unemployed. Direct assault on poverty and undernutrition is needed, especially in the short-run, to strengthen the poverty-reducing impact of employment-oriented growth strategy. They include such measures as food subsidies, narrowly targeted towards the poorest and the vulnerable groups, as well as labour-intensive public work projects.

What is needed is a coalition of political forces or interest groups in developing countries which look beyond the short-term interests of those who own most of the economic assets and wield economic and political power, and are able and willing to adopt reformist policies in favour of the poor. At the same time the organizations of the poor need be promoted or strengthened in order to enable them to effectively utilize the opportunities opened up by the public policies and institutions as well as to serve as a countervailing force where markets are noncompetitive.

International assistance can strengthen poverty and hunger-reducing domestic efforts through either food or financial aid. External assistance for promoting technological progress and infrastructural development should help, first, to reduce the price of food, which is a major determinant of the real income of the poor, and second, to expand employment by strengthening intrasectoral, i.e., farm and nonfarm, linkages. Not all such assistance will go directly and fully to the poor; leakages are unavoidable. They occur in domestic programmes as well as in those internationally financed. International assistance should encourage and support policy reforms, which eliminate discrimination against the poor, including discrimination against the rural and the agricultural sectors where the majority of the poor live in most developing countries.

References

Alderman, H. and von Braun, J., *The Effects of the Egyptian Food Ration and Subsidy System*, IFPRI Research Report 45 (Washington, DC: IFPRI, 1984).
Bautista, R., *Relationship between Agricultural Growth and Food Imports*, IFPRI Seminar on Third World Food Markets (Washington, DC: IFPRI, 1987).
Demery, L. and Addison, T., 'Stabilization Policy and Income Distribution in Developing Countries', *World Development* 15 (1987), pp. 1483–98.
FAO, *The Fifth World Food Survey* (Rome: FAO, 1985).
FAO, *Agriculture Toward 2000* (Rome: FAO, 1987a).
FAO, *Food Outlook, Statistical Summary* (Rome: FAO, 1987b).
FAO, *International Agricultural Adjustment,* 6th Progress Report (Rome: FAO, 1987c).
FAO, *Second Progress Report on World Conference on Agrarian Reform and Rural Development* (Rome: FAO, 1987d).
Garcia, M. and Pinstrup-Andersen, P., *The Pilot Food Price Subsidy Scheme in the Philippines: Its impact on income, food consumption, and nutritional status*, IFPRI Research Report 61 (Washington, DC: IFPRI, 1987).
Hazell, P.R. and Roell, A., *Rural Growth Linkages: Household Expenditure Patterns in Malaysia and Nigeria* (Washington, DC: IFPRI, 1983).
Helleiner, G.K., 'Stabilization, Adjustment, and the Poor', *World Development* 15 (1987), pp. 1499–1513.
Hossain, M., *Credit for Alleviation of Rural Poverty: The Grameen Bank in Bangladesh*, IFPRI Research Report 65 (Washington, DC: IFPRI, 1988).
Kaibini, N.M., 'Financial Facilities of the IMF and the Food Deficit Countries', *Food Policy* 13 (1988), pp. 73–81.
Lipton, M. and Longhurst, R., *New Seeds and Poor People* (London: Hutchinson, 1988).
Mellor, J.W. and Ahmed, R. (eds), *Agricultural Price Policy for Developing Countries* (Baltimore, MD: The Johns Hopkins University Press, 1988).
Mellor, J.W. and Desai, G.M. (eds), *Agricultural Change and Rural Poverty: Variations on a Theme by Dharm Narain* (Baltimore, MD: The Johns Hopkins University Press, 1985), pp. 192–9.
Osmani, S.R., *Controversies in Nutrition and Their Implications for Economics of Food* (Helsinki: World Institute of Development Economics Research, 1987).
Reardon, T. and Delagado, C., 'Policy Issues Raised by Changing Food Consumption Patterns in the Sahel', unpublished paper (Washington, DC: IFPRI, 1986).
Reutlinger, S., Poverty and Hunger: Issues and options for food security in developing countries, (Washington, DC: World Bank, 1986).
Sen, A.K., *Poverty and Famine: An essay on entitlement and deprivation* (Oxford: Clarendon Press, 1981).
Wanmali, S., *Rural Household Use of Services: A Study of Miryalguda Taluka*, India. Research Report No. 48, Washington, DC: IFPRI, 1985).

Chapter 20

Hunger, Famines and Poverty: A Few Considerations of Political Economy*

The triangular relationships between long-term or endemic hunger and poverty (the percentage of the population falling below a minimum level of income or calorie intake), famines (a sudden and steep rise in short-run mortality rates far in excess of trend mortality rates) and human deprivation (inadequate access to health, education and social services) have been and continue to be a matter of considerable debate and discussion, initiated by A.K. Sen and followed subsequently by analytical and empirical verification or elaborations by several others (Dreze, et al., 1995; Dreze and Sen, 1989; Lipton and der Gaag, 1993).

This chapter is an attempt to revisit and re-examine some of the issues raised in the above literature in the context of the South Asian experience. Two main propositions of the above literature are: first, countries which suffer from endemic hunger/poverty have in the past often successfully prevented the recurrence of famines, but a country that makes progress in reducing endemic hunger and spreading access to health and education may fail to prevent famine. Second, among countries with high per capita incomes, some have extended health and education widely among their populations while others have failed to do so. Again, there are countries that have reduced income-related poverty but achieved no significant extension of health and education facilities.

Level And Distribution of Per Capita Food Supply, Entitlement Failures and Famines

There is no unique one-to-one relationship between variations in food production and famine. It depends on entitlements, meaning the ability to command food.[1]

An overall shortfall in food availability is neither a necessary nor a sufficient condition for acute hunger or famine that are characterized by an

* First published in *Asia-Pacific Development Journal*, Vol. 4, No. 1, June 1997.

excessive number of deaths (Dreze et al., 1995). In addition, adequate food consumption is a necessary but not a sufficient condition for a reduction in long-run endemic undernourishment, as anthropometrically measured or as indicated by high infant mortality or low life expectancy. Access to healthcare, especially primary healthcare, the eradication of preventable diseases and sanitation measures, meaning access to clean water and sewage systems, is crucial for alleviating endemic undernutrition.

If 'entitlement protection' schemes such as public food distribution or employment schemes are in place, a collapse of entitlements can be successfully mitigated or alleviated. What is required is a system of early warning and preparedness, as well as an institutional capacity to formulate and implement quickly the entitlement protection schemes. The entitlements of the poor may collapse or decline not only because of a fall in the income of those engaged in food production, but also through a fall in incomes in the non-food sector, including both the agricultural and non-agricultural sectors or through changes in macroeconomic circumstances, meaning fiscal, monetary and trade developments that cause a general inflationary pressure and/or a rise in food prices.

Exclusive concentration on the role of failure in entitlements for explaining famines may underemphasize or fail to recognize two important circumstances. The first is that, in many countries, food production is the predominant source of employment and income and is therefore the most important source of entitlements for the poor. If the majority of the poor earn their livelihood from food production, a shortfall in food production will result in a failure of entitlements. This can no doubt be offset by transferring purchasing power/income to the poor so that they obtain access to foodgrains in the market at a higher price. There will be a redistribution of food consumption to the poor and hungry from the rest of the community. The second is the role of speculation in causing a rise in food prices. A fall in market supplies and a rise in food prices may occur not only because of a fall in production or imports but also because of a rise in speculative hoarding on the part of consumers, farmers and traders, triggered by factors within or outside the domestic food sector. If the hoarded foodgrains could be physically recaptured and distributed directly to the poor and hungry, the failure in entitlements can be offset. However, such a policy is most likely to be inflexible, given the administrative machinery prevailing in most countries. When famine takes place without any speculative hoarding or without any rise in food prices but exclusively because of a failure of entitlements, meaning the, so-called 'slump famine' (Dreze and Sen, 1989), the provision of employment/income in the short run on an urgent basis is the effective way to combat famine without an additional food supply. In all other circumstances, additional supply of food in the market or for direct distribution to the hungry is needed as an effective and expeditious anti-famine policy, not least to dampen or counteract the impact of speculative hoarding and a consequent rise in food prices.

The success of countries in South Asia in combating or preventing famines in general depended on their ability to expand the entitlement protection schemes rapidly through widely known and publicized income supplements combined with an increase in the supply of food released from domestic stocks or imports or a combination of both. Since the food crisis in South Asia and Africa in 1974 there has been an increasing awareness of the need for both an early warning system and domestic capacity and preparedness to meet food crises or threats of famine. Not only has the international assistance for capacity-building necessary in this regard expanded, but food aid, both bilateral and multilateral, to meet emergencies has greatly increased. In addition world trade in food has increased, with considerable diversification of the sources of food supply since the days of the world food crisis in 1974.

Why did Sri Lanka and Pakistan not suffer from famine during the past few decades? When the low per capita food consumption of the poor is combined with a low level of health and nutrition, a sudden collapse in food consumption below what is already a very low level is likely to cause famine. This is not likely in a situation of high average level of food intake and health status, such as that in Sri Lanka, where, given the distribution of income, the poor have a long-run level of food consumption that is adequate to cushion the impact of a short-run or temporary decline in entitlements. Sri Lanka had the highest per capita food consumption in South Asia, followed by India and Bangladesh. In addition, the rate of growth in per capita income was higher in Sri Lanka and Pakistan than in India and Bangladesh (World Bank, *World Development Report 1990*, Tables 1 and 2). At the same time, the pattern of income distribution was not significantly different among the countries (ibid.; World Bank, *World Development Report 1996*, Tables 5 and 6).

During the 1970s and 1980s, Sri Lanka had a higher average level of per capita calorie supply than the other three countries and Bangladesh had the lowest. Furthermore, during this period, the increase in per capita calorie supply was higher in Pakistan and Sri Lanka than in the other two countries (World Bank, *World Development Report 1990*, Table 28). Pakistan and Sri Lanka not only had a higher level of per capita calorie availability but also recorded a higher rate of improvement over time. Under these circumstances, a fall in entitlements in Sri Lanka and Pakistan was unlikely to cause as much distress and deprivation as in the other two South Asian countries. Neither in Pakistan nor in Sri Lanka was there the magnitude or diversity of public or self-employment programmes, as was the case in India and Bangladesh. At the same time, it should be recognized that the overall higher average level of per capita calorie availability in Sri Lanka was reinforced by subsidized or free food distribution for the better part of the two decades; the real value of food stamps provided to the poor in later years was eroded by a rise in food prices. In Pakistan, the distribution of subsidized food, predominantly favouring the urban population, was subject to substantial leakage to the non-poor. This added to the food supply in the market, containing any rise in the market price above the level that might have occurred in the absence of the public food distribution system. Above all, it should be noted that in both of these countries, shortfalls in entitlements or speculation in the food market, in so far as they occurred, were not as severe as in Bengal in 1946 or in Bangladesh in 1974.

Democracy, Free Press and Famines

In a multiparty democracy, with a free and independent press and respect for human rights, a sudden occurrence of acute hunger and famine is widely publicized and the demand for public action is loudly voiced by the political parties and other people's organizations, so that the government is obliged to make the utmost efforts to combat famine (Sen, 1995). This was, it is suggested, the most important reason why recourse to public action to avert famine was so persistent in post-independence India. In China, on the other hand, news about the great famine of the 1960s was suppressed from the public; there was no scope for public outcry for action even if some information filtered through the controlled media or rumours. The government was under no great pressure to take action and therefore could let large numbers of people suffer. It was also possible that the local or provincial functionaries in the famine-stricken areas did not provide correct information for fear of being held responsible for any disruption or shortfall in the food supply. As a result, it was likely that the central government might not have had access to reliable information and could not have assessed the magnitude of the distress. National pride could also have prevented the government from letting the outside world know about the famine. This could have been considered as a failure of Mao Zedong's bold economic and social experiment that was under way especially later during the Cultural Revolution.

To generalize on the basis of the Chinese example about the public policy response of non-democratic governments to economic distress may not be appropriate. Circumstances do differ greatly among countries which otherwise may share the same type of political regime. All autocracies, therefore do not behave in the same way under all situations of economic distress. This was seen in the behaviour of the Democratic People's Republic of Korea during the recent food shortage in 1997. Not only was the government fully aware of the severity of the food shortage and the consequent distress, but it was also very eager to let the outside world know and to publicize the extent of food scarcity and distress widely. Moreover, it requested food aid from the international community. It was the outside world that questioned its eligibility for assistance on two main grounds. First, its persistently inefficient economic policies not only nourished an inefficient and inadequate domestic food-production system but also severely constrained the country's export earnings to enable it to buy food in the world market; second, its large expenditure, including that of scarce foreign exchange, on the purchase of armaments diverted resources away from food imports. In fact, the major food aid donors, (the United States, Japan and the Republic of Korea) made the grant of food assistance conditional upon the Democratic People's Republic of Korea's willingness and readiness to enter into peace negotiations with the Republic of Korea.

The Bangladesh famine of 1974 does not seem to fit in nicely with the hypothesis that under a democratic regime famines in the event of entitlement failures are less likely to occur. When Bangladesh had a famine – two years after independence – it was a democracy with a free and independent press which vigorously and widely reported on the sharp rise in food prices and the dire prospects of worsening distress and conditions of famine. Even though the average annual food output in 1974 was not below that of previous years, the successive floods and droughts in previous years very heavily depleted stocks in private hands (traders and farmers) and in the public distribution system. Therefore, there was great pressure on the part of the traders and farmers to replenish their stocks at the earliest opportunity. As a consequence, availability in the market declined, including the farmers' marketed surplus in 1974. To this were added the exaggerated reports of crop damage caused by seasonal floods in the press and other media. The media thrive on sensational news, and this was an ideal setting for speculative hoarding on the part of consumers, farmers and traders. This led to a rise in prices that, in turn, fuelled further speculation and hoarding. Since the public distribution system had very few or no stocks, something that was also widely reported in the press, it could not distribute food in order to calm down the speculative fever.

Apart from overwhelming pressure by the press and public opinion on the government, there was no lack of a sense of urgency on the part of the government itself to deal with the situation. There were two possible courses of action open to the government under the circumstances. One was to provide more accurate or reliable information about the crop damage caused by floods or droughts and therefore a more reliable and authentic estimate of food production so that it would help to counter or offset the exaggerated estimates and thus to dampen speculation. The available estimates of current food production were often rough and preliminary, the food information system in the early years of independence was inadequate and of insufficient quality. The government was unable to produce more reliable or very different estimates than those reported in the press. In any case, in view of the rise in prices that was already occurring, a more optimistic (or more accurate) estimate would not have achieved much credibility in public perceptions.

The second and more effective policy was to import foodgrains to slow down the rise in prices and thus to dampen the speculative fever. The country's foreign exchange reserves had been depleted to a level scarcely adequate to make any significant purchases abroad; moreover, the sharp rise in world food prices greatly reduced the quantity of imports that could be obtained with a given amount of foreign exchange, apart from the fact that there were serious bottlenecks in arranging quick shipment. At the time, a simultaneous and competitive rush by many countries to buy cereals and secure access to international shipping facilities made it very difficult, if not impossible, for a newly independent country with no established contacts to obtain easy access to world shipping facilities. Therefore, the only feasible option left was to seek food aid, including assistance in transporting food, from the only source able to provide it at short notice and ship or transport it fast: the United States. This was done in right earnest and well ahead of time, but for a multiplicity of reasons recorded abundantly elsewhere, the aid was not available in time (McHenry and Bird, 1977; Faaland, 1981). The traders and stockholders rightly speculated or believed that the government would be unable to implement a suitable stabilizing response to the reported damage to the food crop (Ravallion, 1985). As a result, the speculative hoarding took its full toll in causing a famine.

The lessons of experience of the 1974 famine were learned well by subsequent governments in Bangladesh. In spite of the autocratic military regimes that ruled Bangladesh during 1975–91, and in spite of an actual decline in food production and the associated threat of a steep decline in entitlements in 1979 and 1984, the incipient rise in food prices was quickly arrested and the threat of famine was averted in both instances. This was accomplished through large-scale imports in both years of shortfall, made possible by the availability of adequate foreign exchange resources, on the one hand, and liberal food aid, on the other. Thus, the threat of adverse price expectations fuelled by speculation in the food trade was avoided. In the decades following 1974, every time there was a possibility of a sharp rise in food prices, the government was able to take recourse to imports and built up stocks, often in excess of requirements. Frequently, after such anticipatory food imports in a year of panic, surplus stocks were created, with attendant problems of stock management in subsequent years.

What about the hypothesis that authoritarian regimes are prone to letting people die from famine whereas democratic regimes with an independent press do not or cannot? In fact, authoritarian regimes do not all have the same degree of choice, nor are they totally free from pressure for action. Most authoritarian regimes frequently seek some degree of legitimacy, as was done in Bangladesh. The successive military regimes were convinced that the recurrence of a famine in Bangladesh such as that which had occurred in 1974 and, among other factors, had contributed to the overthrow of the elected government, would have destabilized a military regime as well. They seldom succeed in completely suppressing the press or the media. After all, even in developing countries, with less than complete totalitarian systems, information travels very rapidly, by word of mouth. Rumours are counterproductive; often they accentuate rather than abate destabilizing speculation in the food markets.

Human Development and Level of Income

Experience and empirical analysis confirm that high income is neither a necessary nor a sufficient condition for achieving a high level of human development, as indicated by high life expectancy, literacy or low infant mortality. The oft-quoted examples in South Asia are those of Sri Lanka and Kerala, a relatively poor state in India. At relatively low levels of income, they have achieved rates of life expectancy and infant mortality comparable to high-income countries.

The social indicators in Sri Lanka were higher than those in other South Asian countries as early as the 1960s, as shown in Tables 20.1 and 20.2.

While Sri Lanka and Pakistan have enjoyed higher incomes or higher rates of growth than other countries, Pakistan has done poorly in social indicators compared with Sri Lanka and not much better than Bangladesh. Compared with Bangladesh, India has done better, as can be seen in Tables

Table 20.1 Social indicators in selected South Asian countries

	Life expectancy			Infant mortality			Enrolment in primary education as percentage of the relevant age group		
	1960	*1983*	*1992*	*1960*	*1983*	*1992*	*1960*	*1993*	*1992*
Sri Lanka	62.0	69.0	71.9	71.0	37.0	18.0	95.0	103.0	107.0
Pakistan	43.1	50.0	61.5	163.0	119.0	91.0	30.0	56.0	102.0
India	44.0	55.0	60.4	165.0	93.0	82.0	61.0	79.0	–
Bangladesh	39.6	50.0	55.6	156.0	132.0	108.0	47.0	62.0	77.0

Source: United Nations Development Programme, *Human Development Report* (various issues).

Table 20.2 Public expenditure on health and education, as a percentage of GNP

	Education (1986)	Health (1980)
Sri Lanka	3.6	1.3
Pakistan	2.2	0.2
India	3.4	0.9
Bangladesh	2.2	0.2

Source: United Nations Development Programme, *Human Development Report* (various issues).

20.1 and 20.2. The average figures in India conceal considerable interstate divergences, especially the remarkable performance of the state of Kerala. It is not high income, but high public expenditure on health, sanitation and education that account for high social indicators. High public expenditure on social sectors is not necessarily related to high GNP or high rates of growth in GNP. It was mostly in the 1980s and 1990s that there was rising interest in social development in the South Asian countries. Therefore, emphasis on related public expenditure on the social sectors received momentum.

If it is not the level of or the growth in national income, then what is it that determines the level of social expenditure? Is it the nature of the political system? Is it the nature of cultural and historical traditions unrelated to political or economic institutions? The answers to these questions are not fully known. One possible hypothesis for a low emphasis on social expenditure in the past may relate to the lack of knowledge on the part of policy makers about the productivity-enhancing and growth-promoting effects of health and education expenditure. During the 1960s and early 1970s, there was a preoccupation among both development specialists and policy makers with investment in physical capital as the prime mover of economic growth. It is only in recent years that the growth-enhancing effects of social expenditure have been recognized. At the same time, there is no evidence that the historically high levels of social investment in Sri Lanka and Kerala were stimulated by such an awareness on the part of their policy makers about the productivity-enhancing effects of social expenditure.

It is important to remember that a healthy and educated population does not necessarily guarantee a high economic performance. Kerala, with high levels of social development, did not achieve a high rate of economic growth because other prerequisites of economic growth were missing, such as a system of economic incentives and institutions that encourage savings and investment. The state controls and bureaucratic intervention inhibited private enterprise and investment, and Kerala suffered stagnation in per capita income during the 1980s. Recent experience in Kerala suggests that, in the absence of economic growth, it is difficult to sustain, much less expand, social gains. China presents the opposite example of a country which not only had heavily invested in the past in human capital, meaning the education and health sectors, but also had a high level of domestic savings and investment in physical capital throughout 1960–70. In recent years, starting in the early 1980s, wide-ranging economic reforms have encouraged private initiative and enterprise and expanded as well as diversified the patterns of investment.

The differences in the nature of the political system or of the government do not seem to account for differences in social expenditure. Both Sri Lanka and India were multiparty democracies, representing a wide spectrum of political ideologies, but they performed very differently in respect of social expenditure. It is often suggested that in Kerala the state governments were ruled for decades by leftist parties; in fact, for years it was dominated by Marxist/communist political parties that had a tradition in other parts of the world of emphasizing social expenditure. However, this hypothesis is belied by the fact that priority of social expenditure dates back to periods in history when the local monarchs ruled the states. Even before the days of British rule, the native kings and rulers attached great importance to education and health expenditure. This tradition was continued from the pre-British to the

British and the post-British period. In Sri Lanka, the high priority of social expenditure was continued during the regimes of both centre-right and centre-left political parties that governed the country alternately during the last several decades (Dreze et al., 1995). Thus, it is most likely that social investments in Sri Lanka or Kerala were undertaken because the governments/rulers attached great importance to education and health expenditures as goals or objectives in themselves rather than as inputs to material progress or physical wealth.

There seemed to be a social consensus rooted in culture, history and local traditions in these societies in favour of according high priority to social expenditures. All political parties in Kerala and Sri Lanka emphasized the role of health and education. This could be due to a greater degree of participation by the poor in the political process that articulated a demand for public action in health and education. This does not, however, resolve the puzzle posed earlier, because the effective political organization of the poor or their participation in the political process in Kerala and Sri Lanka was itself a function of high levels of literacy and education. However, it can be argued that, in view of the traditional and historical emphasis on health and educational objectives, the political parties helped to sustain and strengthen their importance in public policy.

In India, Pakistan and Bangladesh, all political parties and governments, in their policy pronouncements, have laid emphasis on poverty and improvement of the social conditions of the poor. However, the effective implementation of their pronounced intentions has lagged behind because the pressure groups in the political arena did not have social issues high on their agenda. Neither did the media play a role in articulating social issues and mobilizing public opinion or exerting pressure on their behalf.

It is remarkable that all South Asian countries, some more than others, adopted with varying degrees of effectiveness not only famine protection measures, including subsidized food distribution, but also poverty alleviating measures such as public employment or credit programmes for the poor. However, excepting in Kerala (in India) and Sri Lanka, they did not substantially expand health and education facilities. This might also partly be due to a lack of understanding, until recently, that adequate food intake needs to be matched by supporting health and sanitation measures in order to yield the expected results in nutritional improvement. For example, the incidence of morbidity and ill health offsets the effect of increased food intake and prevents its efficient absorption. The governments, on the other hand, were quite alert to the need for controlling epidemics and infections the effects of which were, in the short run, sometimes almost as catastrophic as those of famines in causing excessive deaths. While the media were quick to take up sensational issues of famines or epidemics, the endemic long-run effects of lack of health and education did not rouse similar public attention.

Policy Implications

It appears from the foregoing that factors or circumstances that stimulate public action for human development or the alleviation of poverty are many and varied and are affected by social, traditional and political conditions in different countries. What is needed for stimulating public action in this regard is to create a constituency for reforms based on an understanding of political constraints. Common interests between the poor and the non-poor may exist but are not perceived as such. A healthy and educated workforce increases the profits of employers. Higher productivity in small farms and businesses lowers the price of goods and services and benefits consumers as well, both the poor and the rich. An increase in the income of the poor expands the market that benefits the producers/employers.

There is a role of analysis and public information to help improve the perception of common interests. Self-interest is not always unambiguous or apparent. There is a choice between: a) small present and large future gains; b) concentrated small gains and widely spread large gains; and c) small gains that are certain and large gains that are uncertain. Analysis and public discussions help explain the options and enlarge commonality of interest.

Secondly, there may be scope for mutuality of interests and bargains between the rich and the poor that are aided by compensation. For example, the non-poor may be bribed into or compensated for accepting pro-poor policies such as subsidized food, allowing leakages and not insisting on very narrow targeting. The Maharashtra Employment Guarantee Programme in India could obtain the support of urban groups most probably because it reduced migration to the cities; landowners did not object to it because it stabilized the rural labour force and created infrastructure.

Thirdly, the mobilization of the poor by building up their organizational strength enables articulation of their interests and creates pressure for advancing their welfare. The participatory organizations can promote self-help and bottom-up development.

Fourthly, the power in the government is divided between ministries, departments and agencies that do not always have a common interest, such as between the executive and legislative branches or between local and national governments. Each pulls in a different direction and responds to different pressures. The final outcome is the result of various pulls or pressures. This happens more in democratic societies than in others. The political leaders must recognize these centrifugal forces and consciously steer a pro-poor public policy response.

Fifthly, economists, professional groups committed to various standards, churches, action groups and a multiplicity of voluntary organizations can all contribute in different ways, both as 'trustees of the poor' and as 'guardians of rationality'. The middle classes, such as civil servants, teachers, doctors, lawyers and university students, who frequently constitute very vocal and influential political groups and who may sometimes lose from pro-poor policies, may yet on occasion respond to appeals of solidarity and patriotism. They may act as 'trustees of the poor' if they are assured that the burden or sacrifices borne by them are seen to be equally and equitably shared by bureaucrats, politicians and the rest of the non-poor (Streeten, 1995).

In the light of the foregoing it may be worth speculating about the reasons behind the heightened awareness of the importance of public expenditure on social sectors among policy makers in South Asia as well as in other developing countries. Firstly, there is an increased flow of analytical and empirical knowledge on the role of social expenditure in economic development. In the search for reasons as to why some countries develop faster than others, the role of health, education and other aspects of social development has received increasing attention among development theorists and practitioners. Secondly, the discussion of the role of social sectors has spread beyond the domain of professional analysis to the wider public through greater awareness of the subject displayed in the media as well as increased information on the subject disseminated by the media. Thirdly, and most importantly, the driving force in this respect in many countries has been the priority attached to the social sectors by the international donor community, both bilateral and multilateral. Fourthly, the civil society institutions, especially the non-governmental organizations (NGOs) involved in a wide range of social activities, are increasingly playing a role in advancing the cause of education and health services in developing countries.

Note

1 'Entitlements' is not a synonym for income or purchasing power, as some would like to use it, at least as a short-cut definition. Any individuals in a household, for example, women

and children, may not have an income in the usual sense but still have access to food depending on the allocation of resources within the household: also, society may provide free food to the disadvantaged.

References

Alamgir, M., *Famine in South Asia* (Cambridge, MA: Oelgeschlager, Grinn and Hain, 1980).

Basu, K., 'The Elimination of Elimination of Endemic Poverty in South Asia', in J. Dreze, A. Sen and A. Hussain (eds), *The Political Economy of Hunger: Selected essays* (Oxford: Clarendon Press, 1995), pp. 372–99.

Dreze, L. and Sen, A.K., *Hunger and Public Action* (Oxford: Clarendon Press, 1989).

Dreze, L., Sen, A.K. and Hussain, A., 1995, *The Political Economy of Hunger: Selected essays* (Oxford: Clarendon Press, 1995).

Faaland, J. (ed.), *Aid and Influence: The case of Bangladesh* (London: Macmillan and Company, 1981).

Lipton, M. and der Gaag, J., *Including the Poor* (Washington, DC: World Bank, 1993).

McHenry, D.F. and Bird, K., 'Food Bungle in Bangladesh', Foreign Policy, No. 27 (1977), pp. 76–86.

Osmani, S.R., 'The Food Problem of Bangladesh', *Working Paper WP-29* (Helsinki, Finland: World Institute for Development Economic Research, 1987).

Ravallion, M., 'The Performance of Rice Markets in Bangladesh during the 1974 Famine', *Economic Journal*, Vol. 9, No. 377 (1985), pp. 15–29.

Ravallion, M., *Markets and Famines* (Oxford: Clarendon, 1987).

Ravallion, M., 'The Economics of Famine: An overview of recent research', in D.W. Pearce and N.J. Rau (eds), *Economics* (New York: Harwood, 1988).

Sen, A.K., *Poverty and Famines: An essay in entitlement and deprivation* (Oxford: Oxford University Press, 1981).

Sen, A.K., *Resources, Values, and Development* (Cambridge, MA: Harvard University Press, 1994).

Sen, A.K., *Economic Development and Social Change in India and China: A comparative perspective*, The Development Economies Research Programme, No. 67 (London School of Economics, 1995).

Sobhan R., 1980, 'Policies of Food and Famines in Bangladesh', in E. Ahmed (ed.), *Bangladesh Politics* (Dhaka: Bangladesh Centre for Social Studies, Dhaka University).

Srinivasan, T.N. and Bardham, P.K. (eds), *Rural Poverty in South Asia* (New York: Columbia University Press, 1988).

Streeten, P., 1995, *The Political Economy of Fighting Poverty* (Geneva: International Labour Organization, Development and Technical Cooperation Department).

Thomas, T.M. and Michael Therakan, P.K., 1995, 'Kerala: Towards a new agenda', *Economic and Political Weekly*, Vol. 30, Nos. 31 and 32 (1993–2004).

United Nations Development Programme, various issues, *Human Development Report, 1990 to 1995* (New York: United Nations).

World Bank, various issues, *World Development Report 1985, 1987, and 1995* (Washington DC: The World Bank).

PART VII

FOOD SECURITY: INSTABILITY OF FOOD SUPPLIES AND PRICES

Chapter 21

World Food Security: National and International Measures for Stabilization of Supplies*

The objective of world food security in a wider sense and a broader perspective is to assure all people, at all times, the physical and economic access to the basic food they need. Sustained physical availability, the stability of adequate food supplies, and command over purchasing power to secure access to food are all necessary components of food security at the level of individuals, households and nations. Measures are needed at national, regional and global levels, not only to expand food supplies through increasing production, especially in low-income, food-deficit countries, but also to maximize the stability of supplies in the face of production fluctuations and to secure access to food supplies on the part of all, especially the poor people and poor nations.

The three components of food security listed above are not unrelated, even though the attainment of one does not assure the achievement of the other. Higher levels of production do not bring in their train greater stability in year-to-year production; in fact higher yield may be associated – and in many cases *is* associated – with a greater variability of yield. No doubt at a higher level of food availability or consumption an individual or a nation can withstand the effects of shortfalls in food supplies or food consumption without much distress or deterioration in nutritional status, as compared to a situation where the average level of consumption is near the minimum nutritional requirements. At the same time, at a higher level of income and food consumption, an individual or a nation has greater capability and resources to build up stocks to face possible shortfalls in production.

Adequate and stable supplies for a nation do not necessarily ensure that every individual, especially the rural and urban poor, rural landless or urban jobless, would have access to food; they may have neither the resources to produce, nor employment and income to be able to purchase, food. Significant inequalities in consumption levels coexist with high levels of aggregate food

* First published in Lall, S. and Stewart, F. (eds), *Theory and Reality in Development: Essays in honor of Paul Streeten* (London: Macmillan, 1986).

supplies both within and between nations. Moreover, year-to-year stability of food supplies either at national or household levels does not imply that they are adequate to meet the minimum nutritional requirements. In the long run, food supplies should be adequate not only to meet the minimum nutritional requirements of all, but also to meet the effective demand for food from a growing population with rising income levels.

This chapter concentrates on one aspect of the broader concept of food security, namely the problems and possibilities of achieving the optimum stability of supplies in the face of fluctuations in domestic food production, as well as in supplies and prices in world trade. The discussion is limited to the analysis of cereals only. Of the three components of world food security, this is an area where there are still many unsettled questions and continuing controversy – some real, some imaginary, as well as some minor and some important – as to the sources and consequences of instability and as to the feasibility and cost-effectiveness of those measures suggested to reduce instability or offset its effects both at world level and at the level of individual countries.

Nature, Magnitude and Sources of Instability

The evidence indicates that variability of cereal production, measured in terms of average percentage variation from trend, increased during the 1970s compared to the 1960s. The world average variability in yield of cereals per hectare has increased by more than 50 per cent (from 2.26 to 3.36 per cent), whereas variability in production increased by 30 per cent (from 2.57 to 3.22 per cent). The extent of variability in developed countries is higher than in developing countries, both in respect of yield as well as of total production: the largest increase in yield variability is in the USA, from 3.19 to 10.38 per cent, whereas the average for all developed countries increased from 2.43 to 5.65 per cent.

Throughout the last two decades variability in yield and in production in the USA and the USSR was higher than the average variability in developed countries as a whole. During the 1970s the highest variability in yield occurred in the USSR (15.89 per cent), with the USA following (10.38 per cent). During the 1970s the variability in yield (3.36 per cent) and in production (2.29 per cent) in another large cereal-producing and importing country compare with 2.38 per cent and 3.46 per cent respectively for the developing countries as a whole.

The fluctuations in world trade are considerably greater than that in domestic production because only a small proportion of world cereal production is traded. Except for major exporters like the USA, Australia and Canada, the percentage of output that is internationally traded is small.

Moreover, since world trade is dominated by a few exporters, it is the degree of fluctuation in their domestic production that is mainly relevant in determining variations in world supplies. The percentage variation in yield is generally greater in exporting countries, most of whom depend on rain-fed agriculture. For example, percentage variability in yield during the 1970s was 7 per cent in the EEC, 8.64 per cent in Canada, 18.43 per cent in Australia and 10.38 per cent in the USA, as against the average for the world of 3.36 per cent and for the developed countries of 5.65 per cent.

Among the big importers in the world market are the USSR and China. The high degree of yield variability in these countries causes corresponding variation in their import demand, given the increasing tendency on their part to protect domestic consumption from the impact of fluctuations in production. The year-to-year changes in the Soviet cereal imports in absolute amounts as well as a percentage of world trade have been considerable. During the 1970s Soviet imports sometimes increased in one year by 13–20 per cent, and in the next year fell by 11–15 per cent. As a percentage of world trade Soviet imports have varied from 18 per cent in one year to 4 per cent in the next (1972–73 and 19734), again decreased from 18 per cent in 1975–76 to 7 per cent in 1976–77, and again from 12 per cent in 1977–78 to 9 per cent in 1978–79. In 1981–82 the percentage rose to 21 per cent. The cereal imports of China during the 1970s in some cases rose from one year to the next by as much as 170 per cent (from 1976–77 to 1977–78 or fell by 60 per cent between two years (as from 1974–75 to 1975–76) (World Food Institute, 1983, pp. 34–5).

What is important from the point of view of the impact of variations in imports on the stability of world trade is not only the percentage variation from year-to-year in countries' imports, but also the relative importance of these imports in world cereal trade. The larger the volume of imports, the greater is the impact on world trade of a percentage change in countries' imports. The proportion of Soviet cereal imports in world trade averaged between 15 and 21 per cent during the early 1970s, and that of China varied between 14 and 16 per cent during the same period.

There are various factors that contribute to production variability, some of which have become more important in recent years. Yield variations are often equated with the weather, but can also result from input shortages, the use of marginal land, crop disease and poor husbandry. Climatologists have various notions about long-term climatic change, but seen to have no way of predicting weather fluctuations from season to season. What weather does do is to give a skewed yield distribution, with a few sharp yield decreases and many average or slightly, above average years. This unevenness is perhaps more important than yield variance *per se*. Other causes of yield variation are more tractable. The spread of high-yield variety of seeds has often been associated with a reduction in the genetic heterogeneity of seeds and hence increased the risk of simultaneous and similar fluctuations in output in different fields/regions in a country. Input shortages, including the timeliness of input supplies and variations in their price, can be as serious as bad weather. Similarly, the control of pest and diseases rests as much on the organization of pest control measures as on scientific skills. The expansion of crop production to land only marginally suited to its production can magnify the weather-related and other causes of yield variations.

Scientific advance seems to be moving us in two directions – to control variations in yield through disease and pest control and appropriate seed varieties, and to increase variations by contributing to greater genetic homogeneity and by breeding out resistance to disease found in indigenous species as a result of the search for higher yields through new varieties. A choice based on better information on both average levels and variability of yields has to be made between two considerations: to preserve the genetic heterogeneity of crops, or to expend additional resources to control pests and diseases.

The instability of the area under crops, which also contributes to variability in output, seems less well studied. Farmers, owing either to the attraction of other crops or to the general lack of profitability from farming, can at times sharply reduce supply. Where producer-oriented price support systems exist, such fluctuations are rare.

Interrelated Instability in Cereals and Feed Sectors

Instability in perennial crops and in livestock products can be just as troublesome as yield changes in annual crops. These cycles arise from an inevitable feature of the production process in which investment implies, at least in aggregate, a temporary fall in production. Disease problems also often have an effect lasting for more than one year. This suggests that government policies need to be particularly sensitive in such areas. Governments can create investment cycles by their own action if they encourage the planting of perennials or the building up of livestock herds at certain times which then lead to overproduction later in the cycle. Developing country livestock is usually based on short-cycle enterprises such as poultry and pigs, and the use of imported feed will tend to instability generated by the uneven availability of local crops. International feed prices could, however, have significant repercussions of emerging livestock industries and food crop shortages if they in turn generate the occasional slaughtering of the breeding herd, which could destabilize the livestock markets and, in principle, the feedgrain markets.

Is the spread of livestock enterprises based on the utilization of cereals a positive or negative influence on the stability of food supplies? The answer seems to depend on: a) the source of the instability; and b) the reactions of governments to price changes. If the source of instability is in foodgrain (or other staple crop) production, then the availability of a parallel feedgrain marketing system can be an advantage. Even if the type of grain is not directly appropriate for switching to human use, it should be easier to effect a substitution if the buying, storage and transportation systems for feed products is in place. Moreover, the livestock herd itself can be run down at times of food shortages, although at a fairly high cost in terms of instability in the supply of livestock products. A policy to preserve stability in the livestock sector will tend to reduce its effectiveness as a food reserve: a policy of deliberately using the livestock sector as a food security device will place a heavy burden on a small industry and possibly hamper the development of a profitable integrated agricultural system. If the source of instability is from the feedgrain sector, then there is the opposite dilemma: whether to allow the foodgrain sector to help stabilize the livestock activity. Some degree of instability in the market for food crops might then be attributable to the existence of livestock sector. Whether positive factors outweigh the negative depends to a large extent on policy reactions.

Instability arising from outside the agricultural sector has received considerable attention in recent years. In the halcyon days of macrostability, when the three macro-prices – inflation, interest rates and exchange rates– were, at the least, fairly predictable, agriculture could largely ignore monetary matters. Nowadays, inflation can turn remunerative prices into heavy taxes, interest rates can make debt repayment a major cost and stifle new investment and exchange rates can create or destroy foreign markets almost overnight. The problems are essentially short-term: they are the instability of these variables and the uncertainty they create.

Price Instability

The problem of price instability is of broad concern to all trading countries. The fluctuation in the world price of cereals, which is the most important traded food item in world trade, has been aggravated in recent years. World market prices were relatively stable during the 1950s and up to the mid-1960s. The situation changed in the 1970s when prices became highly volatile. Aside from the crisis year of 1973–74, this variability is significant in other years as well. During the 1970s wheat prices varied more than eight times as much as during the 1960s; the price variability of rice exports more than doubled. The coefficient of variation in export prices (in real terms) of wheat was 3.6 per cent during the 1960s and 30.0 per cent during the 1970s, whereas that of rice (in real terms) was 17.5 per cent during the 1960s and 39.0 per cent during the 1970s (Huddleston et al., 1981). Between 1950 and 1971 the largest year-to-year percentage change in US prices of commercial wheat was 16 per cent between August 1977 and April 1979 the US export price of wheat increased by 80 per cent; during the calendar year 1980, corn export prices increased by 50 per cent, primarily in response to reduced crop in the USA and low yields in the USSR.

The sources of increased variability in the prices of cereals are many and various. The major reasons for stability in cereal prices during the 1950s and 1960s were the support policies of the USA and Canada and concomitant carry-over stocks held by them; the stocks were held either directly by the government or under its control. By the early 1970s, the major grain exporters were no longer willing to carry large stocks: they made major efforts to control surpluses by output/acreage reduction programmes. Wheat and feedgrain stocks held by major exporters in 1961–62 were 14.4 per cent of world production, compared to 9.6 per cent in 1970–71 and 8.6 per cent in 1972–73.

Individual countries, both developed and developing – the former more successfully than the latter – adopt policies to shelter their economies from the direct effects of international market instability through border mechanisms which neutralize world price movements or through state trading or direct interventions in domestic cereal/food markets which have the same effect. National policies are directed towards stabilizing domestic prices and domestic consumption levels by varying net trade. This shifts the impact of variability in production on to international markets, which are left free to the play of market forces.

During the late 1960s and 1970s an increasing proportion of the world's production and consumption occurred in nations which pursued internal price stability through managed trade. This was true of the EEC and a few developing countries like India. It was also true of centrally planned economies, whose share in world net imports of cereals increased from 8 per cent in the early 1970s to 30 per cent in the early 1980s. In the Soviet Union, for example, in the earlier period, substantial shortfalls in supply were tolerated; in the later periods, serious efforts were made to reduce or eliminate them. There is no reason to expect that policy-induced international price instability will decrease during the 1980s. If anything, it is likely to increase.

This has at least three disadvantages. First, a volatile world price is a much less reliable indicator of efficient resource allocation. This matters a great deal in a period when developing countries are deciding on incentive levels for domestic agriculture. Secondly, the allocation of burdens and benefits is arbitrary, leading to further tensions in the trade system. If a country is unsuccessful in its insulation policy it will suffer unduly, especially if it is unable to finance the losses from selling high-priced imports on to a stable domestic market. Thirdly, it will destabilize other related markets where domestic insulation is not so complete – in particular the livestock sector.

The problem, then, is to get the right balance between domestic market stability and world market stability. To try for the latter alone is likely to be costly and impracticable.

Sources of Consumption Instability in Developing Countries

Food security at national levels can be achieved only if a degree of stability in consumption levels is attained. Obviously, a complete stabilization of consumption levels is not feasible; adjustment in consumption levels at times of shortfalls in production or supplies up to a limit is expected. The lower the average level of consumption, the lower is the limit to further squeeze on consumption. Moreover, the consumption levels of the poorest need to be protected at times of shortages.

The year-to-year variability in food/cereal consumption in individual developing countries is quite high. Levels of variability of 10 per cent or more are observed in six out of 24 countries. The coefficient of variation ranges from a low of 3 per cent to as high as 20–30 per cent.

At the level of individual developing countries the sources of instability in food supplies and prices are twofold: fluctuations in domestic food production and in import prices. Variability in consumption of cereals is highly correlated with changes in domestic production. Consumption may vary with changes in disposable income. Individual countries try to offset variations in domestic production by imports of foodgrains, but do not necessarily succeed in completely offsetting variations in domestic production and in stabilizing consumption. Foreign exchange for imports of cereals may not be available to enable an offsetting purchase of imports. Moreover, even if foreign exchange for food imports is available, the volume of imports depends on the price of imports – a rise in import price cuts down the volume of imports. In many countries, the food import bill (including cereals and other staples) constitutes a high proportion of foreign exchange earnings, especially in years of low exchange earnings and high food prices, rising to as high as 40–50 per cent of foreign exchange earnings in exceptional cases. Also, marketing and distribution facilities, including port facilities, may not be flexible enough to handle large year-to-year variations in cereal imports.

Import bills of developing countries are therefore dependent on: a) fluctuations in domestic production; and b) variability in import price. During the 1960s, when prices were stable, variability of import volume was more important than that of prices in determining the variability of import bills. In contrast, during the 1970s, in 10 out of a sample of 18 countries, price changes explained more than 60 per cent of the variability in import bills (Huddleston et al., 1981).

National Measures for Stabilizing Supplies

National measures to enhance the stability of food supplies in the face of fluctuations in food availability are needed mainly in three spheres. The first set of measures is directed towards reducing the degree of fluctuations in output. An extension of irrigation facilities, wherever physically feasible and economically justifiable, would help to offset the uncertainty, irregularity and inadequacy in the amount of rainfall; it would also help extend multiple cropping and expand the land area under cultivation, thus both reducing seasonal fluctuations and spreading the risks of year-to-year fluctuations. In the same category are measures designed to ensure a smooth flow of purchased inputs such as fertilizers, improved seeds and pesticides at the right time and in appropriate quantity. Also important are continued research efforts to reduce variability to yield from new varieties of plants, while striving to improve their average yield and to reduce the impact of pests and diseases while efforts are made to develop pest- and disease-resistant varieties of seeds.

The second category of domestic measures relates to the strengthening of the state of national preparedness to meet efficiently and readily the impact of food shortages so that distress and suffering can be mitigated. Foremost in this category is the improvement of physical infrastructure for the marketing and distribution of food, including the improvement and expansion of logistic and transportation facilities, including storage facilities. Many developing countries do indeed suffer from serious logistic and transportation bottlenecks in ensuring the quick delivery of both imported and domestic food supplies. It is necessary to improve national preparedness to meet food shortages in various ways: by establishing or strengthening national monitoring and early warning systems for basic food supplies; by establishing, if necessary, special units of government arrangements to deal with disasters and food emergencies; by compiling a manual of food relief activities and tasks to be undertaken and procedures to be adopted in the event of food shortages. It is important to monitor transportation and logistic factors regularly.

The third category of national measures relates to the food stock or reserve policies in food-deficit, low-income countries. An individual country can hold food reserves or foreign exchange in order to secure access to food supplies at times of shortages.

Holding stocks is more expensive than holding foreign exchange. For one thing there are investment costs in storage capacity as well as in its maintenance and operation. Secondly, food stocks are subject to wastage and spoilage, whereas foreign currency held in banks is not. But there are a number of advantages in holding food reserves as against foreign exchange. First, it is not always possible to get physical access to imports, if a country has command over foreign exchange. Recent experience indicates that embargoes or export restraint can be applied which prevent access to supplies. It is undoubtedly true that embargoes can, at a cost, be circumvented either by obtaining supplies from alternative sources or from re-export through third countries; but this involves extra search or transaction costs on the part of a country cut off from its traditional and familiar sources of food supplies. Secondly, while physical availability at times of world shortages may not be a serious problem, even though a small importing country usually has to pay higher prices under these conditions, it may still face the problem of shortage of shipping space. When the amounts to be transported are small, when there is an extraordinary demand for shipping services for all importers, and when the country happens not to be located on the traditional or convenient international shipping routes, this problem may become acute. For such cereals as rice or white maize, the markets are then at the world level and transactions generally take a much longer time. Domestic stocks in these cases are essential to meet shortfalls in supplies. This is especially true if the size of a country's import demand is large in relation to the total volume of international trade.

Thirdly, even when supplies are available and shipping can be arranged, the kind of food supplies available may not correspond to the consumption patterns of the country. A country which habitually consumes maize may, for instance, have to import wheat. Traditional crops consumed by some developing countries, such as sorghum, millet or roots and tubers, are not commodities that are heavily traded on the world market.

Under all these circumstances, the advantages of holding food reserves cannot be ignored. Domestic food reserves can consist primarily of domestically produced traditional crops. For countries that have attained a high degree of self-sufficiency, and in good years generate export surpluses with deficits in bad years, domestic stockholding offers a way of taking excess output of the market at times of surplus and using the stocks for meeting deficit in bad years. It is difficult for a country that generates occasional surpluses to enter the world market easily without incurring significant marketing costs and without avoiding price concessions. In good years the domestic market-clearing prices may be above the f.o.b. price and in bad years its domestic price will be below the c.i.f. price. If this is the case, then domestic storage capacity would be a viable alternative, particularly if the gap between the low and high prices is wide enough to cover the storage cost.

The most important reason why a minimum amount of domestic food reserves is needed for food security purposes is the time lag in obtaining supplies from abroad. Pipeline or working stocks of traders or households are not likely to be enough to meet food security needs during the time that elapses before imports are physically available for domestic distribution. Most countries consider the expense of holding minimum food reserves to tide over this gap as a necessary and essential cost of relieving human suffering and possible starvation.

Obviously, delays in obtaining supplies vary from country to country, depending not only on location – i.e. whether a country is landlocked or borders on the sea, or is near the traditional international shipping routes – but also on its skill, knowledge and expertise in operating in world markets, including arrangements for shipping and transportation. The difficulties of obtaining supplies are aggravated at a time when domestic shortfalls in production in a developing country coincide with shortfalls in major consuming/producing areas. Major importing countries like India and China already have large stocks, partly for meeting the needs of their public food distribution systems and partly for purposes of price stabilization in the domestic market. A part of this stock is also used for meeting shortfalls in domestic production. The establishment of minimum food security stocks in the sense explained above does not constitute an alternative to trade or a reliance on imports as a means of meeting domestic shortages, but is a necessary supplement to it.

Many developing importing countries do not have adequate skill and expertise in operating in the world cereal markets, in terms of buying the right quantity at the right time in order to obtain the most advantageous terms of purchase. There is a need for assistance to developing countries to augment their skill and expertise. Also, many countries do not have adequate infrastructure for holding and operating pipeline or working stocks to carry surpluses from harvest seasons or surplus areas to lean seasons or deficit areas within the country to even out seasonal variations in supply as well as across the various regions in the country.

In addition, there is a need for training in management and the operation of pipeline stocks, to be held by private traders or public agents, depending on the nature of the food market of a country, storage capacity and distribution facilities. There is a need for public investment or financial assistance to stimulate private investment in these latter.

International Measures for Food Security

Among the international measures to meet shortfalls in food production or to dampen the rise in prices, an important measure that has been discussed in various forums in the post-Second World War period, especially since the World Food Conference in 1974, is the proposal for international food reserves. Various formulations of the proposals for international reserves in the context of world food security have been debated.

It is recognized that stocks are essential for imparting stability to prices in the face of fluctuations in production. If world cereal production and the availability of cereals in world trade out of current output levels fluctuate as described above, then stocks must be held by some countries or by all – stocks that will be 'adequate' and 'available' to act as a buffer to promote market stability. In other words, stocks need to be accumulated at times of low prices and released when prices go up. What level of stocks is to be considered adequate? The level of world stocks of cereals that is to be considered adequate for the purpose of market stability would depend on the price range within which stability is sought, apart from the level of world production and trade in cereals. This criterion for judging adequacy of stocks is different from what is relevant to the concept of food security, i.e. meeting the minimum consumption requirements (at around the level of average consumption of the recent past years) and times of production shortfalls.

World Food Security Reserves

The discussion on world food reserves, their level and adequacy has often been marred by a lack of clear distinction between working or pipeline stocks, on the one hand, and reserve stocks on the other. Carry-over stocks at the beginning of a year consist of supplies from harvests of previous seasons (or from imports) which are available at the beginning of the new crop season. They comprise both working and reserve stocks. Working stocks are those required to assure a smooth and uninterrupted flow of supplies from the farmer, or point of imports, to the final consumer. Working stocks comprise those held by farmers, processors and merchants, including the quantities in transit or afloat, as well as those held by governments for sale through public distribution systems and for intraseasonal supply and price stabilization. Reserve stocks are in excess of working stocks, and include stocks which can be drawn on to meet unexpected deficits in supplies due to crop shortfalls, emergency food shortages and other contingencies.

An attempt has been made by FAO in the past to estimate the level of world stocks of cereals that would be adequate to meet shortfalls in world food production. The objective behind the estimate was to measure the level of reserve stocks needed to maintain the long-term trend in consumption and to cover the world for one year's unexpected and unplannable events such as crop failures due to low yields in various regions. The methodology followed was as follows.

For the exporting countries, an estimate was made of total cereal production and available exports in a bad crop year. In order to establish available exports, an estimate was made of the trend value of domestic output in exporting countries.[1] The trend value of production (based on the utilization of the full production area) is subsequently adjusted downwards by assuming that yields per hectare would be reduced by specified amounts related to either the two standard deviations from trend or to the lowest actual yield level that was achieved in the past, when droughts or plant diseases affected the cereal crop production. In the rice-importing countries, the same assumption was made to derive net imports in a bad year. For imports of wheat and coarse grains, in the absence of data, it was assumed that their net imports would increase by two standard deviations from the trend. The excess of the 'adjusted' import requirements in a bad year over the adjusted estimate of what is available in the exporting countries for exports and stocks in the bad year determines the amount by which carry-over stocks at the onset of the bad crop year need to be drawn down in order to meet the shortfall in world production. Expressed as a percentage of the previous years' annual consumption, this works out at 5–6 per cent. It is to be noted that between the mid-1950s and early 1970s the amount of absolute maximum single-year shortfall in world cereal production works out at approximately 5–6 per cent of 1973–74 world consumption.

An analysis of historical data on stocks for several years for each of the three crops was undertaken, with a view to identifying stock levels in those years when the price situation was more or less equilibrated, i.e., when stocks did not include any 'surplus' element beyond 'working' and 'reserve' stocks. The stocks at the end of the year 1973–74, when there was an extremely tight supply and demand situation and high prices, were assumed to be all working stocks, which exclude any reserves for unforeseen shortages. Deducting the 1973–74 stocks as working stocks, the size of food security stocks was calculated for normal years. It was estimated that the figures shown in Table 21.1 are the normal ratios of working stocks and reserve stocks for different crops, for exporters and importers separately.

Table 21.1 Safe level of stocks as percentage of total disappearance on annual consumption

	Wheat			Coarse grains			Rice
	Export -ers	**Import- ers**	**Com- bined**	**Export -ers**	**Import- ers**	**Com- bined**	**exporters/ importers (combined)**
Total	50	16	(26)	22	10	(15)	15.1
Reserve	(33)	(2)	(9.5)	(8)	(1)	(3.7)	(4.7)

Note: Exporters' stocks are as a percentage of total exporters; disappearance; importers' stocks are a percentage of domestic consumption, and combined stocks are a percentage of total world annual consumption.

Source: *Approaches to World Food Security* (Rome: FAO 1983), p. 34.

The calculation of minimum safe levels of food reserves, including both working and reserve stocks, for the world as a whole and for all cereals, implies first that stocks, irrespective of where they are located, are available to all countries at times of production shortfalls to meet shortages; and secondly, that different cereals are completely substitutable for each other. Looked at from the point of an individual country, e.g., a food-deficit, low-income country, required stocks would be very different, as an individual country has a choice between holding stocks and acquiring supplies through imports. As shown in Table 21.1, what are adequate stocks for importers is different from what is adequate from the point of view of exporters if exporters are to be a reliable and dependable source of supplies for the importers. Moreover, the safe level of stocks was considered to be adequate to meet one year of worst shortfall and not two consecutive years of bad harvests.

The FAO Secretariat's estimate of the minimum safe level of cereal stocks, set at 17–18 per cent of annual consumption (12 per cent of annual consumption is considered a working stock, whereas 5–6 per cent is considered adequate as food security reserves), is used by policy makers/analysts in general for evaluating the adequacy of world stocks from year-to-year, in the absence of any other available yardstick. The year-to-year variation in world cereal stocks in absolute amounts and as a percentage of annual consumption, including their distribution amongst various groups of countries, are shown in Tables 21.2 and 21.3.

The relevant question is whether the separate and independent interests of exporters and importers would result in a size of world stocks that would meet the worst shortfalls in production experienced in recent years. The question has been mooted whether it is necessary, either from the point of view of relieving human distress or from the point of view of cost-effectiveness, to hold stocks which take care of the worst food crisis. In other words should consumption be maintained at the trend level, except in the poorest countries where accommodation through shortfall in consumption would have both serious and long-term adverse consequences for health and undernutrition? Should not a certain level of fall in consumption, let us say 6.5–10 per cent below trend, be allowed in most countries of the world, provided efforts are undertaken internally by different countries to maintain at the minimum level food consumption of the poorest and vulnerable groups during a disaster?

As has been explained above, it is not enough that world stocks are at the safe minimum level; what is equally important is how the stocks are distributed around the world in the exporting and consuming countries. If stocks are concentrated, as they often are in developed exporting countries, there is no

Table 21.2 World carry-over stocks as percentage of total utilization (million tons)

	Wheat	Coarse grains	Rice (milled)
1960–61	34.8	25.1	9.8 *
1966–67	29.4	14.6	6.0
1971–72	23.5	14.2	7.1
1976–77	26.2	11.3	7.6
1978–79	23.8	12.2	10.9
1979–80	18.3	12.3	9.0
1980–81	17.1	11.2	8.3
1981–82	19.5	15.4	7.6
1982–83	21.1	19.3	5.6
1983–84	23.6	14.1	5.1

Note: * = 1961–62.

Source: The World Food Institute, *World Food Trade and US Agriculture 1960–82* (Iowa State University, August 1983) (3rd annual edition).

Table 12.3 Share of different countries in world cereal stocks

	Total world cereal stocks (% of annual cereal consumption)	North America	Others	Among others EEC	Japan	USSR	Developing countries
1972	18	44.2	26.9	6.5	2.5	9.7	28.8
1973	14	37.2	32.4	7.6	2.3	13.4	30.3
1974	15	25.6	38.8	8.3	2.5	20.0	35.5
1975	14	23.1	38.6	11.0	2.0	15.1	38.4
1976	14	26.2	27.7	7.7	3.1	6.9	46.2
1977	18	32.7	27.3	6.0	2.9	9.8	40.1
1978	17	39.2	22.0	5.8	3.7	4.2	38.8
1979	19	34.6	30.0	6.5	3.6	11.0	35.4
1980	18	36.2	25.0	6.2	4.2	6.3	38.7
1981	16	32.5	25.4	6.8	3.8	6.1	42.0
1982	19	43.4	21.0	5.3	2.7	5.1	35.6

Source: FAO, *Food Outlook 1982, Statistical Supplement*, March 1983, p. 21.

assurance that at times of food shortages they are easily accessible to the food-deficit importing countries. Apart from the logistical problem and the bottlenecks in transportation, distribution and marketing, there is the question of bans or embargoes on exporters for various reasons, including the desire to avoid a price rise in the exporting countries themselves. Furthermore, even if the physical availability is not a problem, the price may rise beyond the purchasing capacity of low-income food-deficit countries.

International Grains Negotiations

International negotiations over the six years, 1973–79, have centred on a system of internationally coordinated buffer stocks, designed primarily to stabilize prices within a given range. It was expected that, at the same time, this would ensure that stocks so held would also be able to meet shortfalls in world cereal production. The market will have to bear a part of the shortfall through rising prices. The stocks under international agreement were expected to help iron out extreme fluctuation, as happened in 1972–73.

How are the benefits of price stabilization distributed among producers and consumers, between exporting and importing countries? During a period of price rises, the food import bill of deficit countries goes up, while exporters increase their profits. With price stabilization, importers gain and exporters lose. The net balance sheet of gains and losses for producers and consumers would depend on the assumptions about elasticity of demand and supplies, as well as the behaviour pattern of countries in which production fluctuates strongly (such as the USSR) because of climate and other factors; the latter usually enter and leave the market unpredictably from year to year.[2] Various analyses of the costs and profitability of buffer stocks show that in most cases such stocks are expected to lose money over time, especially if price bands are narrow. This is expected; a buffer stock would protect consumers against unknown events, for example, a crop failure every five years or more. Private traders do not, and are unlikely to, hedge against major crop failures because of private risk aversion and storage costs. There are no profits to be earned in constantly investing in importable prospects. The market is not likely to be stabilized in extreme situations by private traders. The behaviour or the public and private stocks in the USA during the 1960s and early 1970s shows that when public stocks decline, private stocks increase, but the latter increases proportionately much less, and therefore do not offset the decline in the former. When public stocks were generally depleted to negligible amounts in 1975, for example, the amount of private stocks was not comparable to public stocks held in earlier years, and would have been inadequate to meet any serious shortfall in production.

Why is private stockholding in exporting and importing countries not enough to stabilize world prices? Apart from working or pipeline stocks, the private trader or stockholder will hold excess stocks or reserve stocks only if he finds it profitable. It is worthwhile for a private speculator to hold stocks if the expected future price is greater than the current market price plus carrying or storage costs. He does not hold stocks to stabilize prices within a predetermined range. Unless the predetermined price range by accident happens to be equal to the current and anticipated market price, as seen from the point of view of the speculators, it is unlikely that the size of these two sets of stocks would be the same.

Public stockholding has usually been discussed in the context of stabilizing price within a predetermined range. The lower the range of prices, the higher is the size of stocks, and vice versa. Public stockholding to some extent does substitute for private stocks. Apart from stabilizing prices, public stocks may be held to raise prices above the competitive level. This would result in a growing accumulation of stocks, which would eventually be disposed of in the long run, with the result that stocks as a means of raising prices over the long-run competitive level are only a short-term measure. The public authority need not hold all the desired stocks itself; it could provide a subsidy to the private traders or producers to hold stocks on behalf of the public sector, above the levels at which the private traders would otherwise hold for speculative purposes.

The considerations that weigh with the public authorities to take concerted action at an international level to stabilize prices are many and various. First, a greater stability of world cereal prices would help the heavily import-dependent poor countries and save the increase in import costs at times of rising prices. The greater the food import bill, and the greater the proportion of export earnings, constituted by the food import bill, the higher would be the saving in import costs from not having to pay higher prices, for poor food-deficit countries at times of shortfall in world production and high prices.

Secondly, considerable instability of world prices may stimulate national action to follow protectionist policies in order to achieve a higher level of food self-sufficiency and to reduce the level of import dependence, thus resulting in a pattern of world food production which is not compatible with comparative cost considerations and results in misallocation of resources.

Thirdly, instability in world supplies and prices would persuade individual countries to insulate domestic production and consumption from the impact of variations in world prices, by varying net trade with a view, to stabilizing domestic prices. The result would be an intensification of instability in world markets, which in a vicious cycle of action and reaction would lead to attempts at further insulation of domestic markets. This is more true of the developed importing countries, which can afford the costs of domestic support price policies to insulate the domestic market.

Fourthly, faced with high instability of world supplies and prices, national governments in the developing countries that want to embark upon a policy of stabilizing the food consumption of the poor and vulnerable groups would find such measures too costly and may abandon such poverty-alleviating and equity-promoting measures.

The wider the range within which prices are to be stabilized, the smaller is the amount of publicly held stocks and the smaller is the degree of substitution of private stocks by public stocks. The abundant stocks that were held by the USA and Canada in the 1950s and 1960s for domestic reasons of supporting producers' incomes in their own farm economies, held price fluctuation within a very narrow range.

From the late 1960s and in the 1970s the USA and Canada were following different policy measures for stabilizing farmers' income, such as acreage control programmes and deficiency payments or direct income transfers to farmers; the acquisition of stocks with a view to raising or stabilizing farmers' income was no longer the principal consideration of stockholding policy in North America. The downward pressure on prices resulting from an upward swing in production, to the extent that it was not offset by acreage reduction in exporting countries, was partly met through stock-building in the importing countries. In the 1970s importers were increasingly building up stocks.

It must be noted that stocks that would be needed to stabilize prices within a range narrower than that which would result in the absence of public intervention would not pay for themselves. The costs of holding stabilization stocks need to be subsidized. This is what happened to stockholding by the USA and Canada in the 1960s; during the 1970s the public authorities subsidized both publicly held and farmer-owned reserves in the USA. Because there were large grain reserves in the 1960s, prices were held within a very narrow range. During the 1970s publicly-held stocks in the USA were smaller. Moreover, there is no obligation on the part of the US public authorities to sell when market prices rise above support prices; it is permissive, not obligatory. Market prices could, in theory, be given a much greater role in rationing supplies.

The present system of stocks in various countries, both developed and developing, results from domestic policy objectives. There is no consensus as to how high these stocks should be; they were not adequate to prevent prices falling below costs of production in many countries, as in 1982–83, or rising to exorbitant levels with dire consequences for the poorer importing countries, as in 1972–73.

Breakdown of Negotiations

The protracted negotiations for an International Wheat Agreement, which broke down in 1979, were designed to create a framework of international action to promote stability in supplies and prices and to provide some measure of food security for the low-income food-deficit countries.

It was always understood that the publicly-held stocks under the auspices of an international agreement would not be a *net* addition to world cereal stocks, since they were likely either partly to replace private or public stocks of individual nations held prior to the international agreement. They were supposed to operate under a set of internationally agreed rules for their acquisition and release, with specific trigger prices, so that divergent actions would not be taken by countries. Even though the strictly enforced trigger prices were not agreed upon, there is still a need for a greater degree of harmonization of stock targets and policies than exists at present. Individual countries first need to formulate their own national stock policies and targets and let them be widely understood and known; secondly, they need to harmonize so that at times of shortages some countries need not start building stocks and thus aggravate shortages, or vice versa. There needs to be a harmonization of policies in respect of the release or acquisition of stocks; there should be a much greater exchange of information and knowledge of each other's behaviour in the grain market. Speculators in the world grain market, who have accentuated physical scarcity in the past, will have to contend with the fact that there are publicly-held stocks that would intervene to reduce the extreme fluctuation in prices.

The negotiations for an international grain agreement broke down because there was no consensus between the exporters and importers, or between developed and developing countries, in respect of the major economic provisions of such an agreement.

First, there was no agreement between major exporters, i.e., North America on the one hand and the EEC on the other, on the size of the buffer stocks, with the USA arguing for a larger and the EEC for a smaller reserve.

Secondly, there was no agreement on the trigger prices, i.e., ceiling and floor prices, at which buffer stocks would intervene in the market. The exporters were arguing for higher ceiling prices on the grounds that they are essential for offsetting the low profits or losses they incur at times of good harvests and low prices.

Thirdly, there was the question of whether the size of stocks that were discussed, i.e., between 30 and 60 million tons, was consistent with the range of prices within which stability was to be sought. The EEC was seen to be arguing for a narrower price range and a smaller reserve, implying, according to the USA, that somehow exporters would in any case be obliged to hold stocks outside the agreement, which would contribute to the stability of the market.

Fourthly, in any case, there was no clear understanding as to how internationally coordinated reserves would relate to the rest of the stocks held by individual countries outside the agreement for domestic price stabilization or for meeting domestic shortages.

Fifthly, there was no agreement on how the stockholding obligation would be shared among the exporters and importers, among the developed and developing countries. In many other commodity agreements, stocks are held by the exporters rather than by importers and exporters, and are in some cases partly financed by levies on exports. In this agreement, every signatory, importer or exporter as the case may be, was to undertake the obligations to hold stocks at their own expense.

Sixthly, while there was a recognition of the fact that developing countries suffered from a lack of resources, both to build storage capacity and hold stocks, there was no agreement as to how and to what extent developing countries were to be assisted in either building up food security infrastructure or in purchasing or procuring food reserves. The developing countries argued that if they were to hold reserves as part of their obligations under the International Grains Agreement, the latter should provide for financial and technical assistance to enable them to fulfil their obligation. The developed countries, on the other hand, argued that they should seek ways of fulfilling the obligations by seeking assistance from the external resources presently available to them for development purposes, rather than creating additional financial facilities.

Seventhly, developing countries argued that they should have an assured and preferential access to supplies at times of food shortages in view of their limited capacity to buy in the open market at times of shortages and high prices. This was not acceptable to the others.

The foregoing analysis reveals the complexity of issues underlying the negotiation of an international agreement, especially if market stability and achievement of food security, i.e., meeting the urgent import requirements of food-deficit low-income countries at times of food shortages, are mixed together in one package.

The search for an international agreement was more or less abandoned in the face of the unwillingness of the most important exporter, the USA, to enter into any such negotiations, and in view of its clear position that national stocks for the purposes of stability and food security should be left entirely to national decisions without any international coordination, and that all countries should be encouraged to build up stocks in the light of their own requirements and capabilities.

Alternative Measures for Promoting World Food Security

What alternative measures are available to promote world food security, especially to meet the requirements of low-income food-deficit countries? In the absence of an international agreement on grains to promote market stability in supplies and prices, price fluctuations are likely to be greater. The impact of high prices on the low-income food-deficit countries can be moderated if international assistance is available to finance the extra cost of food imports. The recently (1981) established IMF cereal-financing facility, which covers the extra cost of cereal imports in a particular year – over the average cost of cereal imports in the past five years – is a step in the right direction. It covers the extra cost of cereal imports, not only due to a rise in import price but also due to a shortfall in domestic production. The new facility is an expansion of the pre-existing IMF compensatory financing for export shortfalls. Insofar as a rise in export earnings accompanies a rise in cereal import costs, access to finance for the extra cost of cereal imports is reduced by the extent of increase in average export earnings over the past five years. The terms of finance are not concessional; it is a 3–5 year loan, and not for a long period. To date, only a few countries have taken advantage of the facility, partly because world cereal prices have been depressed in the last two or three years due to large surplus, and partly because heavy drawings on the compensatory financing for shortfalls in exports have not left scope for any additional drawings for cereal imports, which are permissible under the IMF quota limitations.

An expansion of the facility in terms of more liberalized terms of access and wider coverage of commodities, in order to include imports of sugar and edible oils, for example, in addition to cereals, has been urged by the developing countries. Under the facility, the shortfall in net export crops is calculated as: a) trend exports *minus* actual exports; *plus* b) actual import costs *minus* trend cereal import costs. The total amount of credit is constrained to 100 per cent of a country's quota for each of the two components (loss of export earnings and excess cost of cereal imports) and 125 per cent of the quota for the combined total.

Another source of stability of food supplies available to food-deficit countries is the Food Aid Convention, renegotiated in 1980, under which food aid donors undertake to provide, irrespective of fluctuations in their domestic production and in the availability of export surplus for commercial sales, a minimum food aid of 7.6 million tons a year, primarily to the low-income, food-deficit countries.

The total food aid flows in recent years, both bilateral and multilateral, are at about 9 million tons, which falls short of the minimum target laid down by the World Food Conference in 1974. Recent estimates of food aid requirements put the figure at 20 million tons by 1985, indicating that requirements exceed actual aid flows by a wide margin.

While the overall flow of food aid up to a certain amount, i.e., 7.6 million tons, has been stabilized, the actual flow above this level fluctuates according to the available export surplus and the multiplicity of policy considerations governing food aid. Under this present agreement, there is no provision for an expansion of food aid when food shortages, scarcities and high prices occur in the world market. Moreover, the year-to-year flow of food aid to individual food-deficit countries is not necessarily related to the fluctuations in the domestic food production of the recipient countries, nor is it related to the fluctuations in world prices of food to help relieve the import burden of the food-deficit countries which rely on commercial purchases, in addition to food aid, for meeting their food requirements. A system of food aid could conceivably be devised by an international agreement under which food aid would be distributed among recipient countries from year-to-year, in such a manner as to stabilize the food consumption within a certain range, let us say 5 per cent, of their trend value of consumption. In order to make such a scheme feasible, its coverage may have to be restricted to the very low-income food-deficit countries and should be limited to compensating shortfalls in trend value of production only to 95 per cent or so. This scheme must leave a part of food aid, the inter-country and inter-temporal allocation of which would be predominantly governed by political/strategic considerations.

Developing countries are concerned that at times of food shortages and high prices in the world market they may be crowded out of the world market. Hence proposals have been put forward for long-term sales and purchase agreements between developing countries on the one hand, and the major exporters on the other, so that they are assured of minimum supplies at times of shortages, while undertaking to buy a minimum quantity from the exporters during periods of low prices. This is the nature of trade agreements for the purchase of wheat that have been concluded by the Soviet Union and China with some exporters such as the USA, Canada and Argentina. Usually, such agreements do not specify prices, which are left to the market forces, but specify the quantities that will be traded. Hence they do not protect the importing countries from the, impact of high prices unless such purchases are combined with price concessions or subsidized export credit or blended credit, as has been done in a few cases.

An alternative proposal to ensure that developing countries would have access to supplies – in a way, preferential access – at times of shortages, is for the developed exporting countries to set aside funds/stocks of cereals to meet the urgent import requirements of the food-deficit countries, especially the low-income countries among them. In recent years, a few countries have done so, foremost among them being the USA, which has set up an International Emergency Food Reserve of 4 million tons, intended specifically to meet its obligations for food aid under the Food Aid Convention, as well as to meet food emergencies in developing countries caused by crop failures or man-made disasters. Where food allocations are made in financial terms and food aid obligations in quantitative terms, financial allocations may or may not be adequate to secure the required quantity of cereals, depending on the market price prevailing at the time of purchase. The food reserves built up by the developed, exporting countries protect food aid disbursements from being eroded by a rise in food prices. So far none of these reserves are intended to meet the demands of commercial purchases by the low-income food-deficit countries. It is worth examining, first, whether many more developed countries should not set up such reserves and, secondly, whether the scope of such reserves could not be extended to cover commercial purchases by low-income countries as well. The concept of such unilateral reserves to be held by developed countries for meeting the food needs of low-income food-deficit countries was incorporated in the International Undertaking on World Food Security, endorsed by the World Food Conference in 1974.

This unilateral action, taken separately by individual countries, does not require any international agreement, and is hence easier to implement otherwise. Each country could do it within the context of national systems and procedures; no uniformity is required, at least to start the system. As experience accumulates with this type of reserve, and as several countries adopt it, it may be appropriate to learn from experience, to improve upon it and to explore the possibility of systematizing it under an international framework or umbrella of some sort.

The search for world food security, in the narrow or restricted sense of stabilizing supplies with a view to stabilizing prices and consumption, within socially acceptable limits and consistent with the improvement of producers' incentives, raises complex and difficult issues. They encompass both domestic food policy and international cooperation in trade and development, for the promotion of stability, growth and equity.

Notes

1 The year for which trend values were established was 1974–75 on the basis of historical data from 1955–56 to 1972–73.

2 If demand is non-linear in the sense that prices rise more when supplies are short, then they fall when supplies increase in equal quantity (the demand curve is linked, the upper section being less elastic than the lower section) and if importers only respond partially this year to price changes (reducing price elasticity of import demand) and if production response to price changes takes place only with a time lag of one year, for example, a price stabilization policy is more likely to benefit the consumer rather than the producers/exporters.

References

Huddleston, B., Johnson, G., Reutlinger, S. and Valdés, A., *Financial Arrangements for Food Security*, IFPRI Research Report, October 1981.

Sarris, A., Abbot, P. and Taylor, L., 'Grain Reserves, Emergency Relief and Food Aid', in W.R. Cline (ed.), *Policy Alternatives in a New International Economic Order* (Washington, DC: Overseas Development Council, 1978).

World Food Institute, *World Food Trade and US Agriculture 1960–62* (Iowa State University, August 1983).

Chapter 22

Instability in World Food Prices in the Post-Uruguay Round Situation: Prospects and Policy Issues*

The prices in world cereal markets are highly volatile, with adverse impact on the poor in developing countries, who spend 40–50 per cent of their income on cereals. Following the Uruguay Round of trade liberalization price instability is likely to persist even though the burden of adjustment to supply fluctuations will be shared widely among all trading nations and domestic markets will be less insulated. As a result of a fall in food supplies caused by a reduction in price support programmes, the world cereal stocks that tended to act as a cushion or buffer to absorb supply fluctuations will fall. Under the circumstances, to deal with the impact of price instability on poor countries, what is needed is international agreement to ensure adequate stocks as well as food aid or financial facility at times of high cereal prices as well as national measures to moderate high price instability without seriously jeopardizing commitment to a liberal regime.

During the negotiations in the Uruguay Round of trade liberalization and their aftermath, much concern was raised about its impact on the future level of world cereal/ food prices and consequently on the cereal import bills of the low-income food-deficit countries as well as on the consumption and nutritional status of the poor in developing countries. Not enough attention was paid to what is likely to happen to the degree of price instability in world cereal markets that in recent years fluctuated as much by 30–40 per cent from one year to the next with significant adverse consequences for the welfare of the poor consumers and producers.

Trade liberalization under the Uruguay Round, if faithfully and fully implemented, will affect food price instability. On the one hand, it will seek to restrain the efforts of individual countries to insulate domestic markets and prices from the variability in world prices and to shift the burden of adjustment to domestic production fluctuations onto the world markets. On the other

* First published in *Asia-Pacific Development Journal* ,Vol. 3, No. 2, December 1996.

hand, it will reduce the magnitude of domestic price support programme, thus discourage domestic surplus and hence the accumulation of stocks that tends to absorb the impact of production fluctuations.

This chapter elaborates on the consequences of the Uruguay Round that are pulling in apparently opposing directions, as illustrated above, as well as on recent changes in agricultural protectionism, especially in OECD countries. It also examines the policy options available to developing countries, as well as the international community, to deal with the impact of price instability in the post-Uruguay Round situation, which will be characterized by a fall in food aid and concessional or subsidized sales, on the one hand, and restrictions imposed by the World Trade Organization (WTO) on the freedom of action of developing countries to pursue domestic stock or trade policies, on the other.

Instability of Food Prices

Instability in food prices (cereals mainly) in world markets arises from instability in food supply and demand. These supply or production instabilities arise from weather variations or incidence of pests and diseases, or in variations in the supply of purchased inputs like fertilizer. Demand variations principally arise from changes in macroeconomic conditions and prices. Supply variations are the most dominant factor in the short-run year-to-year, variations in food prices.

Since food trade, either for individual farmers or for nations, is a small proportion of food production, a given percentage variation in production is associated with a much larger percentage variation in traded volumes (Tyers and Anderson, 1992).

Fluctuations in food prices are aggravated by protectionist policies that insulate domestic markets from the impact of variations in supplies and prices in world trade and transmit to the world market the impact of variations in domestic production. To this end, they follow domestic price support (floor and ceiling) policies that are frequently buttressed or reinforced by domestic stock policies.

The net result of domestic price support and protectionist policies is food surplus and the excess accumulation of stocks in mainly developed exporting countries which, as a consequence, have a depressing effect on world market price.

The fluctuations in major cereal prices (wheat and rice) over the past decades are shown in Figures 22.1 and 22.2. In some years, prices rose as

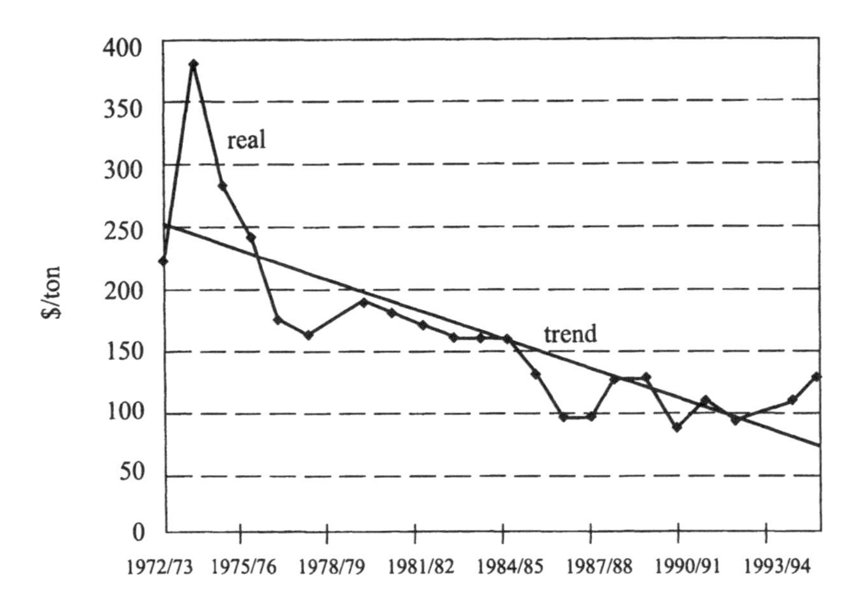

Figure 22.1 Variability and trend in real wheat price, 1972/73 to 1995/96

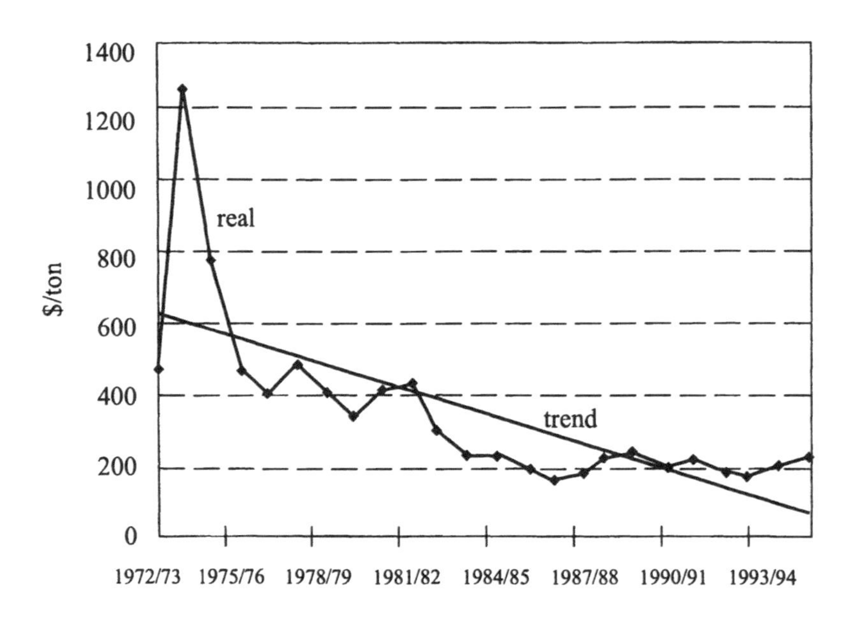

Figure 22.2 Variability and trend in real rice price, 1972/73 to 1995/96

much as 30–35 per cent and in others they fell as much as 27–32 per cent. For example, in October 1995, wheat prices were about 30 per cent and rice prices were 20 per cent higher than their respective levels in 1994. By January 1996, wheat prices rose 37 per cent above the 1994 level while the rise in rice prices was moderate and was 15 per cent higher than in 1994 (FAO, *Food Outlook*, March and November 1995). For more recent periods when variability went up, for example, the variability index of real wheat and rice prices was 46 and 33 respectively during 1970–80, and went down to 19 and 14 for the period 1980–94. For the whole period 1970–94, the variability index of real wheat and rice prices was 44 and 32 respectively.[1] This is shown in Table 22.1.

Table 22.1 Variability of prices of rice and wheat, 1972–73 to 1995–96 (percentage)

Period	Rice	Wheat
1972–73 to 1979–80	43.91	24.98
1980–81 to 1984–85	14.97	2.34
1985–86 to 1989–90	19.10	13.91
1990–91 to 1995–96	74.80	33.64
1972–73 to 1995–96	70.85	27.62

Source: Author's estimates based on annual figures on nominal prices as obtained from various issues of FAO, *Food Outlook*. The deflator used to estimate the real price is obtained from United Nations, *Monthly Bulletin of Statistics*. The deflator used in this context is the unit value index of manufactured exports of industrial countries.

Excessive fluctuations in the prices of cereals have a significant impact on the consumption of the poor because cereals constitute a high percentage – as much as 40–50 per cent – of their consumption expenditure. A rise in cereal prices by 20 per cent may often reduce cereal consumption of the poor by 10 per cent and depress their nutrition, while food producers frequently make large gains. Low prices, on the other hand, favour poor consumers but adversely affect producers who may suffer a depletion in assets with a depressing effect on their long-run income (Pinstrup, Andersen and Garrett, 1996).

As world prices rise, the low-income food-deficit countries are unable to sustain the required level of their cereal imports without an increase in food aid or in international indebtedness. Typically, these countries account for 50–55 per cent of the cereal imports of all developing countries. In normal years, food aid constitutes 10–15 per cent of cereal imports of low-income food-deficit countries.

Effects of Uruguay Round on Instability

The implementation of the Uruguay Round Agreement would result in a reduction by 36 per cent in import protection, and in domestic support programmes by 20 per cent in developed countries and 13.3 per cent in developing countries; there will be a reduction in the volume of exports under subsidy by 20 per cent and in the value of export subsidy expenditures by 36 per cent.

Under the Uruguay Round, payments for acreage set aside reduction programmes in the United States and compensation payments in the European Union are permitted and, therefore, are exempted from commitments for reduction. The provision for reduction of domestic support programmes (aggregate measure of support) is to be considered in the aggregate and is not product-specific so that individual products could be safeguarded by cutting heavily on other products. On the whole, the projected liberalization of production and trade under the Uruguay Round as currently agreed is less than expected and, therefore, its impact on the level and instability of food prices will not be as significant as was originally anticipated.

The new trade regime inaugurated by the Uruguay Round is expected to affect not only the level but also the instability of world food prices in the future. Also, at the same time, the individual countries would be obliged to open up domestic markets and allow food prices to follow the movements in world prices to a much greater extent than before.

Recent estimates conclude that the prices of major cereals by years 2000–05 would be higher by 5–7 per cent than in the absence of the Uruguay Round (FAO, 1994, 1995b). The various modelling exercises undertaken to date about the impact of the Uruguay Round do not quantify the stability of prices. A recent analysis of the consequences of a 5 per cent across the board change in production from the trend (with 1987–89 as the base year) found that there was no significant difference in terms of an impact on price between the pre- and post-Uruguay situations. The extent of a rise in price resulting from a shortfall in production was about the same percentage in the pre- and post-Uruguay situations. For example, a 5 per cent fall in output compared to trend resulted in a rise in price of approximately 25 per cent, 50 per cent and 24 per cent, respectively, for wheat, rice and maize in both pre- and post-Uruguay situations. Similarly, a 5 per cent increase in output above the trend level resulted in a fall of price of about 20 per cent, 25 per cent, and 18 per cent, respectively, for wheat, rice, and maize in both pre- and post-Uruguay situations (FAO, 1995b). This is based on the assumption that the volume of stocks held did not change in the post-Uruguay situation; if stocks were smaller, price fluctuation would be greater. The absence of a difference in price variability caused by a decline in protectionism may be partly due to the limited nature and degree of liberalization undertaken so far.

Future Level of World Cereal Stocks

However, cereal stocks are likely to be smaller than what was traditionally held in the past. This is because food surpluses are in any case expected to decline and continue to do so in the future. Already, prior to the agreement in the Uruguay Round, reforms were instituted in the CAP and in the United States farm laws, under severe budgetary pressure; while they constrained the rise in domestic support price on the one hand, they introduced, on the other hand, programmes for the control or reduction of acreage under important cereal crops. Thus, while, on the one hand, support prices tended to keep up production, the acreage control programmes, on the other hand, tended to reduce surplus production. The 'acceleration' effect of price support was partly dampened by the 'brake' on production through output controls so that food surplus was down.

During most of the period during the 1980s and early 1990s, 60–70 per cent of the world cereal stocks were held in developed countries. Moreover, per capita stocks were considerably and consistently much larger in developed countries than in developing countries. In developed market economies, during the period between 1983–84 and 1989–90, the public stocks as a percentage of total stocks varied between 54 per cent (1986–87) and 39 per cent (1985–86) (FAO, 1990). The cereal stocks, especially publicly-held stocks, were subject to wide variations in developed countries and, hence, played the major role in combating the production fluctuations (FAO, 1995a). Already in recent years, independently of the Uruguay Round Agreement, the ratio of cereal stocks to total utilization in exporting countries declined to 22 per cent in 1992–93 and to about 15 per cent in 1994–95 (FAO, 1994, 1995).

The publicly-held stocks, as analysed above, were associated with producer price support programmes, consumer income support or for meeting sudden disasters or man-made emergencies. The publicly-held stocks associated with domestic price support programmes are likely to be greatly reduced in the future; in the past they partly substituted for private stocks which would otherwise have been held to bridge the difference between current and expected prices, provided the cost of holding such stocks was not higher than the price differential. However, the government price support programmes led to a volume of stocks that were higher than what was warranted by the expected price differential between the current and future period; in fact, government stocks reduced the price differential that otherwise would have obtained. Thus, in the absence of price support programmes, the fall in public stocks would only partly be compensated by private stocks that were displaced in the first place by public stocks and that would now respond to market price differentials over time. In a preliminary exercise undertaken regarding the impact of trade liberalization on cereal stocks, it was found, on the assumption of total liberalization by world as a whole, that the proportion of stocks of cereals (based on average of 1985–89) would fall from 19 per cent (bulk being government stocks) to about 16 per cent (all private) (FAO, 1990).

Impact on Developing Countries

Developing countries will be affected in two principal ways. First, there is likely to be a fall in food aid availability. Second, developing countries have already opened up and are likely in the future to do more so to price signals in world markets. Hence, they will be exposed to a greater extent than before to instability in world markets (Falcon, 1995).

Food aid and concessional food sales in the past to developing countries helped circumvent the impact of high price on low-income food-deficit countries, even though not all food aid flow was fully modulated or regulated systematically or proportionately in response to fluctuations in domestic food supplies and prices in recipient countries. Food aid amounted to about 2 per cent of total production of the major food aid donors, i.e., European Union, United States and Canada (Shaw and Singer, 1995). The new Food Aid Convention of 1995 (effective July 1995 for three years) reduced the minimum aggregate commitment by donors to 5.36 million tons as compared to the previous 7.5 million tons that was guaranteed by the Food Aid Convention of 1986. This situation was due to a fall in commitment by more than half by the United States, the largest food aid donor (Shaw and Singer, 1995). In practice, however, in the past actual food aid generally exceeded the minimum commitment and varied between 8.8 million tons and 13.6 million tons since the mid-1980s. The Uruguay Round provides for the review of the level of food aid required to meet the needs of developing countries during its implementation. The availability of food aid, considering that a certain amount of food aid for the poor countries is permitted under the Uruguay Round, would, however, depend on future prospects of food surpluses in the major developed, exporting countries, that are likely to be reduced in the future. Also, it would depend on the future level of development assistance, that is likely to be curtailed, as well as its distribution between food and non-food aid, since food aid is to be increasingly considered more as a form of development assistance, rather than a way of disposing of food surpluses.

The price fluctuations in world markets will persist due to world production fluctuations that are, after all, the most dominant cause of price instability, aggravated by a fall in world cereal stocks; this may partly be relieved to a yet unknown extent by a further decline in protectionism that may take place in the future. The net result may be that the opposite trends may cancel each other out or may result in either a higher or a slightly lower degree of instability in world markets than in the past. Whatever the net result, if the world price instability is fully transmitted to domestic markets in developing countries, their domestic price instability may rise to a level higher than what prevailed when domestic price stabilization was in place prior to their market liberalization policies. This will pose a difficult problem for low-income food-deficit countries where cereal continues to be a major item of consumption, as well as a major item of income, in view of the dominance of food production in the rural economy.

It may be noted that under the Uruguay Round Agreement, the developing countries are allowed to resort to public stockholding for food security purposes, that is, they can purchase for and sell from public stocks at administered prices, explicitly held for food security purposes. The open market operations are not to be coupled or linked to taxes or subsidies or controls on trade, except within the leeway provided by 'tariff ceilings'. The process of stock accumulation and disposal should be financially transparent. Food purchases by the government will be made at current market prices and sales from food security stocks will be made at no less than the current market price for the product and quality in question (Shaw and Singer, 1995).

To the extent, however, floor/support prices to the farmers are above market prices, this is considered in the category of domestic subsidy. Under the Uruguay Round, the aggregate measure of subsidy should not exceed 10 per cent (as against 5 per cent for developed countries). However, developing countries are allowed to undertake subsidized food distribution to meet the requirements of the urban and rural poor on a regular basis (FAO, 1995b).

Faced with such a predicament, developing countries, while they may move away from interventions in domestic cereal markets by means of large-scale public sstocks due to past experiences of inefficiency, incompetent management and high costs, may yet seek alternative ways for achieving some degree of domestic price stability level based upon world or border prices, by means of taxes and subsidies on trade; this policy allows a large role for private trade and stocks within a sufficiently profitable price band between floor and ceiling prices. Under the new trade regime, the developing countries are constrained to some extent in following a policy of taxes and subsidies on imports and exports. They are expected not only to adopt tariffication in lieu of quantitative controls on imports, but also to 'bind' tariffs at a maximum level beyond which they cannot increase them further, even though they can reduce them. However, in practice, they have been given some freedom of manoeuvre. They have been allowed to fix tariff 'ceilings' above the currently applied rates, so that the majority of them have fixed very high ceilings (called 'dirty tariffication'); they can raise tariffs so long as the level is below the ceiling. Thus, the possible limit of variations up and down, so long as they remain below the 'ceiling' level of bound tariffs, may in many cases allow them to use the instrument of tariffs, to regulate imports and import prices for purposes of implementation of domestic ceiling and floor prices. However, it is not clear whether they can provide import subsidies since, in principle, it is not a trade restricting but a trade promoting measure. New export subsidies, however, are prohibited. In fact, the developing countries are required to reduce over time the export subsidies if they are applied today, except when export subsidies only relate to export marketing and transport. Therefore, subsidy to compensate for the gap between high domestic price and low export prices is not permitted. It should be noted, however, that the period of implementation of the trade reform is extended to 10 years for developing countries as against six years for developed countries. Moreover, as the experience with trade-based stabilization prices throws up problems and challenges in the future, the developing countries can take them up during the next round of negotiations. In view of the limited liberalization of agricultural trade in the current round, the WTO is mandated to continue the reform process by starting new negotiations in the final year of implementation (in six-year term).

Policy Options for the Future

Therefore, in the future, to ensure that the developing countries are able to achieve food security – to enable assured access to basic food for all, especially the poorest – in the face of high price fluctuations, several policy measures need to be considered. These are as follows:

a) to promote international understanding, among developed and developing countries, on measures for assuring adequate level of stocks (public or private, or both) so that excessive price increase at times of acute and large scale food shortage can be contained or excessive falls due to sudden world surpluses be arrested;
b) to assure the low-income food-deficit countries access to food supplies at times of excessive price rise, through food aid under a renegotiated Food Aid Convention and/or financing facilities under the auspices of International Financial Institutions. The Uruguay Round Agreement itself draws attention to these measures;
c) to re-examine the current provisions of the Uruguay Round Agreement to ensure that developing countries are able, if needed, to follow a trade-based policy for stabilizing excessive fluctuations in domestic cereal prices through the use of trade taxes and subsidies. As the experience with the implementation of the Uruguay Round accumulates, these and related policy options need to be taken up in the next round of agricultural trade negotiation expected to start at the end of the current six-year period.

Note

1 The variability is the standard deviation of the percentage differences of actual real prices for the period concerned.

References

FAO, *The Effect of Trade Liberalization on Levels of Cereal Stocks*, Committee on Commodity Problem, CEP: GR 9013, August, 1990.

FAO, *Assessment of the Current World Food Security Situation and Recent Policy Developments*, Committee on World Food Security, CFS 9412, February 1994.

FAO, *Food Outlook, 1994–95*, various monthly issues, 1995a.

FAO, *Assessment of the Current World Food Security Situation and Medium-term Review*, Committee on World Food Security, CFS 9512, February 1995b

Falcon, W.P., 'Food Policy Analyses, 1975–95: Reflections by a practitioner', lecture presented at Twentieth Anniversary of International Food Policy Research Institute, November, 1995.

Pinstrup-Andersen, P. and Garrett, J.L., *Rising Food Prices and Falling Grain Stocks: Short-run blips or new trends?*, 2020 Vision Policy Brief, International Food Policy Research Institute, Washington, DC, January 1996.

Shaw, J. and Singer, H.W., *A Future Food Aid Regime: Implications of the June Act of the GATT Uruguay Round*, Discussion Paper No. 352 (Brighton: Sussex University, 1995).

Tyers, R. and Anderson, K., *Disarray in World Food Markets* (London: Cambridge University Press, 1992).

PART VIII

TRADE AND AID: NATIONAL TRADE POLICY

Chapter 23

Comparative Costs, Factor Proportions and Industrial Efficiency in Pakistan*

The chapter presents, in Section I, new and additional evidence on the comparative costs of manufacturing industries in Pakistan. Furthermore, the findings of the present study are compared with those of earlier studies. Comparative costs in this context are defined as the ratios of ex-factory prices of specific domestic products to c.i.f. prices of closely competing imports. In Section II, it examines whether the tariff rates are an adequate index of the comparative cost ratios; in other words, whether the differences in the tariff rates reflect the differential cost structure of Pakistani industries? We also examine, in Section III, whether the available data provide any evidence on the relationship between the magnitude of cost disabilities of the Pakistani industries and their stage of infancy, i.e., whether and to what extent cost ratios decline with the growing up of infant industries. This chapter also analyses, in Sections IV and V, how far comparative cost ratios can be used as a measure of the relative inefficiency of industries in Pakistan. How far, for example, do the high cost ratios of domestic industries merely indicate that the Pakistani rupee is overvalued? How far are the cost ratios affected by or represent high profits of industries? An attempt is made to adjust for both the overvaluation of foreign exchange and the prevalence of abnormally high profits. Finally, in Section VI, we relate the comparative cost ratios of the manufacturing industries to their factor intensities or their factor proportions in an attempt to explore whether relative efficiency is correlated with the relative intensity of use of the different factors such as capital, labour and skill.

I Evidence on Comparative Cost Ratios

Pakistan's impressive achievement in the growth of the industrial sector has attracted a considerable amount of analysis, especially with respect to the pattern or strategy of industrialization and the efficiency of the industrialization

---

* First published in *The Pakistan Development Review*, Vol. 7, No. 2, Summer 1967.

programme. This chapter is an attempt to analyse the comparative cost structure of the Pakistani industries, based on a direct estimate of the costs of the individual manufactured goods and the c.i.f. prices of the closely competing import products. The data are derived from the published and unpublished reports of the Pakistan Tariff Commission on 115 industries scattered over a 15-year period, i.e., 1951–61.[1] The analysis covers about 359 products. They in no sense, therefore, constitute a random sample from the manufacturing sector as a whole so as to represent various industrial groups. The most important industries, such as cotton textiles, jute textiles, woollen textiles and fertilizers, are not included at all in this analysis since they have not been subject to investigation by the Tariff Commission. For the same reason, there are also important omissions from within such industry groups as machinery, both electrical and otherwise, and transport equipment. The industries covered here, moreover, range over both the large-and small-scale manufacturing sectors. The difference in cost conditions between large and small firms is ignored in the analysis.

The total number of establishments or firms covered in the present analysis, excluding such industries as coir goods industry, washing soap industry, leather footwear industry and *bidi* industry, where small firms predominate, is around 1,164. This number, however, still includes a few small firms in a number of industries such as paints and varnishes, nonmetallic mineral products, engine turbines, etc., which are included in the analysis. Since the number of establishments in the large-scale manufacturing sector as reported in the revised 1959–60 Census of Manufacturing Industries is about 3,800, our coverage is quite large.[2]

Does an analysis of the cost ratios of only those industries which have been investigated by the Tariff Commission yield biased results because only inefficient industries will need such intervention? This is unlikely since the great majority of the manufacturing industries in Pakistan have ex-factory prices higher than the competing world prices, and thereby are prospective candidates for protective tariffs. The reasons why some industries come before the Tariff Commission and others do not are not systematically related to their respective relative efficiency. Thus, for instance, some industries receive protection from foreign competition through high revenue tariffs or quantitative restrictions on imports, none of which require the intervention or investigation by the Tariff Commission. There is likewise no reason to believe that the protective incidence of these measures is systematically related to the comparative cost structure of the industries, since the height of revenue duties or the level of quantitative restrictions is determined primarily without reference to the cost structure of the affected industries. Fiscal and balance of payments considerations are the principal determinants of the levels of these alternative instruments of import restrictions. Therefore, there does not seem to be any strong reason for bias in the sample, one way or the other, with the result that what is true of the given sample of industries in the present study is likely to be broadly true of the whole range of industries in Pakistan.

It is also important to remember that the cost ratios for each industry are the averages of the ratios for the individual products which are included in each industry. There are in some cases important differences in the cost ratios between different products produced by the same industry. While the average cost ratio for the industry as a whole in some cases may be high, individual may have cost ratios which are highly competitive with imports. Moreover, there are instances where differences in efficiency exist between different firms producing the same product. Again, there are differences in the c.i.f. prices of the same product, depending upon the sources of supply. The cost ratios reported in this chapter are, however, averages for a number of firms or represent the cost ratio of the representative firm chosen by the Tariff Commission for its cost analysis. The data relate to the years 1951–66 and the different industries are covered in the different years, limiting the value of an intertemporal comparison. Where cost ratios differ because of differences in the c.i.f. prices from different sources, an average of the cost ratios based on different c.i.f. prices is taken (Islam, 1967).[3]

The average cost ratios have been computed in three different ways for each of the time periods. Firstly, the simple averages of the cost ratios have been computed for the periods, i.e., 1951–55, 1956–60 and 1961–66 respectively. Secondly, for each period the individual industry ratios have been computed as a simple average of the cost ratios for individual products covered by an industry. The output data relating to individual products are not available; in fact, in the majority of the cases, data relating to the value of output of an individual industry are also not available so that they can be used as weights. Accordingly, the number of products produced by each individual industry is used as weights to arrive at an estimate of the average cost ratio for all the industries in each of the time periods. Thirdly, an attempt was made to classify the individual industries into major groups, following the classification used by Pakistan Standard Industrial Classification, which is also used by the Pakistan Censuses of Manufactures. The cost ratios for these major groups have been computed by weighting the individual industry ratios by the number of products covered by them. But then in the next stage, the cost ratios of the major groups of industries are combined together by using as weights the values of gross output of each of the major groups of industries as given in the censuses. This average cost ratio is obtained for each period. The three sets of ratios are given below

Table 23.1 Comparative cost ratios[4]

	1951–55	1956–60	1961–66
		Average values	
(A)	1.56 (1.49)	1.40	1.83 (1.66)
(B)	1.65 (1.57)	1.62	1.83 (1.70)
(C)	1.56	1.33	2.16
		Median values	
	1.43	1.40–1.42	1.64–1.65

The cost ratios are based on ex-factory prices without indirect taxes. The (A) ratios are unweighted, simple averages of individual industry ratios; (B) ratios are based on the number of products produced by each industry as weights; and the (C) ratios are based on the number of products as weights for deriving the cost ratios of each major industry group which are then weighted by the values of output of each major group of industry. Each of the three different systems of weighting has its own limitations. The best weights would have been the value of output of each individual industry to which a cost ratio relates. However, data on the value of output of each individual industry are not available. The first method, in fact, implies an equal weight to every ratio. To the extent that the relative numbers of products produced by different industries diverge widely from the relative outputs of different industries, the (B) ratios may contain a bias in favour of industries with a greater diversity of products. The (C) ratios are based on the appropriate weights insofar as the weights for the cost ratio of each major group of industry are concerned, even though the cost ratio for each major group suffers from the same limitation as that of the (B) ratios. The (A) and (B) sets of cost ratios are roughly comparable. The changes in the cost ratios of all the three sets between the different time periods are in the same direction. They all go down in the second period and go up in the third. The ratios within brackets exclude the extreme values above 3.00 and they are about 4 to 5 items for the whole period. The ex-factory prices, on the average, are about 50–90 per cent higher than the corresponding c.i.f. prices in the different time periods, whereas the median values of the excess of the ex-factory price over the c.i.f. prices range between 40–65 per cent.

It may be pointed out, however, that the cost ratios of three out of 62 industries in the period 1961–66, two out of 29 industries in the period 1951–55 and five out of 24 industries in the period 1956–60 were less than one; this implies that they were highly competitive and their prices were less than the c.i.f. prices of competing products.

An attempt has been made to classify the industries into three groups, i.e., consumer goods, intermediate goods and raw materials and capital goods, and to examine how the cost ratios differ between these three groups of industries, which are given below:[5]

Table 23.2 Cost ratios (without indirect taxes)

	Consumer goods	Intermediate goods	Capital goods
Simple average	1.58 (1.34)	1.78 (1.68)	1.70
Median value	1.40	1.58	1.62
Weighted average	1.60 (1.43)	1.87 (1.77)	1.77

Note: The weighted average ratios used hereafter in this chapter are based on number of products relating to a particular industry as weights.

The cost ratios computed by the different methods for consumer goods are consistently lower than those for the other two categories whereas as between the intermediate and capital goods, both the weighted and simple average cost ratios of the former are higher than that of the latter. But the differences between the cost ratios of the two latter categories virtually disappear when the extreme values are omitted, as indicated in the above table. A more detailed picture of the comparative cost ratios of the three, categories of commodities is seen below (Table 23.3).

The greatest number of consumer goods industries, i.e., 49 per cent of the total, have cost ratios between 1.00 and 1.50 and about 23 per cent have cost ratios between 1.50 and 2.00; in the case of intermediate and investment goods almost equal percentage of industries, i.e., between 33 per cent and 38 per cent have cost ratios ranging: a) between 1.00 and 1.50; and b) 1.51–2.00 respectively. About 20 per cent and 15 per cent respectively of the intermediate and capital goods industries have cost ratios between 2.00 and 2.50, whereas only about 8 per cent of the consumer goods falls in this range. The range of

Table 23.3 Cost ratios for three m

Groups (cost ratios)	Consumer good Frequency distribution	% of frequ
0.50–0.99	7	16
1.00–1.50	21	48
1.51–2.00	10	23
2.01–2.50	1	2
2.51–3.00	1	2
3.01–3.50	–	
3.51–4.00	2	4
4.01–4.50	–	
4.51–5.00	–	
5.01–5.50	–	
5.51–6.00	1	2
6.01–6.50	–	
	43	100

cost ratios derived above seem to compare well with the results of an earlier study (Pal, 1965).[6] The latter estimates the domestic prices of imports which may be used to derive the cost ratios of domestic industries, as shown in the appendix. In spite of the differences in methods and sources of data as well as in the commodity composition of the two samples, the overall comparability of the two sets of cost ratios adds to the degree of confidence in the general level of cost disability of the different manufacturing industries in Pakistan, as evidenced from the present study.

II Comparative Costs and Tariffs

A pertinent question relating to the measurement of comparative costs is whether and to what extent tariff rates measure or fail to measure comparative costs. In a regime of quantitative restrictions the usual assumption is that tariffs underestimate the excess cost of the domestic product over the c.i.f, price of the competing imports. Therefore, the tariff rates on different commodities may not reflect the cost disadvantages of the producing industries.

Another hypothesis suggests that the excess of ex-factory prices (inclusive of indirect taxes) over landed costs (inclusive of excise and sales taxes) is a uniform percentage of landed costs for all commodities. Thus Soligo and Stern (1965, pp. 249–66), in their analysis of effective protection, derive the world prices of the domestic by deflating the domestic prices by tariffs alone, ignoring scarcity margins.[7] They assume that as a group the scarcity margin is the same for goods and investment goods and that domestic prices are at least equal to c.i.f. price of a competing import converted at the official exchange rate plus import duties. Thus where scarcity margins are positive, tariffs alone will understate cost disadvantage, and where scarcity margins are negative, the cost disadvantage of domestic industries will be overstated. The following analysis provides additional evidence on the overall magnitude, both positive and negative, of interindustry differences in scarcity margins. In the majority of the cases ex-factory price exceeds landed cost (67 per cent of the total number of items) whereas in 30 per cent of the cases it falls short of the landed cost. In a very few (about 3 per cent) they are equal. The ex-factory price below the landed costs may imply either that for these products the tariffs are redundant, or that the domestic product fetches a lower price because of inferior quality. Even when rates are determined by the Tariff Commission, at a particular moment of time tariffs and quotas may be out of line with the divergence between the c.i.f. and the ex-factory price and they

Table 23.4 Ratio of actual ex-factory price to landed costs

Year	Equal to one	Below one		Above one	
	No. of items	No. of items	Average ratio	No. of items	Average ratio
1951–55	–	9	0.75	22	1.29
1956–60	2	21	0.72	27	1.42
1961–66	5	47	0.75	120	1.47

need revision which takes place with a time lag. Often, these rates and quotas are fixed without reference to the Tariff Commission and hence without any reference to the divergence between the ex-factory price and the c.i.f. price of the competing product. Moreover, an industry which starts out with its ex-factory price being equal to or higher than the landed cost, both because of high profits as well as of high costs, may after a period not only reduce costs but also earn lower profits owing to increasing competition.

In many cases the available data on the ex-factory prices relate to fair ex-factory prices rather than to actual prices. For the purposes of the estimate of scarcity margins, 'fair' prices have been used only when actual prices are not available so that the former provides an additional evidence on the magnitude of scarcity margins. To the extent that fair prices are lower than actual prices, which is the case in the great majority of the cases, excepting where firms were making losses, the corresponding scarcity margins are an underestimate of the actual scarcity margins. Sixty per cent of the items have positive scarcity margins varying between 43–73 per cent, as seen below.

Table 23.5 Rate of fair ex-factory price to landed costs

Year	Equal to one	Below one		Above one	
	No. of items	No. of items	Average ratio	No. of items	Average ratio
1951–55	1	23	0.80	40	1.43
1956–60	–	12	0.72	12	1.73
1961–66	–	6	0.65	13	1.43

Thus the general range of the positive scarcity margins in the case of both fair price and actual price rise are as high as 30–50 per cent above landed costs except for the period 1959/60, when the margin for 'fair price' went up as high as 73 per cent above the landed costs. In many cases, ex-factory prices are below landed costs, by about 25–30 per cent.

The scarcity margins widely vary as between different commodities as is evident from the coefficient of variation of the ratios of ex-factory, actual prices to landed costs for individual commodities.

Table 23.6 Coefficient of variation of ratio of ex-factory actual price to landed costs

	A[1]	B[2]
1951–55	0.17	0.96
1956–60	0.81	0.56
1961–66	1.51	3.78

Notes

1 The coefficient of variation for the ratios below one.
2 The coefficient of variation for the ratios above one.

As between different classes of commodities, in terms of actual ex-factory price, the highest scarcity margin is in the category of consumer goods whereas in terms of fair price, capital goods have the highest margins. In both cases consumer goods have the largest number of items with negative margins, as seen below.

Table 23.7 Ratio of ex-factory price to landed cost

	Fair 1951–66		Actual 1951–66	
	Below one	**Above one**	**Below one**	**Above one**
Consumer goods	0.72 (23)	2.32 (26)	0.67 (38)	1.53 (35)
Intermediate goods	0.89 (3)	1.39 (12)	0.74 (17)	1.43 (75)
Capital goods	0.78 (15)	1.69 (27)	0.87 (22)	1.39 (59)

III Comparative Costs and Infant Industry

In a young industrializing economy, infant industries start out with high costs which, with the acquisition of skill and experience in terms of management and technical knowledge, are expected to decline over the years. Accordingly, cost ratios may be related to the age of the individual industries. It may be argued that an older industry, irrespective of its nature, benefits more from the growth of external economies and an industrial milieu. The cost ratios of the different industries on the basis of the number of years of their operation can be combined in a few limited number of broad groups as shown in Table 23.8. There does not appear to be any clear relationship between length of life and comparative cost ratios. However, the data presented in Table 23.8 do not provide an adequate test of the hypothesis, partly because the composition of industries in different age groups is different and partly because the number of industries which have operated for a longer period, i.e., above 20 years or so, is very small. Conceivably, different industries have different periods of infancy and some develop competitive efficiency earlier than others. But in the above sample, various industries with widely different characteristics are lumped together for each age group. Table 23.9 below indicates the cost ratios by the three broad categories of industries, i.e., consumer goods, intermediate goods and capital goods as well as by the number of years in operation.[8] A more detailed classification of industries is given in Table 23.10.

The sample of industries in the last two age groups is too small to permit any comparison. Even comparison among the last three age groups does not yield any consistent and systematic behaviour of the cost ratios with increasing years in operation. In some cases, such as the total of the consumer goods or intermediate goods industries; the ratios first increase and then decrease with increasing years of operation; in the case of the capital goods industries, they go down first and then go up. The time pattern of the behaviour of the cost ratios of the individual industries is not very different, as is shown in Table 23.13. If one compares the first and third age groups only the cost ratios in the case of such industries as bakery products, paints, varnishes and polishes, pharmaceuticals, rubber and rubber products, nonmetallic mineral products, electrical machinery, transport equipment, made-up textiles and miscellaneous industries, register a decline. This is not true for the rest of the industries, some of which suffer a rise in cost ratios with an increase in the number of years in operation.

The above analysis, therefore, does not appear to provide any satisfactory or conclusive answer to the problems of the behaviour of the cost ratios over

Table 23.8 Cost ratios of the different i

No. of firms	No. of products	No. of y
41	131	
25	112	
23	49	
2	4	
2	12	31

Table 23.9 Cost ratios by the three bro

Industry	1–5
Consumer goods	1.37
	(12)
Intermediate goods	1.64
	(19)
Capital goods	1.82
	(10)

Table 23.10 Cost ratios of major groups of industries and number of years in operation

Industry	Number of years in operation				
	1–5	6–10	11–20	21–30	31 and above
Canning and preserving	1.41	1.42	–	–	–
Bakery products	1.19	1.23	0.67	–	–
Alcoholic beverages	–	–	–	–	1.33
Footwear	1.44	–	1.63	–	–
Paints, varnishes and polishes	1.55	1.47	–	–	–
Matches	–	–	0.96	–	–
Pharmaceuticals	2.67	1.89	–	–	–
Industrial chemicals	2.71	–	–	–	–
Rubber and rubber products	1.78	– 1.75	–	–	
Petroleum products	–	–	1.40	–	–
Non-metallic mineral products	1.64	1.13	1.25	–	–
Basic metals	–	–	1.00	–	–
Metal products	1.51	1.67	2.07	1.11	–
Non-electrical machinery	–	1.46	–	1.57	1.88
Electrical machinery	2.63	1.62	1.69	–	–
Transport equipment	1.31	1.23	–	–	–
Plastic products	1.13	1.95	–	–	–
Paper and paper products	–	–	1.32	–	–
Made-up textiles	1.40	0.68	–	–	–
Miscellaneous manufacturing	–	1.59	1.42	–	–

time. Partly these comparisons are constrained because each industry group combines a large variety of products and activities whose cost behaviour over time may widely differ. An earlier analysis of the 15 specific industries for which the cost ratios were reviewed by the Tariff Commission indicated an improvement in comparative efficiency in the course of 10 to 15 years. They covered about 40 individual products. For seven industries and 16 products the cost ratios declined by 25 per cent to 60 per cent. The rest of the cost ratios declined by 5 per cent to 24 per cent (Islam, 1967, Table 3, p. 11).

An analysis of the cost ratios over time by itself may not provide an adequate test of the infant industry hypothesis. The cost ratios may turn unfavourable, even when there is an improvement in efficiency and productivity in the domestic industry because the costs of competing imports may fall faster as a result of a more rapid technological progress in the advanced countries. The problem is then one of the speed of technological advance in developed countries and a lag in the developing countries in the process of their 'catching up'. This suggests that a more detailed analysis of the changes in the productive efficiency of the domestic industries over time is necessary; this necessitates an identification of the changes in costs which are not due to: a) changes in wage rates; b) profits; and c) input prices over time, since it is an improvement in efficiency in the use of inputs through a learning process, as distinct from technological advance, that is involved in the infant industry argument. The improvement in efficiency in this sense should be compared with the c.i.f. prices which would have prevailed in the absence of technological advance. The statistical testing of a comparative cost analysis in a dynamic context of technological change thus confronts formidable difficulties.

IV Comparative Costs and Overvaluation of Exchange

The comparative cost ratios in the preceding paragraphs have been obtained on the basis of the official exchange rate. To the extent that the Pakistani currency is overvalued, of which there is some evidence, comparative cost ratios greater than one really reflect the overvaluation of currency so that with an equilibrium rate of exchange the cost ratios all will be equal to each other and equal to one. The Pakistani rupee was devalued in 1955 and the fall in the average cost ratio between the periods, i.e., 1951–55 and 1956–60 may be partly traced to this fact, since with devaluation the c.i.f. price of imports in rupee went up.

One may, therefore, suggest that the comparative cost ratios should be corrected for the relative overvaluation of the currency. The study on the domestic prices of imports as well as analysis of the amount of subsidy implicit in the export bonus scheme tends to suggest that the assumption of a 50 per cent overvaluation in the price of foreign exchange in Pakistan may be reasonable one. It is not easy to estimate what would be the equilibrium rate of exchange in a free market with existing tariffs and domestic fiscal and monetary policies; supply and demand elasticities of exports and imports as well as that of demand and supply of domestic substitutes, etc., enter into picture in a complicated way. The assumption of 50 per cent overvaluation may not be far out, of line for use as an illustration. The adjustment in the comparative costs ratio for alternative estimates of overvaluation of exchange is easily done.

As a result of the correction for the overvaluation of exchange, the comparative cost disadvantage declines and the number of industries which have cost ratios below one and, therefore, are competitive are as follows:

Table 23.11 Number of industries which have cost ratios below one

	Number of industries			
Cost ratios	**1951–55**	**1956–60**	**1961–66**	**Total**
Below one	17 (0.84)	14 (0.73)	26 (0.83)	57
Above one	12 (1.33)	10 (1.27)	36 (1.50)	58

For the period as a whole about half of the total number of industries appear to be competitive with imports while the other half have a higher cost ratio. The overall weighted cost ratios for the three periods after adjustment for the price of foreign exchange are 1.08 for 1951–55, 1.08 for 1956–60 and 1.21 for 1961–66. The weighted ratios for different groups of industries for the whole period are as follows:

Consumer goods	1.03
Intermediate goods	1.24
Capital goods	1.15

The consumer goods industries appear competitive whereas the capital goods industry is 19 per cent more expensive and intermediate goods industry 24 per cent more expensive than the competing imports.

The correction of the cost ratios for the relative overvaluation of the currency, as shown above, is only partial insofar as it corrects the c.i.f. price of the competing imports in domestic currency and does not correct the import component of the domestic ex-factory price in domestic currency. The import component becomes more expensive in terms of domestic currency consequent on a correction for the overvaluation of foreign exchange. The data on the foreign exchange component of all the products are not available. In the case of only 29 industries and their corresponding 107 products, the foreign exchange component (current requirements of imported raw materials and spare parts) is available and the result of adjustment for the value of foreign exchange is given below:

Table 23.12 Cost ratios adjusted for overvaluation of exchange rate

	Unadjusted	**Adjusted**	**Adjusted only for c.i.f. price**
Consumer goods	2.29	1.66	1.53
Intermediate goods	1.82	1.45	1.20
Capital goods	1.79	1.33	1.33
Total	1.86	1.44	1.24

Thus the cost ratio for all industries decline from 1.86 to 1.44, i.e., about 23 per cent. It is important to mention that after either adjustment for the value of foreign exchange, the ranking of the 29 or three broad categories of industries does not change either in terms of the ex-factory price, or the foreign exchange component of the ex-factory price, or the comparative cost ratios.

V Profits, Market Structure and Excess Capacity

One important aspect of the analysis of comparative cost ratios on the basis of ex-factory price is the extent to which the high ex-factory prices may be due to high profits or high factor prices in the relevant industries. If the factor costs do not represent scarcity prices but contain large monopoly or rent elements due to the institutional factors and the imperfections of markets, the high ex-factory price is not an index of comparative disadvantage or inefficiency but represents a transfer from the rest of the community to the factors employed in the industries concerned. The industrial wage for unskilled workers is generally presumed to be higher than the agricultural wage by more than what is accounted for by the costs of movement. Again, wage rate is higher in organized industries than in unorganized, small-scale industries. Even larger firms in the same industry, are found to pay higher wage rates than the smaller firms. One could argue that the kinds of labour employed in large- and small-scale industries or in large and small firms in the same industry are sufficiently different in terms of work, discipline and ability to account for the wage differentials. These are issues on which conclusive answers are neither available nor can they be dealt with within the purview of the present chapter. There is some evidence, however, which permits at least a partial examination of how far excess profits may have contributed to the high ex-factory price.

The cost ratios discussed earlier include mostly actual ex-factory prices which include abnormal profits, if any. In some cases, actual prices are not available and hence 'fair' prices are used for estimating the comparative cost ratios. The 'fair' prices, barring adjustments in a few cases for selling, distribution and overhead expenses, which are considered exorbitant by the Tariff Commission, are different from the actual prices insofar as the latter incorporate abnormal profits. The concept of normal profits expressed in terms of 'mark-up' over the cost of production which the Commission considers fair and reasonable has varied over the years. There are indications, however, in a number of reports that the mark-up permissible has increased over the years. Insofar as the actual percentage of profit on invested capital which is allowed by the Commission in its estimation of fair price is concerned, it is not always mentioned in its earlier reports. During the 1950s, the percentage of profit allowed on invested capital appears to be in the neighbourhood of 10 per cent in those cases where such a mention has been made, whereas in the 1960s, when the reference to the rate of profit is more frequent, it has increased by gradual stages over the years to 121 per cent (1963), 15 per cent (1964), 19 per cent (1965) and 20 per cent (1966).

There are about 39 industries for which data on both fair and actual prices are available; they indicate the prevalence of abnormal profits in the sense that in these industries the actual prices are higher than the fair prices to an extent varying between 8 per cent and 32 per cent (Islam, 1967, Table 7, p. 24). A few have been found to be suffering losses. It is pertinent to point out, however, that abnormal profits earned by the firms under consideration may in fact be larger than what is reported, and reported losses may be misleading insofar as the firms succeed in either misreporting their costs or, in the case of direct investigations by the Commission, succeed in avoiding a careful scrutiny of the detailed cost data. The valuation of the fixed assets on the basis of which the individual industries fix their actual rate of profit or the Commission fixes its fair rate of profit is in any case not an easy exercise. Thus the comparative cost ratios given in Table 23.13 based on fair ex-factory prices may still contain elements of excess profits which are not related to the comparative efficiency of the industries concerned, with the result that the fair cost ratios may indeed be lower than what is indicated in the table.

The comparative cost ratios on the basis of fair ex-factory price are generally lower than those based on actual ex-factory price. The fair cost, ratio in some instances, specially during the periods 1951–55 and 1956–60, is higher than the actual ratio. In some of these latter cases, of course, the firms may be making losses or the two sets of ratios may relate to the different commodities. During 1961–66, the sample of industries covered under each set of ratios is much larger and the commodities covered in the two types of

Table 23.13 Comparative cost ratios [9]

	1951–51		1956–60		1961–66		1951–66	
	Actual	Fair	Actual	Fair	Actual	Fair	Actual	Fair
Consumer goods	1.39	1.53	1.25	1.62	1.83	1.48	1.62	1.30
	(19)	(36)	(15)	(10)	(45)	(11)	(79)	(67)
Intermediate	2.39	1.19	1.91	1.10	1.88	1.68	1.89	1.65
goods	(2)	(3)	(2)	(2)	(82)	(76)	(86)	(81)
Capital goods	2.24	1.60	1.67	1.96	1.82	1.66	1.82	1.70
	(12)	(27)	(30)	(14)	(34)	(12)	(76)	(53)
All goods	1.76	1.61	1.55	1.76	1.85	1.66	1.78	1.62

ratios are more comparable. Over the whole period 1951–66, the fair cost ratio is consistently below the actual cost ratio for all the three categories.

If allowance is made for the fact that actual profits as reported by the Tariff Commission probably underestimate 'true' profits, the fair cost ratio, excluding excess profits, both recorded and unrecorded, would be even lower. The existence of very large profit margins, larger than 10–12 per cent over the cost of production and larger than 12–20 per cent over capital investment is also indicated by the additional evidence which is available on the profitability of industrial enterprises in Pakistan. A recent study of the balance sheets of public limited companies listed in the stock exchange shows that between 1959 and 1963 gross profits as a percentage of capital employed varied between 14 and 15 per cent. This was during the period when normal profits as allowed by the Commission varied between 10–12 per cent (Baqai and Haq, 1967). This is also a very partial evidence and possibly is an underestimate.

Thus if allowance is made for: a) excess profits (to the extent of 10 per cent): and b) scarcity price of foreign exchange or the overvaluation of exchange rate (to the extent of 50 per cent), the aggregate comparative cost ratio is considerably reduced. Thus the cost ratio will be reduced by a factor of 65 per cent, i.e., all the products with the cost ratios which are 1.65 and below will become competitive with the foreign product (ibid., p. 21).[10] The number of industries and products which have cost ratios below 1.50 and below 1.65 are given in Table 23.14. This generally indicates the change in the proportion of industries which become competitive when proper adjustments are made for the above two factors. It appears that between 50 to 60 per cent of industries and between 40 to 54 per cent of the products become competitive as against about 10 per cent of the industries and 5 per cent of the products which are competitive in the absence of such adjustments.

Table 23.14 Number of industries and products which have cost ratios below 1.50 and below 1.65

Years	No. of industries			No. of products		
	Below 1.50	Below 1. 65	Total	Below 1.50	Below 1.65	Total
1951–55	17	21	29	49	68	99
1956–60	14	15	24	32	36	73
1961–66	26	32	62	57	87	179
Total	57	68	115	138	191	351

An important question is to what extent the ranking of industries is affected by either an adjustment of foreign exchange or an adjustment for excess profits. A rank correlation between fair and actual price is found to be very high, i.e., 0.68, and it is significant at 5 per cent level. Similarly, the ranking of industries is undisturbed if cost ratios are adjusted for the overvaluation of foreign exchange. The rank correlation coefficient between unadjusted cost ratios and cost ratios adjusted for the overvaluation of exchange is very high, i.e., 0.96, and is highly significant at the 5 per cent level of significance.

The cost ratios may also be affected by the existence of excess capacity or a structure of market where few firms dominate. There is a considerable excess capacity in the manufacturing industries in Pakistan. About 60 per cent of the industries examined worked below 40 per cent of their installed capacity. The prevalence of excess capacity is not significantly different between the three categories of industries, as is shown in Table 23.15.[11]

Most of the industries have only a few firms. In many instances, the number of firms is larger than justified by the limited extent of the domestic market. Since each produces a substantial excess capacity, a reduction in the number

Table 23.15 Utilization of capacity in manufacturing industries

Percentage of capacity utilization	Consumer goods	Intermediate goods	Capital goods
	(percentage of total number of industries in each group)		
0–20	27.60	34.80	11.10
20–40	34.50	43.50	38.90
40–60	10.30	17.40	22.20
60–80	10.30	–	22.20
80–100	17.30	4.30	5.60

of firms would enable a concentration of output in a fewer firms, each producing at a fuller capacity.

The shortage of imported raw materials has been suggested as an important factor in explaining the existence of excess capacity. An attempt is made to see whether more import intensive industries suffer from a greater excess capacity. No strong correlation is discernible. This is plausible in view of the licensing procedure under which the licensing is a function of the assessed import requirements and there may be in fact a bias towards a more liberal licensing for the import intensive industries, specially if they are successful in the export markets. An analysis of excess capacity and reasons thereof, as well as the opportunities of cost reduction consequent on the utilization of capacity in the industries under investigation has already been pointed out (Islam, 1967, pp. 25–6).[12]

To what extent are high cost ratios due to the prevalence of excess capacity? While there is evidence that the cost of production declines with a greater utilization of capacity, it is difficult to quantify the reduction in costs in the absence of additional data. Moreover, even if costs decline consequent on a greater utilization of capacity, the behaviour of prices depends on what happens in the meanwhile to a number of factors such as changes in demand and factor prices. It also depends upon the pricing policy of the industry, i.e., whether it maximizes profits or not, and upon whether the crucial limitation on the expansion of output is the limited market or the shortage of inputs.

The above argument relates to the price behaviour or changes in the comparative cost ratio of a given product under a changing utilization of capacity. Whether the differences in the comparative cost and price ratios of different industries are related to the differences in the degree of utilization of capacity is another matter and raises additional questions. The differences in the cost ratios between different industries is only partly a matter of differences in the utilization of capacity. An industry may operate at a smaller capacity than another but may still have lower cost ratio because it is more efficient in the use of its inputs even at a lower level of operations than the other at a higher level vis-à-vis its competing imports. Its managerial and technical efficiency may be higher; prices of the factors necessary in its operation may be lower, and taxes on its inputs may be lower to offset the disadvantages of lower scale of operations. Thus an attempt to explore the correlation between cost ratios and percentages of capacity utilization does not indicate any significant results. Nor do the profit rates (P) seem to be related to the number of firms (N), as is seen below:[13]

$$P = 0.096 - \underset{(0.00083)}{0.00098}\, N + \underset{(0.021)}{0.0228}\, T$$

$$R^2 = 0.56$$

The excess profit P is defined as the difference between actual and normal price expressed as a ratio of actual price.[14] T stands for a dummy variable which is taken to be zero for all values of P which are negative (losses) and one for all values of P which are positive. The correlation is not significant and without the dummy variable T the relationship is still not significant as is below:

$$P = 0.035 + \underset{(0.0012)}{0.0007}\, N$$

$$R^2 = 0.0033$$

The relationship between excess profit (P) and capacity utilization (W) is as follows:

$$P = -0.088 - \underset{(0.00063)}{0.00079}\, W + \underset{(0.023)}{0.204}\, T$$

$$R^2 = 0.57$$

where T is a dummy variable as explained above, but without T the relationship is statistically significant.

$$P = 0.104 - \underset{(0.00081)}{0.0029}\, W$$

$$R^2 = 0.\ 14$$

Excess profit seems to be negatively related with capacity utilization. That is, the greater the utilization of capacity, the lower is the excess profit as a percentage of actual ex-factory price. With a larger output, the firms tend to charge a lower profit margin, since a lower profit margin on a larger output may still lead to large absolute profits and, more importantly, to a higher return on capital. But the magnitude of the fall in excess profit in response to a given change or increase in the utilization of excess capacity is very small, as is seen from the small magnitude of the correlation coefficient between excess profit and capacity utilization.

VI Comparative Costs and Factor Proportions

There is a considerable diversity in the comparative costs of different industries as analysed above. A relevant question is whether the comparative costs of different industries can be related to and explained by the diversity in the characteristics of these industries in respect of technology and factor proportions. It is expected that in a labour-abundant country like Pakistan more labour-intensive industries are likely to have lower cost ratios compared with the less labour-intensive industries. Data on labour costs as a proportion of ex-factory price are not available for all the industries under investigation. For a number of industries data are available on direct labour costs and not on indirect labour costs, which are part of the overhead cost, i.e., administrative as well as selling and distribution expenses (Lary, 1966).[15] The regression equation relating the comparative cost ratios (Y) to the direct labour cost ratios (L) is shown below:

$$Y = 2.02 - \underset{(2.11)}{2.47}\, L$$

$$R^2 = 0.019$$

$$\text{Log } Y = 0.252 - \underset{(0.032)}{0.154} \log L$$

$$R^2 = 0.18$$

The double logarithmic regression gives reliable regression coefficients. The correlation is not very high but the regression coefficient is negative and statistically significant at 5 per cent level, implying a reliable relationship between variations in the labour cost ratio and in the comparative cost ratio. A 10 per cent increase in labour cost ratio causes a decline in the comparative cost ratio to the extent of 15 per cent.

Attempts to correlate the comparative cost ratios to the capital-output ratios are limited by a lack of direct data on the capital-output ratios of the specific industries under investigation. The Censuses of Manufacturing Industries in Pakistan provide data on the fixed assets, employment and wages and salaries by the major groups of industries. The comparative cost ratios which relate to specific individual industries may be grouped according to the classification of the major industry groups in the census. The capital-output ratios of the major groups of industries are then related to the cost ratios of the groups of industries (Central Statistical Office, 1955 and 1959/60).[16] The capital-output ratios are available for 1959/60 and 1955. The former is related to the cost ratios for the period 1961–66, and 1956–60 and the latter to the period 1951–55, respectively. The regression equations relating the cost ratios (Y) to the capital-output ratios (K) are given below:

$$\text{Log } Y = 0.078 + \underset{(0.049)}{0.003} \log K \; (1951\text{–}55)$$

$$R^2 = 0.0038$$

$$\text{Log } Y = 0.195 + \underset{(0.075)}{0.021} \log K \; (1956\text{–}60)$$

$$R^2 = 0.0092$$

$$\text{Log } Y = 1.175 + \underset{(0.0207)}{0.704} \log K \; (1961\text{–}66)$$

$$R^2 = 0.87$$

The cost ratios and the capital-output ratios are positively correlated for the period (1961–66) and the correlation is highly significant. The higher the capital-output ratio of an industry, the higher is its comparative cost ratio. The magnitude of the relationship is also high, implying a 7 per cent rise in cost ratio in response to a 10 per cent rise in capital-output ratio. However, for the other two periods, i.e., 1956–60 and 1951–55, the relationship between cost ratio and capital-output ratio, though positive is not significant. The simple linear relationships are less reliable than what is shown in the above logarithmic equations.

The above two exercises relating the cost ratios to the labour cost-output ratio and the capital-output ratio as possible explanations of the variability in the cost ratios between industries, tend to indicate that the industries which use more labour per unit of output and less capital per unit of output are more advantageously placed in respect of their comparative efficiency. Admittedly, the evidence on the effects of differences in the capital-output ratio is not very conclusive, firstly because, in two of the three periods for which the hypothesis is tested, the regression coefficients are not significant, and also because of the general limitations of the data on which the exercise is based.

A low capital-output ratio is consistent with a high capital-labour ratio. A more capital intensive product (a higher proportion of capital to labour) may result in a low capital-output ratio if capital is very productive in the sense that it raises the productivity of labour more than the increase in capital intensity. An attempt is made to relate the comparative cost ratios to the skill and capital intensities of the industries, i.e., the relative proportion of skilled to unskilled labour and the proportion of capital to labour. Under the simplest of assumptions one can hypothesize that, the higher the ratio of wages and salaries per employee, the higher is the proportion of non-wage value added per employee, the higher is the proportion of capital to labour. This assumes that the differences in wage rate per employee between industries are entirely due to the differences in skill and labour of the same skill receives the same wage in different industries. Similarly, the assumption involved in treating the non-wage value added per employee as an index of a higher capital-labour ratio is that the return on capital is the same irrespective of the variation in the amount of capital used, and that the return on capital is the same as between different industries. Under constant returns to scale and perfectly competitive conditions, this is a correct assumption. The results of the following exercise have to be viewed in the context of the restrictive nature of the above assumptions.

In the following analysis: a) the wage and salaries per employee (T); and b) non-wage value added per employee (S) are related to the comparative cost ratios (Y) of different industries (Basevi, 1966; Krueger, 1966).[17] In all the periods the simple linear relationships do not yield reliable results.[18] The double logarithmic relationships yield reliable results for all the periods except for wages and salaries per employee in 1961–66, as is seen below:

$$
1951\text{–}55 \quad \left\{ \begin{array}{l}
\text{Log } Y = 0.046 + \underset{(0.037)}{0.127} \log S \qquad R^2 = 0.11 \\[2ex]
\text{Log } Y = 0.452 + \underset{(0.070)}{0.327} \log T \qquad R^2 = 0.18
\end{array} \right.
$$

1956–60

$$\text{Log } Y = 0.726 - \underset{(0.042)}{0.268} \log S \qquad R^2 = 0.360$$

$$\text{Log } Y = 5.216 - \underset{(0.547)}{2.575} \log T \qquad R^2 = 0.244$$

1961–66

$$\text{Log } Y = 0.48 - \underset{(0.030)}{0.110} \log S \qquad R^2 = 0.072$$

$$\text{Log } Y = 0.291 - \underset{(0.200)}{0.022} \log T \qquad R^2 = 0.0006$$

The regression coefficients, except for T in 1961–66, are all significant at the 5 per cent level. But the results for 1956–60 and 1961–66 contradict those for the period 1951–55. The data for 1951–55 indicate that the more skill intensive and capital intensive industries have higher cost ratios, whereas the data for 1956–60 and 1961–66 indicate the reverse. The strength of the relationship in all the cases as indicated by the correlation coefficient is very small, i.e., the coefficient of determination varies between 0.11 and 0.18 in 1951–55 and between 0.36 and 0.24 in 1956–60 and is 0.072 in 1961–66. Therefore, a very small proportion of the variance or the variations in the interindustry comparative cost ratios are explained by the variations in factor proportions in the sense defined here.

A more appropriate relationship to explain interindustry variations in the comparative cost ratios would be a joint or multiple relationship between the cost ratios and the differences in skill and capital intensities, between the individual industries. A simple correlation does not reveal the joint influence on the comparative cost ratios and may even yield biased results. A multiple correlation yields a better result in terms of the explanation of the variance or the interindustry variations in the cost ratios as given below:

1951–55

$$\text{Log Y} = -0.424 + \underset{(0.098)}{0.284 \log S} + \underset{(0.048)}{0.030 \log T}$$

$$R^2 = 0.185$$

1956–60

$$\text{Log Y} = -0.356 - \underset{(0.045)}{0.415 \log S} + \underset{(0.127)}{0.679 \log T}$$

$$R^2 = 0.548$$

1961–66

$$\text{Log Y} = -0.132 - \underset{(0.046)}{0.217 \log S} + \underset{(0.117)}{0.403 \log T}$$

$$R^2 = 0.111$$

The multiple correlations yield more reliable results and provide a better explanation of the variations in cost ratios. All the multiple correlation coefficients are much larger than the simple correlation coefficients obtained earlier and are statistically significant at 5 per cent level; all the regression coefficients, excepting that of T (wage and salaries per employee) in 1951–55, are statistically significant at 5 per cent level. Generally speaking, the skill-intensive industries as indicated by the higher wages and salaries per employee seem to have high cost ratios, except in 1951–55 when the coefficient is not reliable. This is also confirmed by one of the two simple regressions which yield reliable results. Again, excepting in 1951–55, the capital intensive industries seem to be associated with the low cost ratios. This is also corroborated by the results of the simple regressions.

The earlier results indicate a significant positive correlation between cost ratios and the capital-output ratios for the periods 1956–60 and 1961–66 but not for 1951–55. This is consistent with a significant negative correlation between the cost ratios and the non-wage value-added per employee which is yielded by both simple and multiple correlations in two of the three periods. An increase in the non-wage value added per employee, implying a higher ratio of capital to labour, may lead to a high productivity of capital and thus a decline in the capital-output ratio. The evidence on the positive relationship between wages and salaries per employee and cost ratio between the time periods is less conclusive, insofar as the results of all the simple and multiple regressions are not consistent, even though multiple regressions which yield a positive relationship provide, in general, a better explanation of the relationships examined. While, as is seen earlier, on the one hand, a higher ratio of labour cost to output is associated with a low comparative cost ratio, higher wages and salaries per employee, on the other, hand, are associated with a higher cost ratio. This would imply that skill-intensive industries, in the sense of a higher proportion of skilled to unskilled workers, are associated with a lower ratio of labour cost to output. It is only when an increase in wages and salaries per employee is associated with a faster increase in output per employee that the labour cost per unit of output will decline with an increase in skill intensity. Thus high labour cost-output ratio is associated with a low proportion of skilled to unskilled labour. This indicates that the industries with a low skill intensity but with a high proportion of labour costs to value of output are more competitive with imports from abroad. Though more skill-intensive industries have a lower labour cost per unit of output, their total costs and ex-factory prices vis-à-vis the prices of competing imports are higher than those of the less skill intensive industries. The labour cost ratio, though it does not distinguish between unskilled and skilled labour as used in this study, includes only direct labour and excludes the managerial labour and the labour component of the overhead and distribution expenses, which may include a higher proportion of skilled labour than is included in the category of direct labour cost. Moreover, while interpreting the interindustry variations in the comparative cost ratios with reference to wage and non-wage value added per employee, it is important to note that the use of these ratios as the indices of capital and skill intensities is subject to a number of limitations which have been discussed earlier. But subject to the limitations of data and methodology, the above analysis provides some evidence, though very tentative at this stage, that the industries with a high component of skill and with high capital-output ratio tend to have high comparative cost ratios. This seems to provide a confirmation of the common-sense view that technical skill is a very scarce factor and high-priced in Pakistan at our present stage of development. Until a wide diffusion of technical knowledge and skill takes place, consequent on increased investment in this direction, the industries which need high levels of skill will tend to suffer from high cost disadvantages in relation to competing imports. Similarly, the limited evidence on the positive correlation between the capital-output ratio and the comparative cost ratio may be attributed to the learning costs associated with the use of capital in the early stage of development.

Conclusion

The manufacturing industries in Pakistan suffer, on the whole, from a high cost disadvantage vis-à-vis the competing imports. The weighted average cost ratios vary between 1.50 and 1.90, i.e., the ex-factory prices are 50 to 90 per higher than the c.i.f. prices. Thirty per cent of the industries examined in this study have ex-factory prices 51–100 per cent higher than their corresponding c.i.f. prices; about 16 per cent of the industry have prices 100–200 per cent higher than the c.i.f. prices (Islam, 1967, Table 2, p. 8). This range of the comparative cost ratios compares well with those obtained from the earlier studies based on an analysis of the domestic prices of imported goods, allowing for the differences in the methodology and commodity composition of the two studies. The tariffs do not seem to provide an appropriate measure of either absolute or relative cost disadvantage of the different industries in Pakistan. On the one hand, because of quantitative restrictions, there are positive scarcity margins over the landed costs of the competing imports for a great majority of the commodities examined here; on the other hand, there are many commodities for which the ex-factory price is below landed cost. For example, in 30 per cent of the cases, *actual* ex-factory prices and in 40 per cent of the cases, *fair* ex-factory prices fell short of the landed costs. Positive scarcity margins usually vary between 30–50 per cent except in one year when it rose to 73 per cent. In those cases in which ex-factory prices fall below the landed costs of competing imports, they are as much as 25 to 30 per cent lower.

As is well known and is further corroborated by the present analysis there is widespread underutilization of capacity, engendered partly by a lack of coordination between industrial investment licensing and the licensing of imports of raw materials and spare parts (ibid., fn. 18). Even though excess capacity contributes to high cost in a particular industry, there is no evidence that interindustry differences in cost ratios are explained by the differences in excess capacity which are substantial in almost all the industries anyway.

High comparative cost ratios of the Pakistani industries, however, do not necessarily measure the inefficiency of industrialization in Pakistan to an equivalent extent, if allowance is made for: a) an overvaluation of exchange rate; and b) the existence of high or excessive profits in many industries generated by a lack of competition, either internal or external, in the domestic market. An adjustment of the c.i.f. price of the competing imports as well as of the foreign exchange component of the ex-factory price, on the assumption of 50 per cent overvaluation, in the case of a sample of 29 industries and 170 products, shows that the aggregative comparative cost ratio declines from 1.84 to 1.44, i.e., by 23 per cent. If one adjusts only the c.i.f. price and not the foreign exchange component of the ex-factory price, 50 per cent of the industries under examination become competitive. Again, if allowance is made for the fact that ex-factory price contains elements of excess profits, the comparative cost ratio will be further reduced by 10–11 per cent. When, however, both adjustments are made for overvaluation in the few cases when the data permit them to be done, the industries with the comparative cost ratio of 1.33 and below become competitive in the world market.

In so far as the growth of the infant industries into adulthood is concerned, the evidence examined in this chapter on this subject is quantitatively very small. The performance of only a few industries, about 15 in all, has been reviewed by the Tariff Commission subsequent to the grant of protection by it. The evidence indicates an improvement in comparative advantage and a decline in cost ratios; an attempt to correlate the cost ratios with the length of the period of operation of the corresponding industry groups does not provide any conclusive, systematic and consistent evidence relating to the development of infant industries. While there are a few cases of a fall in the cost ratios with an increase in the number of years of operation and an accumulation of experience, it is not true for all, nor is there any evidence that the cost ratios are a smooth and a continuously declining function of an increase in the number of years of operation. A satisfactory examination of this problem is, however, inhibited by the limitation of the data.

Insofar as the differences in the comparative costs among the individual industries are concerned, the industries with a higher labour-output ratio have a greater comparative advantage. However, the gain in terms of comparative costs resulting from the choice of industries with a higher labour cost does not appear to be appreciable. At the same time industries with a high component of skilled to unskilled labour tend to have high cost ratios, though the evidence on this is not conclusive. There is some evidence that industries with a high capital-output ratio have also high comparative cost ratio. Contrariwise, the industries which have higher non-wage value added per employee, i.e., with a higher ratio of capital to labour tend, however, to have lower comparative cost ratios, implying that with a greater application of capital to labour, productivity of capital goes up and cost disadvantage declines. However, it is necessary to point out that these conclusions are very tentative and need to be further verified on the basis of additional, more reliable and comprehensive data and for different samples of industries and different years.

Notes

1 All the data relating to cost ratios, unless otherwise specified, are from Tariff Commission, *Reports*, various years. The specific reports have not been cited and in some cases the reports have not been published so far. The cost data relating to sugar, cement and paper are taken from IBRD, 1966, pp. 121–8.
2 Since data on the value of output of industries covered in this analysis are not available it has not been possible to estimate the proportion of the total industrial output which the industries covered by the present study contribute. However, an attempt is made to compare the number of establishments in specific industry groups which are covered in the present analysis with the number of establishments in corresponding major industry groups which are recorded in the Census of Manufacturing Industries 1959/60 in Appendix B. This comparison only includes those industry groups from the Census, which include commodities investigated by the Tariff Commission.
3 There are in many instances indirect taxes in the form of sales and excise duties on the manufactured goods produced at home. The indirect taxes are not relevant for an analysis of comparative costs and efficiency of the manufacturing industries; however, they affect the prices facing the domestic consumers and the relative use or consumption of the different products is affected by their relative prices, including taxes. The c.i.f. prices of competing imports on which comparative cost ratios are based include the costs of transportation, a change which will affect the c.i.f. prices and cost ratios. This need to be borne in mind while using cost ratios as an indicator of the cost disability or the relative inefficiency of the Pakistani industries.
4 The weights for the 1951–55 are obtained from the Censuses of Manufacturing Industry, 1955 (Central Statistical Office, 1955) and those for 1956/60 from Census of Manufacturing Industries 1959/60 (Central Statistical Office, 1959/60). The weights for 1961–66 are obtained from the unpublished Interindustry Table (Stern and Tims, n.d.).
5 The ratios within brackets exclude extreme values and are weighted averages, weights being the number of products in each industry.
6 Unweighted averages in Table 23.4 are simple averages of cost ratios and weights used in the estimation of weighted averages are the values of imports of individual items or of the category of commodity to which the item belongs.
7 The landed costs in Tables 23.4 to 23.7 include both the tariffs as well as excise and sales taxes on imports and the ex-factory prices likewise include indirect taxes. The ratios are weighted ratios, weights being the number of products to which each price ratio relates. Each industry ratio is a simple average of the ratios for the different products and the ratios for different industries are weighted by number of products of each industry in order to obtain an overall ratio.
8 The cost ratios are weighted cost ratios, weights being the number of products. The figures in brackets of Table 23.9 indicate the number of firms relevant to each period.
9 The figures in the brackets relate to the number of products.
10 This estimate, however, does not make any adjustment for the foreign exchange component of the domestic product which has the effect of raising the cost ratio. Making this double adjustment for overvaluation, where permitted by the data, would reduce the cost disadvantage by 23 per cent, and then all industries with cost ratios of 1.33 and below would be made competitive.

11 Excess capacity is defined as the difference between the actual output and the installed capacity as assessed by the technical investigations of the Tariff Commission. The number of shifts implied in the assessment of installed capacity is not clearly indicated. The observations relating to excess capacity relate to different industries in different points of time.

12 The ex-factory price of 15 industries may be reduced anywhere between 8 per cent and 15 per cent depending upon the nature of the industry.

13 All the regression equations in the present study are as 'weighted' equations in the sense that each variable is given a weight equal to the number of products it relates to.

14 Excess profit is used here in the sense of abnormal mark-up or the excess over the 'normal' or fair mark-up on ex-factory price as used by the Tariff Commission.

15 In this exercise the comparative cost ratios and the labour cost ratios of the individual products are not combined to arrive at the average ratios for each industry. This increases the number of observations which are conducive to a satisfactory statistical testing. There are differences, often significant, between the labour cost ratios of individual products in a given industry.

16 For each industry group, the cost ratio is the weighted average (weights being the number of products) – cost ratios of the individual industries included in the group. The weights are the number of products relevant to each cost ratio. This method of estimating the capital-output ratios for the industries under examination suffers from the limitation that in many groups of industries, the census contains a much wider variety of individual industries than is included in the present list of industries. Often the cost ratios relate to a small fragment of the major group of industry or which capital-output ratio is used. The limitation of the present set of data is added to the limitations of the data on fixed assets as they are reported in the census. They represent book values of fixed capital assets, depreciated by each industry following its own method of depreciation, which does not necessarily reflect the physical life of the capital equipment. In addition the depreciated book value does not indicate the cost of replacement, and does not reflect the changes in the price of the equipment since the time of its acquisition.

17 The data on non-wage value added is subject to error in the sense that it includes: a) depreciation allowances; b) rent; and c) some miscellaneous expenses. As in the case of the exercise on capital-output ratios, these two ratios for the major groups of industries are obtained from the Censuses of Manufacturing. Industries are related to the combined comparative cost ratios of industries which fall in the same group but which in some cases cover only a small fragment of the major industry group or cover only a few selected individual industries. Moreover, the value added (wage and non-wage) ratios for 1955 are related to the cost ratios for the period 1951–55 and those for 1959/60 are related to the cost ratios of 1956–60 and 1961–66.

18 1951–55

$$\left\{\begin{array}{l} Y = 1.54 + \underset{(0.0006)}{0.0008\,S} \qquad R^2 = 0.020 \\ Y = 1.09 + \underset{(0.0012)}{0.0053\,T} \qquad R^2 = 0.15 \end{array}\right.$$

1956–60

$$Y = 1.64 - 0.0003\,S$$
(0.0006)
$R^2 = 0.0016$

$$Y = 1.68 - 0.0005\,T$$
(0.0008)
$R^2 = 0.0053$

1961–66

$$Y = 1.93 - 0008\,S$$
(0.0004)
$R^2 = 0.0185$

$$Y = 1.92 - 0.0009\,T$$
(0.0013)
$R^2 = 0.0020$

References

Baqai, M. and Haq, K., 'A Study of Savings in the Corporate Savings in Pakistan, 1959–63', paper read at the Pakistan Institute of Development Economics Seminar on *Current Economic Problems of Pakistan*, held Karachi, 28–30 January 1967.

Basevi, G., 'The U. Tariff Structure: Estimates of effective rates of the US industries and industrial labour,, *Review of Economies and Statistics*, May 1966.

IBRD, *The Industrial Development of Pakistan* (Washington: IBRD, 1966).

Islam, N., *Tariff Protection, Comparative Costs and Industrialization in Pakistan*, mimeographed Research Report No. 57 (Karachi: Pakistan Institute of Development Economics, January 1967).

Krueger, A.O., 'Some Economic Costs of Exchange Control: The Turkish case', *Journal of Political Economy*, October 1966, pp. 466–80.

Lary, H.B., *Exports of Manufactures by the Less Developed Countries* (New York: National Bureau of Economic Research Inc., June 1966).

Central Statistical Office (Pakistan), *Census of Manufacturing Industries* (Karachi: Central Statistical Office, 1955 and 1959–60).

Tariff Commission (Pakistan), *Reports* (Karachi: Manager of Publications, various years).

Pal, M.L., 'Domestic Prices of Imports in Pakistan: Extension of empirical findings', *Pakistan Development Review*, Vol. V, No. 4, Winter 1965.

Soligo, R. and Stern, J.J., 'Tariff Protection, Import Substitution and Investment Efficiency', *Pakistan Development Review*, Vol. V, No. 2, Summer 1965, pp. 249–70.

Soligo, Ronald and Stern, J.J., 'Some Comments on the Export Bonus, Export Promotion and Investment Criteria', *Pakistan Development Review*, Vol. VI, No. 1, Spring 1966, pp. 38–56.

Stern, J.J. and Tims, W., *Inter-Industry Input-Output Tables for Pakistan, 1962–63*, unpublished Mimeographed Report (Karachi: Pakistan Institute of Development Economics, n.d.).

Appendix A

A Comparison with Earlier Studies

The comparative cost ratios derived above may be compared with the results of an earlier study, which estimated the domestic prices of imports; domestic prices include tariffs, other indirect taxes and scarcity margins on c.i.f. prices of the imports (Islam, 1967, fn. 7), as a result of quantitative restrictions on imports. On the assumption that the imported and the domestic products closely competing with them fetch the same wholesale prices in the domestic product, the ratio between the c.i.f. price and the domestic wholesale price is given in Table A-1 for the three main categories of commodities.

A comparison made between the cost ratios derived from the present study and those derived from the study of domestic prices of imports is given below. The import study relates to the year 1964/65 and hence its findings should appropriately be compared with the cost ratios of the period 1951–66; the cost ratios for the whole period 1951–66 are also given in Table A-2. The average cost ratios (weighted and unweighted) for all commodities as obtained from the import study are generally higher than the cost ratios derived from the direct comparison of ex-factory prices (including taxes) and c.i.f. prices. However, when the commodities are grouped into three categories the ratios for consumer and capital goods appear to be distinctly higher in the case of import study than in the present study, whereas the ratios for intermediate ratios are lower in the import study than in the present study. The critical assumption for the use of the domestic prices of imported products to represent comparative cost ratios for the domestic industries is that the ex-factory price is equal to the domestic price of their competitive imports. This assumption may not hold for a number of reasons. In the first place an industry may no longer be an 'infant industry' and may have developed competitive efficiency, in which case the domestic price, therefore, falls below the import price plus tariffs and scarcity margins. Tariffs, therefore, become redundant. Secondly, the domestic prices of imports may be higher than the ex-factory prices of domestic competing goods because of consumers' preferences for the imported products. In the extreme case one can conceive of a highly differentiated market for import goods catering to a special clientele. This is specially true for commodities which are the products of international firms and are noticeable in the field of luxury items of consumer goods as well as in the case of drugs and pharmaceuticals and chemicals, etc. Thirdly, the domestic producers may

Table 23.A1 Cost ratios based or

	Karac
	Unweighted
Consumption goods	2.85
Intermediate goods	1.97
Capital goods	2.42

Table 24.A2 Cost ratios

	Present study With indirect tax unweighted *1951–66*
Consumer goods	1.79
Intermediate goods	2.03
Capital goods	1.89
Total	

not charge the maximum scarcity price which the comparable and competing imported products may fetch in the market, even though there are only a few domestic producers. Various considerations which induce monopolistic producers not to maximize short-run profits but to take into account various long-run factors affecting their profits may induce the domestic producers not to exploit the domestic scarcity to the maximum. Moreover, the comparison of the results of the present study with those of import study is also inhibited by the fact that the commodities in the two studies are not all comparable. The earlier study includes a number of imported commodities for which no domestic production exists. Hence their domestic prices cannot be said to represent what would have been the prices if the commodity is domestically produced, since tariffs and import controls in these cases bear no relation to the need for protecting an existing domestic industry.

In order to obviate this difficulty an attempt has been made to select a few specific items which are common to both the studies (Table A-2). It appears from the table that the unweighted simple average cost ratios for the selected items, which are comparable as between the two studies, are about the same, i.e., 2.12 and 2.05 respectively. The domestic wholesale prices which constitute the basis of the cost ratios in the import study include the normal profit margin of the wholesaler whereas the ex-factory prices which are the basis of the cost ratios in the present study do not include this element, excepting where the producing firm is itself the wholesaler. The indirect evidence suggests that this normal mark-up may be about 12 per cent. The unweighted average cost ratio after deducting the 12 per cent mark-up for all the items listed above comes to about 1.92, as against 2.05 in the present study.

The average ratios conceal significant differences between the individual cost ratios. If we assume that the c.i.f. prices of a commodity are the same in both the studies, the differences in cost ratios indicate the differences between the ex-factory prices of the domestic products and the wholesale prices (with and without mark-up) of the competing imports. Thus, on this assumption, the ex-factory prices of caustic soda, electric meter, G.I. pipe, batteries and sugar appear significantly higher than the domestic prices of competing imports. This may imply that the above-mentioned industries are underprotected and that the domestic and imported items sell at different prices. In the case of cement, safety razor blade, transformer and bicycle the relative position is reversed with domestic prices being lower than the prices of competing products. In these cases, it is probable that importers do not charge, the maximum price that the market will bear. In the case of the rest of the items, cost ratios are comparable.

Table 24.A3 Cost ratios of selected items

	1964 import study	Import study with mark-up 1964/65	1964/65 without mark-up	Present study (with taxes)	Year
Free list items					
Nylon twine and monofilament		1.92	1.72	1.97	1965
Caustic soda	2.72	1.73	1.55	2.34	1964
Sodium bicarbonate	2.15	2.17	1.94	1.90	1965
Soda ash	2.73	2.19	1.95	2.32	1966
Acetic acid	2.48	1.73	1.54	1.89	1966
Cement		2.45	2.19	1.67	1965
G.I. pipe	2.24	1.53	1.36	2.21	1963
Brass sheets		1.65	1.47	1.50 [a]	1962
Aluminium sheets		1.62	1.44	1.93	1965
Licensed items					
Wheat flour		1.39	1.24	1.32 [b]	1962
Safety razor blades	3.27	3.20	2.85	1.24	1964
Electric lamp	2.20	1.98	1.76	1.64	1964
Leather belting		1.98	1.77	1.31	1962
Electric meter		1.96	1.75	2.93	1963
Transformer		2.66	2.38	1.44	1963
Batteries		2.35	2.09	3.02	1965
Bonus items					
Sugar	2.90	2.89	2.58	4.55	1966
Bicycle	2.74	3.33	2.97	1.84	1963
Simple average		2.12	1.92	2.05	

Notes

a Brass ingots and strips
b Wheat flakes etc.

Appendix B

Table 24.B1 The coverage of the present study*

Industry	No. of establishments 1959/60 Census	No. of establishments covered in this analysis
Food manufacturing	460	112
Alcoholic beverages	4	1
Silk and artificial silk	233	2
Manufacture of textiles, n.e.c.	57	8
Footwear	101	(Number not known but very large including both small- and large-scale industry)
Paper and paper products	27	6
Rubber and rubber products	36	8
Chemicals		
Basic industrial chemicals	42	29
Paints and varnishes	36	42 (including small scale)
Medicines and pharmaceuticals	63	184 (including small scale)
Perfumes, cosmetics and soaps	80	(Includes only soap and number is large including small-scale)
Matches	20	19
Nonmetallic minerals	83	–
Glass and glass products	29	2
Concrete products and nonmetallic mineral products, n.e.c.	39	100 (including small-scale)
Manufacture of basic metal	96	31
Metallic products	467	–
Heating, cooking, lighting apparatus	27	5
Cutlery	8	4
Utensils	132	57
Safes, vaults and trunks	13	37
Metal products, n.e.c.	99	7

Table 24.B1 cont'd

Industry	No. of establishments 1959/60 Census	No. of establishments covered in this analysis
Machinery, non-electrical	277	–
Engines and turbines		103 (including small-scale)
Textile machinery and accessories	39	27
Pumps and compressors	11	2
Electrical machinery	104	38
Manufacture of transport equipment	118	6
Manufacture of plastic products	56	218 (including small-scale)
Manufacture of pens and pencils and related products	23	3
Miscellaneous manufacturing industries	57	60

* *C.S.O. Bulletin*, May 1965, pp. 1095 to 1098.

Chapter 24

Export Incentive and Effective Subsidy in Pakistan: An Evaluation*

Introduction

This chapter seeks to evaluate the effects and the efficiency of the existing system of export incentive schemes in Pakistan. In view of the widespread prevalence of similar export incentive schemes in many parts of the developing world, it is hoped that the analysis may provide a few lessons for the other developing countries.

Since the beginning of 1950s Pakistan superimposed a regime of quantitative restrictions on imports over and above a system of import tariffs, which were continued, even though there was a devaluation of about 45 per cent in 1955, into the 1960s, with varying degrees of intensity. Tariffs registered an upward movement throughout; the average rates of import duties on the various categories of imports, which ranged between 23 per cent and 99 per cent during the period 1955/56–1959/60, increased and ranged between 17 per cent and 142 per cent and between 34 per cent and 180 per cent in the years 1963–64 and 1965–66 respectively. The scarcity margins on the imported goods, i.e., the difference between the landed costs (inclusive of duties) and the domestic wholesale prices of imports, an index of the restrictive impact of the quantitative controls on imports, varied between 7 per cent and 47 per cent during 1964–65 and between 7 per cent and 52 per cent during 1965–66 (Lewis and Guisinger, 1966; Thomas, 1966, p. 532; Alamgir, 1968). The implicit rate of exchange for imports, defined as the official rate of exchange, which has been Rs 4.76 per dollar since 1955, multiplied by the percentage excess of the domestic prices of imports over their c.i.f. price (converted at the official rate) ranged from Rs 7.78 per dollar to Rs 19.93 per dollar during 1963–64 and between Rs 9.12 and Rs 22.69 during 1965–66.

While, on the one hand, the implicit rates for imports were on, the increase during the 1950s as well as the 1960s, the implicit rates for exports started to

---

* First published in *Bulletin of the Oxford University Institute of Economics and Statistics*, Vol. 31, No. 3, 1969.

increase to any significant extent only during the 1960s. The implicit rate of exchange for exports is the total rupee receipts at the official rate per dollar worth of exports, reduced by the percentage rate of export taxation or increased by the percentage of export subsidy provided directly or indirectly under various export incentive schemes. Towards the end of the 1950s, and especially during the 1960s, Pakistan adopted a number of export-incentive schemes such as the export bonus scheme and the export performance licensing scheme and introduced a number of fiscal concessions for exports. Under the export bonus scheme, the exporters receive a fixed percentage of their exports in terms of the entitlements to imports, which are transferable and can be freely bought and sold for the imports of a specified list of commodities. The export bonus scheme, which was introduced in 1959, excluded the major primary commodities such as jute, cotton, tea, hides, skins and wool; the latter constituted in the early 1960s about 80 per cent of Pakistan's total export earnings. The rest of the primary commodities, which in 1959 did not constitute more than 6 per cent of the total exports, and all the manufactured exports were entitled to receive the import entitlements to the extent of 20 per cent and 40 per cent respectively of their export proceeds, with the exception of the jute manufacturing industries which were eligible for the same rate of import entitlement as the minor primary commodities, i.e., 20 per cent.[1]

Since 1962–63 an additional measure of export promotion in the form of export performance licensing has been in existence, under which many export industries are entitled to receive import entitlements at varying percentages of their exports, as under the export bonus scheme. In contrast to those under the bonus scheme, these entitlements can only be used for the imports of raw materials, spare parts and machinery required by the industry concerned.[2] The subsidy element in the export performance licensing arises from the fact that there is an excess demand and a consequent scarcity for most of the imported raw materials in the domestic market, especially in view of the widespread prevalence of excess capacity in the manufacturing industries (Hecox, 1969).[3]

During the mid-1950s, i.e., after devaluation of 1955, the implicit rate for exports varied between Rs 4.09 and Rs 4.15 per dollar, due to the export taxation, as against the official rate applied to the minor primary exports, with the exception of a few which were eligible for the very restricted export performance licensing introduced during 1956–58 and hence were entitled to a rate of exchange of about Rs 5.12. However, the average implicit rates for major primary exports improved to Rs 4.62 per dollar between 1960–61 and 1966–67 with a progressive reduction in the export duties; the minor primary commodities under the export bonus scheme earned between Rs 4.78 and Rs 6.19 per dollar depending on: a) the varying percentage rates of bonus; and b) the rates of premium on the bonus vouchers in the different years. Insofar as the manufactured exports were concerned, not more than 1–2 per cent of the exports of manufactures prior to 1959 were affected by the various import entitlement schemes in effect during the 1950s.[4] The structure of import and export rates in 1965–66, which are the combined result of export incentives, on the one hand and the import substitution policies, on the other, is given in Table 24.1.

Comparing the export rates for manufactured goods, which are of primary interest in this chapter, with the import rates, it appears that about 68 per cent of the manufactured exports earn between Rs 5.90 and Rs 7.47 per dollar, a range which lies below the lowest rate paid by the manufactured imports.[5] About 49 per cent of the imports pay between Rs 9 and Rs 14 per dollar and about 5 per cent pay between Rs 12 and Rs 19 per dollar. The highest rates for miscellaneous manufactured exports, constituting one-third of the total manufactured exports, is Rs 9.52 per dollar. Only three manufactured exports, of minor significance, earn about Rs 11–12 per dollar. A considerable increase in the implicit import rates encouraged and speeded up the process of import-substitution in Pakistan during the 1950s. A rise in the implicit import rate without a corresponding rise in the implicit export rate involves a discouragement to the expansion of exports. Firstly, it increases the relative returns from investment and production in the import-substituting industries compared to those in the export-promoting activities. Secondly, the costs of the inputs into the export industries, which are supplied by the import-competing industries, go up as they replace the imported inputs by the high-cost domestically-produced inputs. Thirdly, the cost of the imported inputs in the export industries also goes up insofar as the restrictions and tariffs on the imports designed to encourage the import substitution raise their domestic price. Fourthly, the import substituting industries, earning high profits in the protected domestic market, bid up such factor prices as urban wages above the opportunity cost of labour, thus raising the wage costs of the export industries. An import substitution policy which discourages exports foregoes the advantages of economies of scale and of competition provided by a participation in and by an exposure to the world market.

Table 24.1 Implicit exchange r

Category of imports	P
1	
A Licensed imports	
1 Capital goods	
2 Raw materials	
3 Consumer goods	
B Bonus imports	
1 Capital goods	(Percentage distribu
2 Raw materials	bonus im
3 Consumer goods	approximately th
	as licensed i

Note: The export rates include the effects of a
which were subject to export taxes in 1
66. The imports are overwhelmingly m

Source: Islam, 1968a, Appendix B, pp. 27–31

Measurement of Nominal and Effective Export Subsidy

While the excess of the rate of exchange for imports over the rate of exchange for exports does indicate a greater attractiveness of production for home market as compared to production for export miarket, an appropriate measure of the differential incentives, in terms of the allocation of resources between import substitution and export expansion, is however, a comparison between the rate of effective export subsidy and that of effective protection. Just as the protective effect of tariffs and quantitative restrictions is reduced by the imposition of tariffs and restrictions which raise the prices of inputs, so also the impact of the export subsidy is reduced to the extent that the domestic prices of the imported inputs used in the export industry or the inputs supplied by the domestic industry fetch higher prices than their corresponding international prices. The magnitude of subsidy to the export industry should thus be considered in terms of effective subsidy, i.e., subsidy received by 'the value added' in the export industry. The overall impact of the various export incentive schemes can be expressed in terms of a total amount of subsidy per unit of export, i.e., as a percentage of the f.o.b. value of exports, which is thus defined here as nominal subsidy. The rate of nominal subsidy can be derived as follows. The total receipts per unit of export sale (R) is

$$R = X(1 + bP + EP') + D(t_m.m) + D(td.d) + Dt$$

where X = F.O.B. export price;
D = domestic market price;
tm = the rate of duty on the import component of the output (in domestic market price);
m = import component of the output (in domestic market price);
td = the rate of indirect taxes on the domestic inputs;
d = the domestic input component per unit of output;
t = the indirect tax on output;
P = premium on the bonus vouchers;
b = percentage of export bonus;
E = percentage of export performance licensing;
P′ = premium on export performance licensing.

Hence the export price which the exporter could quote if he were to equalize his profits per unit of sale at home and abroad (R = D) would be

$$X' = \frac{D\,(1\text{-}tm\,.\,m - td\,.\,d\text{-}t)}{1 + bP + EP'}$$

In practice he may not be able to sell abroad at that price, since the world price which governs the actual export price (f.o.b. export price) may be less than X′ so that X′>X. The nominal export subsidy as a proportion of f.o.b. export is, therefore,[6]

$$\frac{R}{X} - 1 = Pb + EP' + [D(tm\,.\,m) + D(td\,.\,d) + tD]\frac{1}{X}$$

The effective export subsidy is defined as $\frac{V_x - V_w}{V_x}$

where V_x = value added in the export market
= $R - I_d$
Id = the value of the intermediate inputs (domestic and foreign) for exports; this is the same as the value of intermediate inputs in domestic prices, since the export subsidies which are given in the form of remissions and exemptions of taxes on the inputs (domestic and foreign) used in the export industry are already converted in terms of the total subsidy to the gross output as indicated earlier and therefore, are not treated again in order to avoid double counting;
$V_w = X - I_w$
I_w = value of intermediate inputs in world prices.[7]

The relationship between the nominal and effective export subsidy is analogous to the relationship between the nominal and effective protection as seen below. Thus the effective rate of protection is

$\frac{V_d - V_w}{V_d}$ whereas the effective rate of export subsidy is

$\frac{V_x - Vw}{V_x}$[8]

If V_x, which is inclusive of export subsidies, equals V_d which includes the effects of the import restrictions and tariff protection, effective subsidy is the same as the effective protection. This would be the case if the receipts per unit of export including subsidies (R), equal the receipts per unit of domestic sales (D). V_d is different from V_x only insofar as the receipts per units of sale in the domestic market are different from the receipts per unit of sale (including export subsidies) in the export market, since the value of intermediate inputs in both cases is the same.[9] The available data on the differential between the international and domestic prices suggest that the receipts, including export subsidies, per unit of sale in the export market are in some cases less than what is obtained by selling at home. In these cases, exports are not as profitable as domestic sales.

This chapter estimates the magnitude and structure of the nominal and effective export subsidy in Pakistan for the various categories of exports, especially manufactured exports, resulting from the multiple export incentive schemes. It further seeks to examine their impact on the expansion of exports and the degree to which they succeed in counteracting the relative attractiveness of the import substitution policy pursued since the 1950s. It evaluates the prevailing system of the multiple rates of effective export subsidy in terms of the efficiency of use of the domestic resources in the different sectors and industries.

The structure of nominal and effective export subsidy for the different manufactured industries in Pakistan for the year 1963–64 is given in Table 24.2.

The effective subsidy was generally higher than the nominal subsidy and had a much wider distribution or a larger variance. In 1963–64, 60 per cent of the industries, contributing about 78 per cent of the value added in the manufacturing sector, received between 26 and 50 per cent of the export price (including subsidy) as *nominal subsidy* and only about 38 per cent of the industries, contributing about 17 per cent of the value added of the manufacturing sector, received between 51 per cent and 75 per cent of the export price (including subsidy) as nominal subsidy. On the other hand, 36 per cent of the industries, contributing about 35 per cent of the value added of the manufacturing sector received an effective subsidy varying between 51 per cent and 75 per cent; whereas 25 per cent of the industries, contributing about 28 per cent of the value added of the manufacturing sector, received subsidy between 76 per cent and100 per cent, and 18 per cent of the industries, contributing about 12 per cent of the value added, received subsidy above 100 per cent. A few industries, contributing only about 3 per cent of the total value added, received a subsidy of 200 per cent and above. On the other hand, there were a number of industries, which, while receiving nominal subsidy, were subject to effective export taxation but contributed only about 4 per cent to the value added of the manufacturing sector. The highest effective export

Table 24.2 Nominal and effective export subsidy for manufacturing industry

	Nominal subsidy	**Effective subsidy**	
		Unadjusted non-traded inputs	**Non-traded inputs in value added**
	1963–64	*1963–64*	*1963–64*
	(1)	*(2)*	*(3)*
Percentage distribution of industries (unbracketed) and percentage of value added (bracketed)			
Rates of subsidy			
0–25	2	0	5
	(4.02)		(3.84)
26–50	60	12	35
	(77.83)		(20.19)
51–75	38	36	40
	(16.85)	(34.80)	(47.15)
76–100	–	25	10
	–	(27.55)	(0.01)
101–125	–	6	–
	–	(6.08)	–
126–150	– 4	–	
		(1.38)	
151–200	–	2	–
	–	(2.08)	
200 and above	–	6	2
	–	(2.75)	(6.68)
Rates of taxation			
0–25 –		2	2
	–	(0.52)	(64.00)
26–50 –	–	–	
	–	–	–
51–75 –	4	–	
	–	(2.52)	–
76–100 –	–	–	
	–	–	–
101–125 –	–	2	
	–	–	(0.01)
126–150 –	–	2	
	–	–	(13.91)

151–200 –	–	2	
	–	–	(0.41)
200 and above –	–	–	
	–	–	–

Note: The figures within brackets indicate the percentage contribution made by the industries in each group to the 'value added' of the entire manufacturing sector in Pakistan, as given in the input–output table for the year 1963–64. The value added figures as given in the input–output table have been adjusted to exclude two-thirds of the inputs from the service sector to correct for the over-valuation of the service sector (see Lewis and Guisner, op. cit). The value added proportions in column (3), indicating the estimation of effective subsidy inclusive of the non-traded inputs in the value added, has been computed inclusive of non-traded inputs in the value added to maintain comparability. The percentage distribution of the industries is based upon treating those industries which receive export performance licensing twice, since the effective as well as the nominal subsidies are calculated for those industries twice, once including the effects of export performance licensing and again, excluding the effects of export performance licensing. This method is adopted to overcome the difficulty raised by the fact that export performance licensing relates to the specific products or branches of various industries whereas the data on the other variables used for the estimation of effective subsidy relate to the major industries or groups of industries. The sum of the proportions of value added for each group does not add up to 100 since the estimation of nominal subsidy does not cover all the industries for lack of data, a number of industries such as cotton ginning, jute baling, salt and tea, etc., have been omitted as they are not treated as manufactured exports for the purposes of the present study. The nominal subsidy in column (1) is calculated as $\frac{R - X}{X}$ i.e., the difference between total receipts including subsidy per unit of R export and f.o.b. export price (X) is expressed as a percentage of the former in order to make the nominal subsidy comparable in terms of method of calculation to effective subsidy which expresses the difference between value added in export market including subsidies (Vx) and the value added in world prices, (Vw) as a percentage of value added in the export market, i.e., $\frac{Vx - Vw}{Vx}$

The estimates of the effective subsidy in column (2) are based on the use of undeflated nontraded inputs. In other words, the domestic value of the non-traded inputs, i.e., the inputs from construction, gas, electricity and all other services as recorded in the Pakistan inputoutput table of 1963–64 is assumed to be equal to their value in world prices. The use of undeflated non-traded inputs tends to raise the rate of effective subsidy since the value added in ,s-orld prices is lower than otherwise it would be if the opportunity cost of the non-traded inputs is found to be below their domestic market value. The inclusion of non-traded inputs in the value added for the purpose of estimating the effective protection has been proposed on the ground that they measure, in some sense, along with value added, the contribution of the economy to the value of the product and hence may be said to represent the domestic resource cost of the activity engaged in import substitution or in export expansion (Corden, 1966). Accordingly, in the column (3) effective subsidy has been calculated including non-traded inputs in value added. The appropriate treatment of the non-traded inputs in any attempt to measure the value added in world prices of any economic activity is, however, not universally agreed upon. One measure of the social opportunity cost of the non-traded inputs would be to break down the components of the gross output of the non-traded sectors into: a) intermediate inputs from the other traded sectors, including imports which can then be evaluated in world price; and b) primary inputs or value added. It is the evaluation of the latter in world prices which poses the problem and an approximate method of solution may be to assume that the weighted average of the nominal protection to the traded products measures the excess of the value added in the non-traded products over their social opportunity cost. This is on the assumption that the excess of the opportunity cost is primarily due to the protection granted to the traded sectors, which raises the profitability of the domestic production of the traded commodities and hence raises in turn the demand for non-traded inputs necessary for the production of the traded commodities.

subsidies were received by edible oils and fats (255 per cent), with negative value added in the export market, rubber manufactures (123–219 per cent), matches (157 per cent), fertilizers (130 per cent), and metal products (125 per cent).[10] In a few cases such as silk products, coal and petroleum products, one variety of non-metallic mineral products and of electrical machinery and plastic products, the effective subsidy was lower than nominal subsidy. The rates of nominal and effective subsidy were not correlated. The rank correlation coefficient between the industries, ranked according to nominal and effective subsidy respectively, was small and statistically insignificant at the .05 level.

The level of effective subsidy declined in 1965–66. The percentage of industries receiving effective subsidy of between 25 and 50 per cent increased to 29 per cent in 1965–66 from 12 per cent in 1963–64; moreover, they contributed 25 per cent of the value added of the manufacturing sector as against 12 per cent in 1963–64. The percentage of industries with effective taxation increased to 12 per cent but they did not contribute a larger percentage of value added in 1965–66 than in 1963–64.

When effective subsidy was estimated in terms of value added including non-traded inputs, the general level of effective subsidy declined; the vast majority of the industries received subsidy below 75 per cent and a large percentage of industries received subsidy between 26 per cent and 50 per cent. Moreover, a larger proportion of industries suffered from effective taxation. The general lack of correlation between the ranking of the industries by nominal and effective subsidy holds true.

Import Substitution and Export Expansion: Relative Attractiveness

The relative attractiveness of import substitution as against export expansion can best be evaluated in terms of an industry-wide comparison of the effective rates of protection and export subsidy for the year 1963–64 as given in the summary Table 24.3.[11] The industries which received an approximately equal degree of effective protection and effective export subsidy account, on an average, for about 42 per cent of the total manufactured exports during 1964–67. They include the second most important export industry, cotton textiles, and other relatively important export industries such as leather and leather goods, sports goods, paper and paper products, all of which together constitute 32 per cent of the total value added of the entire manufacturing sector.[12] A few industries such as footwear, matches, and rubber manufactures which accounted only for 6 per cent of the value added and for about 4 per cent of exports during 1964–67 received effective subsidy greater than effective protection. If total exports excluding cotton and jute textiles are considered, 84 per cent of the manufactured exports earned an effective subsidy greater than 90 per cent of the effective protection. The jute industry which supplied about 44 per cent of the total manufactured exports during 1964–67 received an effective protection greater than effective subsidy. This seems to indicate

Table 24.3 Relative rates of effective protection and subsidy

(Distribution of industries in terms of their percentage contribution to total value added of manufacturing sector and to total manufactured exports)

	Percentage of value added		Percentage of exports			
	1963–64		*1961–64*		*1964–67*	
	A	*B*	*A*	*B*	*A*	*B*
Effective protection > effective subsidy	46	16	58 (jute 54)	59 (jute 54)	48 (jute 44)	46 (jute 44)
Effective protection = effective subsidy	32	32	31 (cotton17)	29 (cotton17)	42 (cotton24)	39 (cotton24)
Effective protection < effective subsidy						
Subsidy	6	37	3	5	4	8

Notes

A Includes undeflated non-traded inputs.
B Includes non-traded inputs in value added.
The columns do not add to 100 because industries such as cotton ginning, etc., are not included in the list of industry for which effective subsidy and protection have been estimated but they are included in the total value added for the manufacturing sector as a whole.

that the higher rate of effective protection has in this case provided high or abnormal profits to the jute industry which in addition is oligopolistically organized in the domestic market.

The industries which received effective subsidy either roughly equal to or greater than effective protection expanded exports at a faster rate than those which received effective subsidy less than the effective protection. Their contribution to total exports increased from 34 per cent during 1961–64 to 46 per cent during 1964–67. Their contribution to total exports excluding jute and cotton textiles, increased from 77 per cent to 84 per cent.

The above general conclusions remain unchanged if the effective subsidy is estimated inclusive of non-traded inputs invalue added. Exports by the industries receiving effective subsidy by this definition roughly equal to or greater than effective protection constituted 47 per cent of the total exports during 1964–67 as against 34 per cent during 1961–64. They contributed 77 per cent of the total exports excluding the exports of jute and cotton textiles during 1961–64 as against 92 per cent during 1964–67. Excluding jute and cotton textiles, roughly about equal proportions of industries in terms of their contribution to the export earnings receive subsidy greater than and less than effective protection. The industries which received roughly equal percentage of effective subsidy and protection judged by both the definitions of effective subsidy and protection (whether non-traded inputs are included in the value added or not) contributed between 60 and 70 per cent of these exports.

A pertinent question to explore is whether a commodity is exported if the rate of effective protection exceeds effective subsidy, since this implies that the value of output in export prices (including subsidies) is less than the value of output in domestic prices. It is conceivable that under certain circumstances the exporter does not equalize his profits per unit of sale in the domestic and export markets so that exports take place – even when the effective protection exceeds the effective subsidy. Firstly, there is the possibility of price discrimination between the home and foreign markets, caused by the quantitive restriction on trade; in view of the imperfections of competition in the domestic market, on the one hand, and of Pakistan's small share in the world market, on the other, resulting in different elasticities of demand in the two markets, price discrimination appears profitable in many cases (Gilbert, 1940, para. 1, p. 70; Kindleberger, 1962, pp. 258–80). Secondly, the manufacturers may operate on the basis of a conventional profit margin rather than maximize short-run profits; so that if the conventional profit margin is earned in the export market, they are willing to export, even though the domestic market is more profitable. Moreover, a short-run maximization of profits may be moderated if a policy of persuasion, exhortation and pressure is pursued by the government to encourage the manufacturers to enter the export market. Entry into the export market is one of the most effective ways of winning the favour of the government, which exercises considerable direct and indirect control over and commands substantial patronage in respect of the operation of the private enterprises. The export policy of the government has in recent years been buttressed by the compulsory fixation of export quotas, non-fulfillment of which incurs the displeasure of the government. Lastly, there are considerable differences in costs and efficiency between the individual firms in an industry, so that for the more efficient intra-marginal firms the difference between the world price and costs is more than met by the magnitude of the export subsidy. These firms are able to forego some profits without making losses. It is usually the more efficient and the larger firms which engage in the export trade.

An important feature of the export policy has been a considerable fluctuation over time in the magnitude of export subsidy. The two most important components of the export policy, i.e., the export bonus scheme and export performance licensing, have been frequently changed in terms of their coverage, percentage of import entitlements and in terms of the permissible list of imports. The fluctuations in the magnitude and pattern of export subsidy, originating from export bonus and performance licensing, are given in Table 24.4.

Table 24.4 Export subsidy as percentage of f.o.b. exports

Rates of subsidy	**Percentage distribution of industries**		
	1963–64	*1965–66*	*1967–68*
0–24	1.78	1.61	3.23
25–50	5.36	62.91	62.90
51–75	58.93	35.48	33.87
76–100	5.36	–	–
101–125	28.57	–	–
126–150	–	–	–
151–200	–	–	–
200 and above	–	–	–

As has been indicated earlier, by 1965–66 there was a decline in the effective export subsidy but no decline in effective protection, judging from the differentials between domestic and foreign prices of imported goods and tariff rates (Alamgir, op. cit.); the result was a diminution in the attraction for exporting as against import substituting. Assuming that the rates of effective protection of 1965–66 were the same as in 1963–64, and comparing them with the effective rates of subsidy, one finds that in the majority of the cases the effective protection exceeded effective subsidy; moreover, the magnitude of the difference between the two correspondingly widened.

The rationale of the reduction of the export incentive scheme and corresponding reduction in the effective export subsidy is complicated. The export subsidization policy after 1963–64 became more discriminatory as between different commodities, since there was a reduction in the magnitude of the less discriminatory of the two principal measures, i.e., the export bonus scheme, and a widening of the coverage, without raising the rate, of the more discriminatory export performance licensing. It was not clear whether the export bonus scheme was conceived by the authorities less as a device for the correction of the overvaluation of the exchange rate for export, given the depreciation of the effective import rate owing to the tariffs and quantitative restrictions on imports, and more as a temporary measure of assistance to the infant manuf acturing industries in Pakistan. More often than not it was conceived as a temporary measure of assistance for infant industries; it was frequently argued that the cost disadvantage should not be entirely offset by an adjustment of the exchange rate for in exports via an export subsidy: it should be partly offset by a reduction in costs. It was expected that with a gradual improvement in efficiency and/or a reduction input prices, coupled with a temporary, moderate measure of export subsidy, the economy would grow into the official exchange rate over time.

It was not clear whether a lower exchange rate was to be associated with an increased effort at lowering costs, and if so, how. If, on the one hand, the effective export subsidy at a lower rate than the effective protection was adequate to ensure the competitiveness of the manufactured exports in the world market the higher effective rate of protection was redundant and it resulted only in high rents or profits for the manufacturing sector. So long as a higher effective rate of protection restricts markets and contributes to monopoly profits, the expected improvement of efficiency of the manufacturing industries from an increased participation in the export market tends to be thwarted. However, there have been subsequent increases in export subsidy in 1967–68 (following the devaluation of sterling) owing to changes in the export bonus scheme involving a rise in the percentage bonus and in the premium on the bonus vouchers. But at the same time there has also been a rise in the tariff rates and an enlargement of the bonus import list, both of which have the effect of raising the degree of nominal protection. Given the close positive correlation between the nominal and the effective protection, as evidenced in a recent study, there has presumably been an increase in the degree of effective protection, along with an increase in effective export subsidy (Lewis and Guisinger, op. cit.).

Multiple Effective Rates, Efficiency Criteria and Export Earnings

What has been the rationale for the differentiation between commodities in terms of nominal and effective rates of export subsidy? Firstly there is discrimination between the agricultural and the manufacturing sectors and secondly there is discrimination between individual commodities in each group. The most major agricultural exports suffer from effective export taxation. Discrimination between different commodities is usually sought to be justified with reference to differential elasticities of demand and supply. In addition, a higher rate is justified for those industries which suffer from high costs of infancy and which are expected to reduce costs over time by 'learning by doing'. When an infant industry faces a more elastic export demand curve, the case for a higher exchange rate is reinforced. The lower rate of exchange for agricultural commodities is sought to be justified with reference to their relatively low supply and demand elasticity. However, excepting raw jute, of which Pakistan supplies a major share of the world market, the elasticity of export demand for most of the other commodities is likely to be high in view of Pakistan's small share in world trade. Even the assumption of low elasticity of demand for raw jute needs reconsideration if the new synthetic substitutes such as polypropylene as well as the entry in the world market of the potential marginal producers like Thailand, Brazil and African countries is taken into account. Reliable empirical evidence on the differential demand elasticities for exports is difficult to come by (Alam, 1968, pp. 170–80; Hussain, 1964; Harberger, 1957; Shinkai,1968; Kreinin, 1967).[13]

The presumption that, view of Pakistan's small share in the world market for the manufactured exports, the elasticity of export demand is likely to be high, needs to be qualified since international trade in many manufactured goods is imperfect in the presence of product differentiation and of marketing costs involved in breaking the barrier of consumer's preference for the established products.[14] However, Pakistan's miscellaneous manufactured exports suffer from poor quality and lack of standardization; furthermore, the selling efforts and non-price factors involved in entering the new markets with whom the traditional trade links have been relatively undeveloped play a crucial role in export sales. This tends to reduce the price elasticity of demand. It appears doubtful in the light of the factors considered above whether the considerable degree of discrimination between the major agricultural exports, on the one hand, and the manufactured exports on the other, is justified only on the grounds of the differential demand elasticities. The reductions in export taxes and the inclusion of wool under the export bonus scheme during 1967–68 constitute a belated recognition of the validity of this argument.[15]

Given the differences in the demand elasticities for exports, the appropriate rate of exchange for the different exports, with a view to maximizing real income or net foreign exchange earnings (net of domestic resource cost) would depend on the differences in supply elasticities. Thus if the supply of agricultural exports is highly inelastic compared to that of manufactured exports, other things remaining the same, a higher exchange rate is called for for the latter. While the aggregate supply elasticity of the agricultural sector may be low, the supply elasticity of the individual export crops is high because of the high degree of substitution in production between the different export crops as well ns between the agricultural raw material exports, on one hand, and food crops, which are intended primarily for the domestic market, on the other. If the aggregate agricultural supply is inelastic, a high exchange rate for raw material exports may tend to divert resources from the production of food to the raw material export sector; high rents and high prices will be earned by the food producers which may result in a rise in the urban money wages and hence in the costs of production of the manufacturing sector. In these circumstances an exchange rate for the agricultural export sector lower than that for the manufacturing sector may be in order. Moreover, a higher rate for the manufactured exports can also be justified on the basis of infant industry argument but this is intended to be temporary. Thus a certain degree of discrimination in terms of effective exchange rate between agriculture and the manufacturing sector in Pakistan is justified on the above grounds; but the present degree of effective taxation of the exports of the agricultural raw materials may discourage production, especially since the introduction of new technology in food production, mainly in terms of the high yielding seeds, and favourable price incentives encouraged by a system of subsidies, reduces the relative profitability of export crops. An improvement in the exchange rate for the major agricultural raw material exports appears necessary to meet the increase in their opportunity costs.

The wide discrimination between the individual manufactured exports, ranging from high effective subsidy to high effective taxation, seems to be unrelated to any plausible range of diffences between them in terms of elasticity of demand or of infancy. While higher rate for the newer manufacturers as a whole (excluding jute and cotton textiles), in consideration of their relative infancy in production and export marketing, may be warranted for a temporary period, the extent to which they should be higher is critically dependent upon: a) the extent of competition from synthetic substitutes; and b) costs of entry into new and unfamiliar markets. But a considerable degree of discrimination as between the individual newer manufacturers, as is the case now, is difficult to justify in view of an inadequate knowledge about the differential elasticities and infancies of the individual industries; in fact the present system runs the risk of providing a higher exchange rate and higher nominal subsidies to the higher cost industries or industries having greater difficulty in their sales effort in the export markets. It is noteworthy that in view of a lack of close correspondence between the rates of nominal and effective subsidy, the discrimination in terms of nominal subsidy does not result in discrimination in terms of effective subsidy, either in the desired direction or degree. While *nontinal* subsidies for the textiles products in general are lower than those for the newer manufactures, many newer manufacturers have lower *effective* subsidies and in some cases even negative effective subsidy, which is contrary to the assumption of the higher demand and supply elasticities as well as the greater infancy of the newer manufacturers.

One way of judging the efficiency of the differential structure of export incentives is to relate the rates of effective subsidy to the marginal cost of earning additional foreign exchange from the individual manufactured exports. Since, on the basis of the available data, it has not been possible to estimate the domestic niarginal cost of the additional foreign exchange earning, a few alternative estimates of the average domestic cost of average foreign exchange earnings have been made. Whichever index of the average domestic resource cost is used, the rates of effective subsidy for different industries are unrelated to their domestic costs of earning foreign exchange.[16] The absence of any significant correlation between the average domestic costs of earning foreign exchange and the structure of effective subsidies tends to confirm that the differential structure of effective subsidies is not designed to minimize average costs of earning foreign exchange by granting higher subsidies to the lower cost exports.

How have the exports responded to the export subsidies? The manufactured exports of all kinds received subsidies and expanded rapidly during the 1960s. Between 1957–58 and 1966–67, among the various exports enjoying export subsidies, the minor agricultural exports increased annually at a compound rate of 15 per cent, the jute and cotton textiles expanded at 22 per cent and the minor manufactured exports at 38 per cent. The high rates of growth of the minor exports, both agricultural and manufactured, are partly statistical phenomenon, since the value of these exports was very small in the late 1950s. The growth of cotton and jute textiles has been partly at the expense of the exports of raw cotton and raw jute. The growth in the exports of tanned leather and leather goods has been partly at the expense of hides and skins. The average net foreign exchange earnings of the four major manufactured exports, i.e., f.o.b. export value minus the import component and exportable domestic raw materials incorporated in the manufactured exports, as a percentage of f.o.b. export value, are as follows: jute textiles, 27 per cent; cotton textiles, 26 per cent; woollen textiles, 30 per cent; and leather and leather goods, 20 per cent, as of 1963–64 (Islam, 1968b).

In the present context the more pertinent question is whether the differential performance of the individual exports is related to the magnitude of the differential incentives. The rates of growth of individual manufactured exports during 1960–67 are not related to the differential export subsidies, either nominal or effective, judged by the rank correlation between them. The relative importance of the different exports, i.e., the average ratio of the the individual exports to the total manufactured exports during 1960/61–1963/64 or during 1964/65–1966/67 is not correlated with the rates of effective subsidy of 1963–64 or of 1965–66. Moreover, there is a very poor correlation between the rate of effective subsidy and the proportion of export to the output of each industry. This is true of both the estimates of the rate of effective subsidy and for both the years, i.e., 1963–64 and 1965–66. Undoubtedly the relative performance of the individual exports is also a function of changes in the supply and demand situation both at home and abroad, so that the isolation of the influence of other factors is necessary in assessing the effect of differential export subsidies on the differential performance of exports. However, as pointed out earlier, the major proportion of the total manufactured exports (excluding jute and cotton textiles) was contributed by those items which received effective subsidy greater than or roughly equal to the effective protection implying that in these cases the attraction of production for the domestic market was offset by the export incentives. But the differences in the export performance as between individual items as measured by the rates of growth of the individual exports or by the proportion of the total manufactured exports constituted by each individual item is not directly in proportion to the relative magnitude of effective subsidy vis-à-vis effective protection.

Conclusions

Since the early 1960s Pakistan adopted a large number of export promotion measures which constituted a substantial amount of nominal and effective export subsidy to the manufactured exports and to a very few minor agricultural exports. The major agricultural exports were subject to effective taxation, though to a decreasing degree over time. The effective export subsidy was generally higher than the nominal subsidy and had a much larger variance indicating a considerable degree of discrimination between the individual commodities and industries. There were a few items which received effective subsidy less than nominal subsidy; in fact, a few manufactured exports items were effectively taxed. The nominal and effective subsidy were unrelated, in contrast to nominal and effective protection which were highly correlated. Effective subsidy did offset or nearly offset the effective protection in the majority of the cases, thus counterbalancing the relative attractiveness of production for the domestic market, created by the high effective protection throughout the 1950s. However, the reduction of export subsidy in the years following 1963–64 again increased the relative attractiveness of the domestic market. The magnitude of nominal and hence effective subsidy has been subject to frequent changes. The dual role of multiple exchange rates in offsetting the overvaluation of the exchange rate, as evidenced by the high and rising exchange rates for imports, and in providing assistance to the nascent infant industries in their efforts to compete in the export markets, was not often clearly distinguished.

The discrimination between the individual manufactured exports in terms of effective export subsidy does not appear to be related to or justified by any plausible range of variation in the elasticities of and demand or in the relative degree of infancy of the different industries. The rates of effective subsidy are not related to the domestic cost of earning foreign exchange, conceived in terms of average costs and average earnings. While a lower rate for agricultural exports is justified, the present high degree of discrimination between the agricultural and manufactured exports needs to be reduced by raising the effective subsidy for the former. Exports have expanded considerably during the 1960s but the differential rates of growth are not related to the differential export incentives. In other words, the manufactured exports in the aggregate have expanded in response to export incentive schemes at a rate higher than the growth in world trade in manufactures, but not in any direct proportion to the differential export incentives.[17] While a once and for all upward adjustment of the exchange rate will reduce the administrative burden as well as the opportunities and the inducements for evasion, it may be argued that in the present circumstances, with a complicated structure of effective export and import rates, it will be difficult to determine the appropriate amount of devaluation which can replace the existing structure of multiple rates. The generalization of the export bonus scheme, the least discriminatory of the existing export incentive schemes, to include all imports and exports, provides a readily available mechanism for introducing a flexible exchange market; the resultant rate of exchange which emerges out of the operation of the bonus market will provide a guide to an appropriate degree of once and for all official adjustment of the exchange in the future. Discrimination between agricultural and manufactured exports, though at a much reduced scale can be accomplished by taxes.

Even if there is one exchange rate for the agricultural exports and one for the manufactured exports, the effective rates of subsidy on export will be different for the different manufactures, depending upon the relative inputs of the agricultural and manufactured sectors in the production. Under these circumstances, it is necessary to have different nominal rates to establish a desired structure of effective rates with the minimum necessary discrimination. Though there are a few valid cases for discrimination between commodities in terms of differential export or import rates of exchange, there hardly seems to be any case in the same industry for discrimination in favour of the domestic market and against the export market, excepting when there is a monopoly power to be exploited in the export market and when the social evaluation of risks and uncertainties of selling abroad depress the returns from exports below that of the domestic sales (Bhagwati, 1968). On the one hand, if the social evaluation of the risks of the export and domestic sales is the same, but at the same time, private evaluation of risks of the export sales is higher than the social evaluation, an effective export subsidy higher than the effective protection, instead of the other way round, seems warranted. A case in point is when a pioneering firm in an industry incurs costs in introducing its product and establishing itself in the market, which it cannot fully recover owing to the late comers or 'follow up' firms reaping the benefits of export sales without having to incur any of the costs.

Notes

1 In addition, some invisible items such as the earnings from shipping, ship and aircraft repairs, and salvage operations were entitled to a bonus at the rate of 20 per cent. The hotel

industry and private remittances from abroad were subsequently included in the scheme, both of these items being entitled to a 20 per cent bonus rate.

2 The differences as between the different industries, in terms of the extent of subsidy received by them, were thus related to scarcity of the imported inputs which were specific to the industry concerned.

3 The imports may be used either for a larger domestic production destined for the domestic market or for outright sales at high profits to other similarly situated and suffering from excess capacity. Moreover, a greater utilization of capacity is expected to reduce unit costs and thus to increase profits per unit of sale.

The additional import entitlement or export performance license received by the export industries in the initial years was substantially in excess of the import component of exports, especially since the import entitlements were at 100 per cent of the f.o.b. value of export. Until 1964 only 43 industries were eligible for the additional import entitlement, and the import entitlements were a uniform percentage of exports. In the subsequent the number of eligible industries has been expanded; they were 250 in 1965–66. At the same time, the percentage of import entitlements was reduced. Moreover, the import entitlements of each industry were to be fixed separately in the light of its own import component, but in no case was it to exceed 50 per cent of the f.o.b. value of exports. Thus the degree of differentiation between commodities increased and the rates of subsidy accruing to the different industries were widely different.

4 Even though the exemption from the indirect taxes on the domestic intermediate inputs as well as the imported inputs was introduced as early as 1956, the procedures for exemption were not worked out satisfactorily and the administrative delays and complexities limited their usefulness. It was only during the 1960s when the export drive became one of the main components of commercial policy in Pakistan that the implementation of the measures of tax exemption was pursued vigorously. The exemption of exports from the domestic indirect taxes was, however, in force during the mid-1950s. The rebate of income tax on exports was introduced only in 1963–64 and the export performance licensing in its generous form which is quantitatively the most important incentive, next to the export bonus scheme, was introduced as late as 1962–63.

5 Twenty-nine per cent of total exports, which are all manufactured, receive between Rs 5.90 and Rs 7.47 per dollar; all manufactured exports constitute 43 per cent of total exports and hence the former constitutes about 68 per cent of the manufactured exports.

6 In the above derivation of the amount of export subsidy, the rebate on income tax on profits or income from exports is not included. The percentage of rebate on income tax is related in a progressive manner to the percentage of output exported, i.e., a higher percentage of rebate is allowed with a higher proportion of export to the output. During 1963–67, the industries which exported between 10 per cent and 20 per cent of its output were entitled to a rebate of 10 per cent on the income tax payable on profit or income from exports; the rebate was increased to is per cent in 1967. Those exporting above 30 per cent of output received the maximum rebate of 20 per cent, which was raised to 25 per cent in 1967. Assuming profits on the export sales to the extent of 25 per cent of the sales price, and income tax at an average rate of 30 per cent, the subsidy works out between 0.7 per cent and 1.5 per cent of the f.o.b. value of export during 1963–67, which increased to a value of 1.0 per cent and 1.8 per cent in 1967.

7 In cases where data on x were not directly available, they were, indirectly derived as follows:

$\frac{X = D}{P_w (1 + .05)}$ where D is the domestic wholesale price and Pw is the ratio of the domestic wholesale price (including indirect taxes) to world price which is in most of the cases based on the c.i.f. prices of the competing imports. The c.i.f. import price is assumed to be 5 per cent more than the export price.

8 In this chapter V has been derived indirectly from $\frac{V_d - V_w}{Vd}$ which is given in Lewis and Guisinger, op. cit., and V_d is given in the input–output table of 1963–64.

9 If V_x and V_w are always positive, then a positive $\frac{V_x - V_w}{V_x}$ always implies an effective export subsidy and a negative figures implies export taxation. However, when V_x and V_w are negative the interpretation of $\frac{V_x - V_w}{Vx}$ with regard to its sign is no longer simple. If V_x is negative, a positive figure implies an export taxation rather than export subsidy, whereas with a negative V_x, a negative figure implies an export subsidy rather than export taxation. However, with a positive V_x, a positive figure implies subsidy and negative figures implies taxation of exports.

10 See Appendix, Table 24A–1.

11 The comparison is made in terms of the average export subsidy in the case of those industries forwhich two rates of effective subsidy are computed. Two rates of subsidy for one industry are computed because one of the major export incentive schemes, i.e., export performance licensing, relates only to specific products or branches of the industries on which relevant data are available for the purposes of the present analysis. One rate of subsidy is computed for the industry excluding the export performance licensing whereas another estimate is made on the assumption that the entire industry is entitled to the scheme. Detailed industrywise comparisons are given in Appendix I.

12 See Appendix.

13 Some fragmentary evidence on the demand elasticity for agricultural exports is given below. The elasticity of export demand for raw cotton in the US market is estimated at 14.5 whereas that of the Pakistani skin in the same market is estimated at 2.3 during 1955–64. The export demand elasticity for fine rice is estimated at 3.14. However, fine rice is a highly differentiated product and its sales are limited to Middle Eastern markets. These are estimates of elasticity in particular export markets or regional markets, which import considerable quantities of Pakistan's exports. However, if the world market as a whole is considered, elasticity is likely to be higher.

14 A few studies of demand elasticity for the manufactured exports of the developed countries tend to indicate that the elasticity of demand for the textile products is smaller than that for other products such as chemicals, basic metals, metal products, and machinery, etc., even though there are differences within each group, with a few items having a very low elasticity.

A recent estimate of the elasticity of substitution between the Pakistan and Indian jute bags in the United Kingdom markets puts it at 13.24 and that between Pakistani and the competing cotton fabrics in the US market at 6.38. The elasticities of export demand, on the other hand, are respectively -11. 0 and -5. 0; Pakistani's share in the UK and the US markets for these two products is about 88 per cent (Alam, op. cit.).

15 Export tax on a raw material like raw cotton in conjunction with export bonus on the exports of cotton textiles has also been sought to be justified with a view to increasing the relative proportion of exports of the cotton textiles. This is related to the general argument

for industrialization and diversification such as the expansion of employment opportunities, the generation of 'external economies', not otherwise available through an expansion of cotton exports, and the possibility of a higher rate of capital accumulation in the manufacturing sector.

16 The alternative estimates of domestic costs are: 1) direct domestic resource cost, i.e. $\left(\frac{D+V}{X-m}\right)$ 2) direct and indirect domestic resource cost, i.e., $\left(\frac{U}{X-M}\right)$ 3) direct and indirect wage cost, i.e., $\left(\frac{W}{X-M}\right)$ and 4) $\left(\frac{V_d}{V_w}\right)$ D is the domestic intermediate inputs and V is the value added in domestic prices, excluding indirect taxes on output; X is the f.o.b. export value, in is the direct import component, excluding indirect taxes on imports, U is the direct and indirect value added per unit of output excluding indirect taxes, W is the direct and indirect wage costs per unit of output, M is the direct and indirect import component, excluding indirect taxes on import components, V_d is the direct value added in domestic prices and Vw is value added in world prices. If it is assumed that the domestic intermediate inputs in a given activity can be either exported directly or be used in other export or import substituting activity, then the opportunity cost of domestically traded inputs is the value of the intermediate inputs in world prices and hence the net foreign exchange earnings will exclude not only import components but also the value of domestic inputs in world prices. The relevant domestic costs would be the value added in domestic prices. If the costs of the intermediate inputs in domestic prices are included in the domestic costs, the relative inefficiency of the supplying industries is reflected in the costs of earning or saving foreign exchange by the specific activity or industry in question. The last measure of cost, i.e., $\left(\frac{V_d}{V_w}\right)$ is derived from the effective protection which is $\left(\frac{1-V_w}{V_d}\right)$ The ranking of industries by $\left(\frac{V_d}{V_w}\right)$ or $\left(\frac{1-V_w}{V_d}\right)$ is the same.

15 The evaluation of the relative performance of the individual manufactured exports by means of the individual rates of growth, especially in the case of a large number of new manufactured exports each of which are very small value, is subject to a serious limitation in the sense that the lower the base year, the higher the rate of growth.

References

Alam, N., *The Experience of an Overvalued Currency: The Pakistan Case*, unpublished dissertation (Yale University, 1968).

Alamgir, M., 'The Domestic Prices of Imported Commodities in Pakistan – A Further Study,' *Pakistan Development Review*, Vol. VIII, No. 1, Spring 1968.

Bhagwati, J., 'The Theory of Practice of Commercial Policy: Departures from unified exchange Rates', *Papers in International Finance*, No. 8, January 1968, Princeton University, USA, pp. 11–14.

Corden, W.M., 'The Structure of a Tariff System and the Effective Protective Rate', *Journal of Political Economy*, June 1966.

Gilbert, M., Temporary National Committee, Monograph No. 6 (Washington, DC: Government Printing Office, 1940).

Harberger, A.C., 'Some Evidence on International Price Mechanism', *Journal of Political Economy*, December 1957, pp. 506–22.

Hecox, W.E., *The Use of Import Privileges as Incentives to Exporters in Pakistan*, Research Report No. 30 (Karachi: Pakistan Institute of Development Economics, 1969).

Hussain, M., 'Export Potential of Fine Rice from Pakistan', *Pakistan Development Review*, Vol. IV, No. 4, Winter 1964.

Islam, N., *Imports of Pakistan: Growth and Structure, A Statistical Study* (Karachi: Pakistan Institute of Development Economics, 1967).

Islam, N., 'Export Policy in Pakistan', in *Essays in Honour of J. Tinbergen* (Amsterdam: North Holland Publishing Company, 1968a).

Islam, N., 'Commodity Exports Net Exchange Earnings and Investment Criteria', *Pakistan Development Review*, Winter 1968b.

Islam, N. and Malik, I.O., *Comparative Costs of the Manufacturing Industry in Pakistan – A Statistical Study*, Research Report No. 58, June (Karachi: Pakistan Institute of Development Economics, 1967).

Kindleberger, C., *International Econoinics* (Homewood, IL: Richard D. Irwin, Inc., 1962).

Kreinin, M.E., 'Price Elasticities in International Trade', *Review of Economics and Statistics*, November 1967, pp. 510–16.

Lewis, S.R. and Guisinger, S.E., chapter on Pakistan, in B. Belassa (ed.), *The Structure of Protection in Developing Countries*, unpublished manuscript (IBRD).

Lewis, S.R. and Guisinger, S.E., *Measuring Protection in a Developing Country: The case of Pakistan*, Research Memorandum No. 6 (Williamstown, MA: Williams College, 1966).

Shinkai, Y., 'Price Elasticity of the Japanese Exports: A cross-section study', *Review of Economics and Statistics*, June 1968, pp. 268–73.

Thomas, P.S., 'Import Licensing and Import Liberalisation in Pakistan', *Pakistan Development Review*, Vol. VI, No. 4, Winter 1966.

Tims, W., *Analytical Techniques for Development Planning* (Karachi: Pakistan Institute of Development Economics, 1968).

Appendix

Table 24.A1 Effective protection and

Industry	N Effec subsi 1963- *(1)*
Canning of fruits and vegetables	
Bakery products and confectionery –	
Sugar	-0.69
Edible oils and fats	-2.54
Alcoholic beverages	
Tobacco products	
Cotton textiles	.86
	.93
Wool textiles	
Jute textiles	.64
Silk and silk textiles	.03
Knitting	
Thread ball	.52
	.81
Manufacture and repair of footwear	.45
	.68

Table 24.A1 cont'd

Industry

Wearing apparel

Manuf. of cork and wood products

Manuf. of paper and paper products
Articles of paper and paper board
Printing, publishing and allied industries
Tanning and leather tanning

Manuf. of rubber

Manuf. of fertiliser
Paints and varnishes
.8184
Perfumes, cosmetics and soap
.7960
Matches
Chemical produces, n.e.s.
.6862
Coal and petroleum products

Table 24.A1 cont'd

Industry	N Effec subsi 1963- *(1)*
Manuf. of nonmetallic	.60
mineral products (1)	.76
Manuf. of nonmetallic	.44
mineral products (2)	
Manuf. of metal products	1.24
	1.11
Machinery, except electrical (1)	.66
.8235	
Machinery, except electrical (2)	.57
Electrical machinery (1)	.77
	1.21
Electrical machinery (2)	.56
Manuf. of transport equipment	-0.41
Photographic and optical goods	
Plastic products	-3.24
Sport and athletic goods	.43
Pens, pencils and related products	.38
	.62

Sources: Lewis and Guisinger, n.d., Islam and Malik

Chapter 25

The New Protectionism and the Nature of World Trade: Comments*

Professor Jones has presented a very stimulating and wide-ranging paper replete with considerable insights into various aspects of both trade theory and policy. Most of what he says I have no disagreement with; what I will have to say will mostly be in the nature of clarification and amplification.

The focus of Professor Jones's paper is on the rapidly growing trade in intermediate inputs and its implications for standard trade theory and protectionism, including regional trading arrangements or free trade areas. While noting the rapid growth in recent years in world trade in intermediate inputs, it should be emphasized, however, that a large part of this trade in intermediate inputs is in the nature of intra-firm trade, often within the multinational or transnational corporations. About 30–40 per cent of world trade in manufactured goods is in the nature of intra-firm trade. To the extent that this is so, trade becomes closely interlinked with international investment, i.e., location across countries of enterprises owned by multinational or transnational companies. The factors which govern the pattern of international investment under these circumstances have a direct influence on the composition of trade, especially trade in intermediate inputs.

As Professor Jones suggests, the pattern of specialization is no longer a simple function of the relative endowments of primary factors of production but depends upon a host of other considerations, such as availability of social overhead capital, as well as transport, communications, electric power, nature of the financial and taxation systems, and rules and regulations governing the operations of foreign corporations, etc. – considerations which also influence the flow of foreign direct investment. Moreover, the cumulative advantages enjoyed by a firm or a country that happens to be first in the field, that aggressively undertakes investment in research and development and becomes a pioneer in capturing the economies of a scale, assume an important role in determining the pattern of specialization in trade. However, it should be emphasized that this is relevant not only to trade in intermediate inputs, as

* First published in *The Pakistan Development Review*, Vol. 32, No. 4, Winter, 1993.

the article by Professor Jones seems to convey, but also that in finished products.

The strategic trade theory which, in any case, interjects the theory or principles of industrial organization or strategy into the theory of trade, suggests that in an imperfectly competitive industry in which firms located in different countries are engaged in monopolistic or duopolistic competition, taxes and subsidies could be used by one of the partner countries in order to obtain a greater share of profits and of the world market. This theory, to the extent that it is valid, applies to monopolistically competitive enterprises in general, whether they produce final or intermediate products. In other words, this applies to competition in the aircraft industry, i.e., between Boeing and Airbus company as much as in the semiconductor industry.

What are the implications for developing countries of the emerging pattern of trade in intermediate inputs including close interlinkages between international trade and investment? They certainly have a much wider range of choice in manufactured goods that they can produce or specialize; they can specialize in 'production blocks', as Professor Jones calls them, so to speak, linked across countries through global communications and information technology. This choice is not necessarily always linked with foreign direct investment; components or inputs produced in a particular country can be used in enterprises in another country, which are not owned by a transnational/multinational corporation, through contractual arrangements with enterprises abroad. What is needed is a high level of organization and managerial ability and greatly developed infrastructure, i.e., transport and communication systems within and across nations. Under these contributions, a major part of the components needed for the production in B of a final output which is imported by Country A may have been produced in Country A; similarly, a major part of the components of an export product from Country A to B may have been imported in the first instance from Country B. The definition of the national origin of the final product under these circumstances becomes very blurred.

Should the government pick a 'winner', i.e., an industry or a part of an industry which it perceives to have the potential to compete in the world market and then provide it temporarily, i.e., for a short period, assistance which ranges from protection in the home market to credit and export subsidies? In this scenario a protected import substitution industry eventually gains sufficient competitive strength on the basis of successful operation in the home market and is thus able to graduate to a profitable export industry. There are examples of this kind in different countries.

Two objections are often voiced against this strategy. First, the government is unlikely to be able to identify or decide ahead of time which industry will in the long run succeed in the competitive struggle in the world market. In this view, what is needed is the provision of infrastructure, appropriate regulatory framework needed for a well functioning market, investment in human resource development, including training, education, etc. The free play of market forces would do the rest, i.e., pick the 'winner', provided a liberal trade and exchange rate regime neutral between the production for the home market and that for the export market exists. It is argued that private enterprise, be it foreign or domestic, is more conversant than the bureaucracy with the details of market possibilities, be it for components or for finished products; they are always better 'pickers'. If private enterprise fails, it goes bankrupt and losses imposed by the market on private enterprises do not impose a cost on the government. Second, even if one justifies a temporary provision of subsidies or protection, they tend to become permanent. The vested interests receiving subsidies become entrenched and powerful enough to ensure their continuation long past the period when they are considered necessary. Therefore, they impose substantial social costs in the long run.

Professor Jones's objections to the government's picking the 'winner', is based on the second set of arguments, i.e., reluctance of the recipients of special privileges or subsidies to surrender them, even after they succeed and earn high profits or rents.

In this respect, the experience of the East Asian countries may be relevant. Frequently, the interpretation of the East Asian experience seems to lie in the eyes of the beholder. Both interventionists and free marketeers quote Korea in support of their case. Increasingly, a substantial group of observers of the Korean scene, including most Korean economists, seem to have reached the consensus that Korea successfully picked the 'winners' and provided them with a variety of assistance, including subsidies and protection, until they established themselves in the world market. Many successful export industries started out as import-substituting industries under protective walls; they received export aid of various kinds as they entered the export market. What was more important was that Korea succeeded in removing special assistance, i.e., protection or subsidies, as the industries became successful in the world market.

Korea had a lower average level of tariffs – lower than most developing countries today – and the tariff rates or degrees of trade restrictions were more uniformly distributed or much less widely dispersed across activities/industries. Protection was provided with an advance notice of their intended decline over time and this policy was enforced effectively. On the whole, policies were skewed more in favour of production for the export market than for the home market.

How did the Koreans succeed in picking the winners and withdrawing support as they became successful? It is suggested that Korea, unlike many developing countries of today, was not a 'soft state'; the government could overcome pressures from entrenched interests, enjoying rents from protection, and could withdraw assistance when it was no longer needed. Secondly, whenever mistakes were made in 'picking' the winners either because world market conditions did not justify the establishment of such an industry in the first place, or because circumstances in the world market subsequently changed so that the initial success could not be sustained, the Korean government followed the changes in market signals and let the losers go down; it moved on to pick and support the next in the line considered to be the potential winner. The response to market signals was implemented with considerable skill and speed; losses incurred due to past mistakes were not allowed to accumulate. In other words, the government of Korea was a 'nimble' government demonstrating considerable pragmatism, vigilance and flexibility in the design and implementation of policies.

It is alleged that the success of this policy depended on the discipline and efficiency of a competent bureaucracy and a political/administrative leadership devoted to the cause of Korea's development. The present day developing countries, it is suggested, do not have the kind of efficient and development-oriented administrative and political leadership which Korean had. This is a proposition which is considered plausible. The underlying circumstances, historical and cultural, which led to the emergence of a development-oriented bureaucratic and political leadership need further analysis. In recent years, many developing countries have come increasingly to accept advantages of a liberal trade and exchange rate regime. They have sought membership of the GATT – long considered a 'rich man's' club and accepted the rules and disciplines of GATT in their trade regime. They have become active participants in the on-going GATT negotiations under the Uruguay Round. Advantages of competition and scale economies emanating from export-oriented development strategy are being gradually recognized. Many of them have undertaken unilateral liberalization of trade while at the same time developed countries continue to persist in a high degree of protectionism.

Professor Jones suggests that trade in components tends to discourage protection for high cost input-supplying industries since such protection puts the producer of the final or finished product at a competitive cost disadvantage.

There is indeed an incentive for the producer of the finished/final product to oppose the protection of input-supplying industries. However, both of these groups – the final product and the input supplying industries – are small in number and often have adequate organizational strength to secure protection for both of them at the expense of consumers who are diffused and unorganized. The producers of final product seek compensation against protection for input-supplying industries by raising the rate of protection on their products so that it offsets the high cost of inputs. What matters to them is the rate of effective protection rather than that of nominal protection. Similarly, export industries which rely upon protected domestic inputs seek exemption through tax rebates and subsidies to offset the cost disadvantage. Many countries seek to ensure that export industries obtain inputs at world prices by providing them with drawbacks on duties/taxes on imported components as well as tax rebates/ subsidies to offset the high cost of domestic components. Thus, the tension between the competing interests of producers of final products and of those producing inputs and components is often partially resolved by raising the general level of protectionism and subsidies across the economy. This resembles the case of appropriate degree of protectionism for agriculture vis-à-vis industry. Industrial protectionism hurts agriculture by raising the costs of inputs to agriculture and by diverting resources away from agriculture to the manufacturing sector. The agricultural pressure groups, instead of demanding reduction or elimination of industrial protectionism, seek even higher degrees of protection for themselves.

Professor Jones's example of restrictions on rice imports in Japan leading to high wage costs and, therefore, high costs of manufacturing industries is not quite comparable with the example of corn laws in England in earlier centuries. This is primarily because the relative importance of corn in the consumption expenditure of the average British worker in eighteenth-century England was much higher than the importance of rice in the consumption expenditure of the average Japanese worker. At the present time, the contribution of high rice price to the high wage costs in the manufacturing industry in Japan is unlikely to be substantial. Similarly, the case of the expensive Japanese rice leading to higher wage costs in Japan is not comparable to high costs of components/inputs which place the manufacturer of a finished product at a competitive disadvantage in the world market. The high input costs affect only the industries which use a high proportion of the protected inputs, whereas expensive rice, to the extent it is an important element in the determination of wages, affects wage costs in the whole range of manufacturing exports. While it is true that the elimination of protection to rice production would divert resources from rice production to other sectors, especially the industrial sector, agriculture uses a very small proportion of resources in Japan. Its overall impact on the Japanese economy is unlikely to be significant.

Professor Jones is right to emphasize that the trade in inputs provides a protective bias to the regional trading arrangements since it has a trade diverting effect on trade in, inputs. If, for example, cheaper inputs from outside the free trade area are surrendered by Country A in favour of higher priced inputs from the partner Country B, the cost structure of the Country A, producing the finished product, goes up. Since in a free trade area the external tariffs or trade restrictions of the partner countries are not uniform, this provides incentive to the Country A, faced with higher cost of inputs and, therefore, higher costs of its finished product, to seek trade restrictions on imports of finished products from the third countries. However, since the free access of Country A's finished product to the partner country B remains unchanged, its costs advantage within the free trade area remains undiminished if the high cost of input does not offset the price advantage Country A enjoys in Country B because of its free entry in B's market.

The advantages of free trade area, however, go beyond static gains derived from trade creation offsetting trade diversion; it facilitates economies of scale, efficient division of labour and competitive efficiency, all of which liberal trade regimes offer. Moreover, as income growth is stimulated in the partner countries through a more efficient allocation of resources within and between the partner countries, their imports from third countries may go up as well. Regional free trade arrangements could be a first step in the way towards a global trade liberalization for countries concerned. As the countries within a free trade area are exposed to competition within a larger regional market beyond their own domestic borders, they gain competitive strength and are able to venture into the world market.

Much has been said about the relative advantages and disadvantages of developing countries having free trade arrangements with a dominant developed partner country rather than among themselves. This is the case with North American Free Trade Area. It is argued that this type of encouragement offers greater advantages for developing countries. Firstly, there is a possibility of greater trade creation because potential divergences in economic structure between developed and developing partner countries are greater than among poorer developing countries with similar economic and trade structures. To date, developing countries have not undertaken free trade arrangements among themselves to any significant extent. Hence, experience so far is very limited to evaluate their effects. Furthermore, the access to larger markets of developed partner countries provides economies of a scale to an extent not otherwise available in a free trade arrangement among developing countries, each with a relatively small domestic market. Thirdly, free trade arrangement with developed countries facilitates access to technology and external capital, much needed for their development. Not only direct investment from the developed partner country may move to the partner developing country to take advantage of lower wage costs, but also third countries outside the free trade area may invest in the developing country to gain access to the larger domestic market of the developed partner country.

In the particular case of Mexico, an additional incentive for joining the North American Free Trade Area was to ensure that outward looking trade and economic liberalization policies, introduced by Mexico after many years of protectionist policies followed in the past, become irreversible by tying them within the Free Trade Area arrangement with the USA.

PART IX

TRADE AND AID: ISSUES IN DEVELOPMENT ASSISTANCE

Chapter 26

Recent Trends in the Theory of International Investment*

This chapter is an attempt to discuss a few recent developments in the theory of international investment.

The development of economics as an empirical science has been marked by the interaction of theory and fact. The consideration of the practical problems and issues of the day has often led to the re-examination of the old theories and formulation of the new ones – more so in the field of international trade and investment since the days of Ricardo. In recent years, the economic development of underdeveloped economies has provided an important context for a re-examination and extension of the traditional theory of international investment. Traditionally, the centre of the stage was occupied by the discussion of the mechanism of transfer and the balance of payments adjustment – a discussion which was sharpened by the short-run Keynesian income analysis, with its emphasis on macroeconomic aggregates and consequently on the role of foreign investment as an employment stimulating and stabilizing factor. However, in the post-Second World War period, the discussion of foreign investment has been associated with three inter-related problems: 1) cure of the dollar shortage; 2) the procurement of sufficient supplies of raw materials and primary products for the advanced industrialized countries; and 3) the economic development of backward economies.

In connection with the latter problem, estimation of foreign capital requirements for backward economies has engaged a great deal of attention. This has been posed from two angles: saving-investment method and balance of payments method. Considerable analytical refinements have taken place in the latter method, especially in analysing the impact of the domestic investment on the balance of payments. Distinction has been made between: a) investment effects; b) direct operating effects; and c) indirect operating effects insofar as the impact on the balance of payments is concerned. The investment effects include: 1) the purchase of machinery and equipment from abroad; and 2) the multiplier effect of investment on income and imports. Direct operating effects

---

* First published in *Pakistan Economic Journal*, 1956.

include: 1) output of a commodity which increases exports or is a substitute for imports; 2) imports (direct and indirect) for the production of the given commodity; and 3) reduction of import requirements for the production of commodities for which the given commodity is a substitute. The indirect operating effects would include: 1) multiplier effect of inflationary financing of consumption; and 2) multiplier effect of a change in import or export surplus. The investment effects, the 2) of direct operating effects and 1) of the indirect operating effects, would have negative effects on the balance of payments. The 1) of the direct operating effects would tend to counteract it. The 3) of the direct operating effects and 2) of indirect operating effects can have either positive or negative influence on the balance of payments (Chenery, 1953).

An important analytical problem in recent years has been the study of the absorption of foreign capital – a systematic treatment of the factors involved in an effective and productive utilization of foreign capital in the furtherance of economic development. The problem of absorption of foreign capital in an undeveloped economy includes and extends beyond the problem of mere initial transfer, which has been the most discussed problem in the traditional theory. It is not merely the process of creation of import surplus via price and/or income effects in the borrowing and lending country that should engage attention but, more specifically and importantly, the nature and composition of the import surplus, the methods of utilizing it in terms of the various sectors of the economy – all being examined in the light of their effect on the economic development of the borrowing country, which should warrant closer examination. The concept of absorption, therefore, embraces the problem of productive utilization of foreign capital in its microeconomic aspects. In a narrower sense, this involves an enquiry as to whether a particular project financed by foreign capital yields expected returns from the point of view of the investor, which might be private or public lending agency from abroad. The criteria of profitability would vary as between different types of lenders. In a broader sense, this would lead on to an examination, beyond the specific profitability of a particular piece of investment, of the several economic effects in the generation of external economics and stimulus to domestic enterprise and saving in the long run. Once it is realized that the productive application of foreign capital is a function to a considerable extent of the availability of supplementary and cooperating factors, the nature of the economy, the character of the resource pattern, the nature and form of economic organization and social institutions, the long-run overall effects of the introduction of foreign capital on the above factors would, in turn, determine the rate of absorption of the subsequent flows of foreign capital. It is evident that absorption of foreign capital in backward economies involves economic, organizational, technical and even sociological considerations.

Once the problem of absorption of foreign capital vis-à-vis economic development is put in its proper perspective, certain aspects of the transfer mechanism, not sufficiently emphasized in the traditional discussion, assume added significance in developing economies. In an environment of economic development in which capital imports and internal investment react on each other cumulatively, price and income effects reinforce each other in the process of transfer. In the context of persistent inflationary bias in the undeveloped economies and consequent balance of payments deficit, the problem is no more how to create an import surplus in order to effect, the transfer but to limit the import surplus to the total amount of foreign capital available – a limit which it tends continuously to overstrip unless inflationary forces are kept under control. The mechanism of transfer is also related to the forms of investment. Direct investment, insofar as it involves direct shipment of capital equipment from abroad, bypasses the foreign exchange market, whereas local expenditure associated with it, and also portfolio investment, involve in the first place a net addition to the foreign exchange resources of the country. The resultant credit-creating potential may be used in the imports of luxuries rather than necessities for the people engaged in the investment projects, as the purchasing power created by the banks may fall into the hands of the income group other than the workers employed. The inflationary impact of the introduction of foreign capital brings into focus the nature and role of supplementary savings. The utilization of foreign capital can be so inflationary as to lead to balance of payments difficulties if the expansion ratio which is defined by Polak (1943) as investment financed by foreign loans plus investment financed by banks.

Foreign loans exceed the maximum which is reached when the amount of investment, directly financed by foreign loans together with the investment financed by the banks on the basis of their augmented resources, is such that the available foreign exchange resources are sufficient to cover all the direct and indirect imports. The extent of the expansionary finance which the banks would be tempted to undertake would depend upon the resources left over after the initial expenditure on imports, associated directly with foreign financial investment, and their reserve ratio. It is likely that at least part of the indirect imports would be obtained only after a time-lag so that, for a period in the beginning, the reserves of the banks would be very large, and this may induce optimism and encourage expansionary policy. The individual banks have no means of foreseeing in quantitative terms the ultimate effect on the balance of payments and exchange reserves of the whole banking system, of their expansionary policy, which would depend on factors such as the character of the investment programme, the resource pattern of the economy, the marginal propensity to save and to import, etc.

They have neither the means nor the incentive as individual bank to take an overall view of the situation and to limit their loan policy accordingly. It would require continuous and efficient watch by the monetary authorities over the investment programme, their nature and effect on the balance of payments and the forecast of the necessary propensities on macro and micro levels on the part of the authorities.

Any investment programme has its domestic component, i.e., expenditure on local resources which can be financed by domestic savings, inflation or by foreign exchange provided by foreign borrowing. Even in the last case there would be a rise in price which would induce imports to be financed by borrowing. In a market economy relying on price and market mechanism large-scale diversion of resources into the expanding sectors would involve rise in prices. In underdeveloped economies a relatively large price rise is necessary to accomplish a given reallocation of resources in the face of institutional barriers' rigidities, influence of customs and traditions – all resulting in a smaller degree of responsiveness to price incentives. It is not always easy to distinguish between sectional and general price rise and the tendency towards sectional rise may become pervasive. The general lack of mobility acts on another level. Because of the high cost of transport there exist large price differentials between regions. Moreover, the margin between the landed cost of imports and the cost to the final consumer is usually very high because of high mark-ups of the monopolistically organized few large import firms and expensive middlemen-riddled system of distribution and marketing. The combination of these factors makes the supply of consumers' goods inelastic so that local price rises substantially before imports appear on the market on a sufficient scale. Such an inflationary rise associated with foreign capital would shift the distribution of income in favour of the wealthier classes and the shift may conceivably exceed the increase in real income of the more numerous lower income groups, if the rate of money income of the latter is relatively sticky, and may cause a deterioration rather than improvement in the standard of living of the latter group. Such an eventuality has led Polak to suggest the existence of a theoretically optimum rate of capital inflow: 'the point at which the current loss in real income of the low income group exceeds the discounted prospective increase in real income which may be expected from the new investment' (Polak, 1954, pp. 418–19).

The analysis of international investment vis-à-vis economic development has led to greater attention being paid to a disaggregative analysis of foreign capital inflow in terms of its different forms and also different fields of application. Conditions and criteria relating to their profitable utilization are different. The importance of this differentiation is derived from the fact that there is no smooth substitutability or free mobility between different forms and fields of investment in the sense that foreign capital which is not flowing in one form or field would flow naturally in another form or field. Portfolio investment as distinguished from direct investment presupposes the existence of institutions and agencies specializing in dealing in foreign bonds. The growing importance of corporate forms of international investment is probably the result of increasing accumulation of surplus or undistributed profits in the large corporations, their relative independence of financial accommodation from the banking system, and concentration in the hands of large corporations of technological leadership in the methods of production. The motivations and factors determining the flow of different types of capital inflow are different. The decision on portfolio investment is based on a calculation of return, i.e., the rate of interest versus risk as compared with alternative opportunities. The risks of investing it foreign bonds are greater because of the ignorance of the conditions abroad, which prevents correct evaluation of the creditworthiness of prospective borrowers, and because of uncertainties associated with exchange rate changes, and political factors. Direct investment is governed by profit opportunities and not by existing rate of interest on money loans in underdeveloped economies. The considerations affecting the calculus of profit in the case of direct investment are quite complex, encompassing business motives of all types. The direct and immediate profitability of a specific undertaking may not be considered separately but only in terms of its anticipated contribution to the overall profitability of the parent concern. Considerations such as the maintenance of sales volume or preservation of market abroad, the necessity of procuring from overseas essential raw materials or ingredients required by the domestic industries and the ease or difficulty of directing the operations of a corporate enterprise from a distance, are important. Branch factories consist essentially in branching out from a domestic nucleus and they would be discouraged in the fields in which such domestic nucleus of private corporations is absent. A creditor country usually undertakes direct investment in those fields in which it has developed specialized skill, technique and experience. A technique of production developed in and suited to the standardized mass market of an advanced economy fits ill with and acts, therefore, as a possible discouragement to investment in an underdeveloped economy where the market is less homogeneous and more fractionalized owing either to the absence of a wide distribution of medium-sized incomes or the persistence of specialized luxury consumption of the rich or regional or social differences in consumers' tastes. The profitability of investment depends on the adequacy of internal market and existence of social overheads. Tariffs can create demand, by shutting out imports, with advantage only where cost conditions eventually prove competitive. In a 'footloose' or market-oriented industry for which the country offers cost advantages capable of being developed in a short time, a temporary and small tariff wall may provide the marginal incentive to foreign investors.

The traditional concentration of foreign investment in the export sector has been sought to be explained greatly by the existence of large and expanding export market in the past century. The case or difficulty of transferring debt service payments may have been important. Though the replacement of imports may as well provide foreign exchange, the low level of imports in the early stages in an undeveloped economy, owing partly to the inadequate level of income and demand and partly to inadequate organization and facilities for foreign, trading and insufficient familiarity with foreign products, discourages foreign investment. The volume of imports, even when replaced, may not be large enough for domestic production on an efficient scale or for any large-scale financing of debt service. The inadequate development of well-diversified export products in the early stages adds to the anticipated risk. The foreign trade of a country, both exports and imports, develops only as a country develops. The safest course for foreign investment under such circumstances is to concentrate and develop the export sector. However, mere concentration of foreign investment in the export sector or in import replacement may not secure either a strong balance of payments position or a sufficiency of exchange earnings. In a developing economy attention has to be paid to other investments which are simultaneously taking place and they may either increase the demand for import or result in the domestic absorption of export. Moreover, a lack of price flexibility or an inflationary policy or a wage policy which disrupts the relation between productivity and wage rates in the export- and import-competing industries is likely to cause pressure on the balance of payments, even if foreign investment is earmarked for the export sector. Substitution may be technical and/or price substitution. Imports may be replaced not only by technically similar goods produced at home but also by technically dissimilar goods because of the possibility of price substitution. The inter-relations between demand and supply of different goods are not always direct and simple but roundabout in many cases. It is also relevant that in the long run, forces governing the existing export markets, i.e., demand and supply conditions, including changes in technology, may change and develop new exports or reduce the importance of existing ones. The pattern of foreign trade changes as the borrowing countries diversify their economies and advance on the road to industrialization. The servicing of foreign debt under such circumstances would be dependent upon a readjustment of the pattern of trade between the lending and borrowing countries. Moreover, under the system of multilateral trade the transfer of investment service payments from the debtor country is effected not necessarily by means of a favourable trade balance with the creditor country, but also by means of export earnings out of its trade with the third country. This provides the debtor country with a wider range of choice in adjusting the various items of its balance of payments – especially important when exports of the debtor country are widely distributed among a number of countries. It facilitates adjustment to the changing pattern of trade by distributing the impact of change over a wider range of countries.

While on the subject of debt service it is important to remember that while portfolio investment has a definite rate of amortization, direct investment has none. Repatriation of direct investment can take place by the sale of enterprise to the domestic investors, influenced by the changing conditions in the given industry, lack of confidence in the monetary and economic policy and situation of the country or political and social factors. The withdrawals via failure do not influence the balance of payments. The timing of the repayment of foreign investment, and hence the necessity of generating export surplus can be regulated by the manipulation of the rate of inflow of new capital in relation to the average realized rate of return on foreign investment.

From the long-run point of view, the capacity to repay and to service investment, however, depends not only upon an increase in foreign exchange earnings but crucially upon the generation of an increase in income by the application of foreign capital and the attainment of a higher level of savings. The increase in total income is not enough, as it may not imply an increase per capita income if the population is increasing. The increase in the level of saving has to be high enough to allow debt service payments consistently with the acceleration of the rate of domestic capital formation. To accomplish such an end, either the initial inflows of capital have to be sufficient to cause large increase in income and savings or the foreign investor has to wait until, through the successive ploughing back of the marginal savings, a high level of income and saving and investment is reached. A small and once for all inflow of capital giving rise to certain increase in income may well be eaten up by an increasing population or an increase in the existing standard of consumption leaving the debtor country unable to service the debt. The inflows of capital over a period of years have to be sustained and large enough to enable a sufficient rise in per capita income, out of which both the servicing of debt and a higher rate of domestic saving and investment can be attained. In a country where average savings ratio is already high and the gap between the existing savings and investment requirement is small, a once for all capital inflow can probably increase the average savings ratio sufficiently high enough to allow a self-sustaining process of growth.

References

Chenery, H.B., 'The Application of Investment Criteria', *The Quarterly Journal of Economics*, February 1953.

Polak, J.J., 'Balance of Payments Problems of Countries Reconstructing with the help of Foreign Loans', *The Quarterly Journal of Economics*, February 1943.

Polak, J.J., 'Conceptual Problems Involved in Projection of the International Sector of Gross National Product', in *Long Range Economic Projection Studies in Income and Wealth*, Vol. 16 (Princeton: Princeton University Press, 1954).

Chapter 27

International Economic Assistance in the 1970s – A Critique of Partners in Development

Introduction

Foreign economic aid is at the crossroads. There is an atmosphere of gloom and disenchantment surrounding international aid in both the developed and developing countries – more so in the for than in the latter. Doubts have grown in the developed countries, especially among the conservatives in these countries, as to the effectiveness of aid in promoting economic development, the wastes and inefficiency involved in the use of aid, the adequacy of self-help on the part of the recipient countries in husbanding and mobilizing their own resources for development and the dangers of getting involved, through extensive foreign-aid operations, in military or diplomatic conflicts. The waning of confidence on the part of the donors in the rationale of foreign aid has been accentuated by an increasing concern with their domestic problems as well as by the occurrence of armed conflicts among the poor, aid-recipient countries, strengthened by substantial defence expenditure that diverts resources away from development. The disenchantment on the part of the recipient countries is, on the other hand, associated with the inadequacy of aid, the stop/go nature of its flow in many cases, and the intrusion of non-economic considerations governing the allocation of aid amongst the recipient countries. There is a reaction in the developing countries against the dependence, political and economic, which heavy reliance on foreign aid generates. The threat of the increasing burden of debt-service charge haunts the developing world and brings them back to the donors for renewed assistance and/or debt rescheduling.

Foreign aid is under attack from the 'radical left' in both the developed and the developing world on the grounds that aid strengthens the hands of the political groups who are in power in the poor countries and who often happen

---

* First published in *The Pakistan Development Review*, 1956.

to be conservative governments composed of a small oligarchy. They inhibit necessary socioeconomic reforms and thwart efficient growth and create considerable inequities. It is alleged that aid serves as a link between the established vested interests of the rich countries and the reactionary elements in the poor countries, blocking progress in both towards a more acceptable and viable socioeconomic and political world order. Aid, according to this line of reasoning, should be curtailed or eliminated until the necessary sociopolitical changes in attitudes and institutions take place in both the rich and the poor countries so that the poor all over the world can share the resources more equitably in the context of a just world order.

Standing between the anti-aid groups is the liberal who is neither convinced of the unchanging and ultimate depravity of human beings nor believes in violent revolution as a means of social reform and economic progress. The liberals want more aid – greater commitment of the international community towards the development of the poor; furthermore, they want an improvement in the terms of aid and a change in the pattern of aid towards greater flexibility in its composition with a view to increasing its effectiveness in development. They expect private capital in the advanced country to play its role in development; they expect national states to curb their sovereignty in the interests of the 'village community' – a state to which the twentieth-century development in communications has reduced the world.

It is in this context that the Pearson Commission was assigned to study the experience with international economic assistance during the past two decades and to map out a strategy of aid for the next decade. The Commission, in its Report *Partners in Development* (1969), strives to lead the middle group – the liberals – and to strengthen its voice in the world community which today, more than ever before, has at its disposal resources and technology to relieve poverty and distress in the underdeveloped two-thirds of the world. The Commission's recommendations are very wide-ranging and cover almost all the ideas now familiar in the fields of world trade and investment. A necessary casualty of this approach is a certain lack of priority in their recommendations. Since its main aim is to make 'reasonable' and 'acceptable' recommendations, it often makes them weak; small improvements on all fronts rather than dramatic improvement in some may reflect a judgment on the part of the Commission of what is feasible. Only time can reveal the relevance or accuracy of this judgment.

The following discussion concentrates on only a few from amongst the wide range of issues raised by the Pearson Commission in international economic assistance. The points of view of both the donor and the recipient countries are considered, with an accent on the latter. We examine the rationale of public internationalism from the points of view of both sets of partners in development, the principles of allocation of aid, various methods of increasing the effectiveness of aid in promoting growth; we make, in addition, a few observations on foreign private investment which the rich countries, faced with a shortage of public aid funds, have increasingly tended to emphasize as an important engine of economic growth of the poor nations.

Rationale of Aid – Developing Countries' Point of View

Why should developing countries seek foreign aid? The now widely familiar two-gap analysis provides a rationale; i.e., developing countries in the uphill task of raising the level of investment suffer from deficiency of domestic savings, as well as from that of foreign exchange. Domestic savings, even if they are adequate for meeting the demands of the target investment rate, cannot always be converted into foreign exchange through either net import substitution or export expansion. An expansion of domestic output requires fixed inputs of foreign-imported capital equipment and, in many cases, imported raw materials; the exports which are predominantly agricultural and primary commodities cannot be expanded to meet the growing import needs. Foreign aid meets both deficiencies and in order to do so the amount of aid inflow needs to be equal to the more dominant of the two gaps so that tile target growth is achieved. Foreign aid thus contributes to growth by enabling the recipient country to overcome the critical bottlenecks, i.e., inadequacy of saving and of foreign exchange.

Two-gap analysis assumes that foreign aid is entirely used to supplement domestic savings or to augment foreign exchange availability for investment imports. Aid is expected not only to augment the supply of these two critical inputs necessary for development but also to finance the transfer of technology from abroad and the inflow of foreign managerial and administrative talents under various technical assistance programmes, increasing the efficiency of use of resources. Moreover, aid acts as a catalyst in persuading the recipient countries to improve the allocation of resources by the pursuit of sensible or efficient economic policies in the field of tax policy, interest policy or exchange rate policies, etc.

> Economic policy aimed at growth requires a certain boldness and willingness to experiment. The growth process is still mysterious, and no decision-maker can

> be absolutely certain that any particular change in policy will produce the effects planned nor can any expert or adviser so guarantee. Results will depend on many variables. Moreover, developing countries live so close to the margin of subsistence that even minor economic reverses have major political and economic consequences. A cushion of foreign resources makes it possible to pursue bolder policies and take steps to accelerate development that might otherwise not be possible. (Commission on International Development, 1969, p. 521)

What is the experience in the past? Has economic growth been positively correlated with foreign aid? Has foreign aid always supplemented domestic savings or foreign-exchange earnings for development? Has aid succeeded, most cases, in inducing efficient economic policies?

It is difficult to find strong and significant statistical correlation between growth rates and aid flows either on the basis of cross-country studies or on the basis of historical studies. This conclusion remains valid whether aid as a proportion of total imports or total GNP is related to the rate of growth of GNP (ibid., pp. 49–50; OECD, p. 183). Some of the reasons for the failure to obtain significant statistical correlation are obvious. A considerable part of aid in the past has been granted on non-economic considerations for military and political purposes, without any reference to the objectives of growth or the country's ability to utilize aid. It was also directed in many cases for promoting exports of the donor countries without particular reference to growth objectives in the recipient countries. The experience with aid for development is short and mistakes have been made in the past. As the Commission points out, its effectiveness is expected to increase (in the future) as a result of past experience with the use of aid in both the donor and the recipient countries.

To some extent aid is expected to be negatively associated with the rate of growth in a *cross-country analysis*. This may indeed be a measure of the success of aid effort. Countries which have attained a high rate of growth should be dispensing with aid in an increasing degree and should be becoming self-reliant. The higher the rate of growth a country achieves as a result of successful application of aid in the earlier period, the higher is the proportion of investment and imports financed by domestic savings and its own export earnings. Moreover, a large part of aid is directed to physical infrastructure, investment which takes place now, while the effects on growth are felt later. Similarly, aid' which is devoted at present to social overheads, such as educational and training facilities and the development of human resources, will have an impact on, growth in the years subsequent to the period of high aid flow.

The factors affecting economic growth are complex and they are often interrelated. A simple statistical correlation between aid and growth is unlikely to be able to distinguish the effects of aid unless all the factors are explicitly included and their effects eliminated. After all, aid has contributed in the past no more than 15–20 per cent of total resources for development and the variation in the domestic resources as well as various policies relating to the composition of investment and efficient utilization of resources are the more crucial determinants of economic growth. The fluctuations in these latter factors are often more pronounced than those in the volume of aid. Economic growth has, however, been found to be correlated with import capacity which is the combined total of export earnings and foreign aid. A few studies which have distinguished between the effect of export earnings and that of foreign aid have found positive correlation between aid and economic growth in a cross-country analysis. The relative contribution of these two factors differs between the different studies; in one study an extra dollar of foreign aid contributes less to the growth of GNP[1] than an extra dollar of exports and in another study the reverse is true (Cohen, 1968; Hagen et al., 1969).[2] The contribution of foreign aid is positive in both cases but is very significant in one and not so significant in the other. Similarly, studies which have used earlier explanatory variables such as an index of natural resources (in terms of ratio of primary exports to GNP) have found a positive correlation between growth and aid (Papanek and Jakubiak, 1970, 11, pp. 10–12).

While analysing the effect of aid on growth, the implicit assumption has been that aid is devoted entirely to investment. This is an unduly restrictive assumption. Aid in the first place supplements GNP and is a component of total resources available to a country for distribution between investment, consumption and, for that matter, augmentation of the foreign-exchange reserves of the recipient country. There is no *a priori* reason why aid should be entirely devoted to investment. The choice is a function of time preference of the individual country between present and future consumption at the margin; of the marginal returns on investment of foreign-aid resources, the interest cost of borrowing (excepting in the case of grant) and the time pattern of repayments. If the objective of economic policy is not only to maximize growth of income within a definite planning horizon but also to attain a certain increase in the rate of consumption within the planning period, it is optimum and rational policy for the aid-recipient country to devote a part of aid to consumption.

In fact, unless one assumes that output maximization is the only objective of policy, there is no rationale for exclusive use of aid for investment purpose. Similarly, capital coefficients of output consist of domestic and foreign-capital goods, which are not substitutable, then as soon as the supply of domestic-capital goods is exhausted, the investment effect of foreign aid will be zero and the entire aid has to be devoted to consumption if it is to be absorbed at all. The macroeconomic models of Chenery and Bruno (1962) and Chenery and Strout (1966), etc., which conceive the role of foreign aid entirely in terms of raising investment, either through supplementing domestic-resource component (savings) or foreign-exchange component (raw materials and capital goods necessary for investment), allow the use of aid, for consumption purposes only in the limiting case when, with a rising level of investment, all the labour is fully employed. Under these circumstances, the shortage of labour puts an absolute limit to investment and additional aid is entirely consumed. A statistical study of a large number of countries on the basis of historical time series reveals that average propensity to allocate aid to consumption and to investment is 0.45 for both (Arskoug, 1969, p. 76), while another cross-sectional study, finds propensity to consume out of aid is 0.24 (Rahman, 1968a).

Once the possibility of the use of foreign aid for both investment and consumption expenditures is admitted, it is obvious that, given the identity of total resource use and resource availability, i.e., $Y + F = C + I$, where Y is GNP, F is foreign aid and C and I are aggregate consumption and investment expenditures, foreign aid is inversely related with saving. As GNP in the period in which aid inflow takes place remains unchanged, an addition to consumption out of foreign aid in the same period reduces saving because saving defined in the national accounting framework is $(Y - C)$. The proportion of GNP saved, i.e., $(Y - C)/Y$, also declines from the previous level, i.e., as compared with no-aid situation. The distribution of aid between consumption and investment uses, as already pointed out, is independent of the commodity composition of aid-financed imports so long as the composition of domestic output is flexible and resources are shiftable between the consumption and investment goods producing sectors in the economy. Various alternative situations are conceivable in terms of the effects of aid on domestic saving; it is to be noted, however, that it is never suggested that aid may, in fact, reduce the domestic investment below the level realized in the pre-aid situation. In the case of private foreign investment, it has often been alleged that the foreign enterprise with a greater command over resources and technology may discourage domestic private investment, if the infant domestic enterprise is unable to survive the competitive struggle. In the case of foreign public aid, the worst possibility is that the entire aid is devoted to consumption in which case domestic investment remains unchanged, but savings fall; if domestic consumption increases by more than the amount of pre aid investment, there will result negative saving. It is conceivable that investment increases by more than foreign aid if the latter is entirely or even partially devoted to investment and if at the same time the domestic savings are stimulated by the inflow of foreign resources. If the savings in the pre-aid situation are frustrated owing to dearth of the foreign-exchange component of investment, foreign aid, by increasing the supply of imported inputs, can increase the investment beyond the level of pre-aid investment by more than the amount of foreign aid.

An increase in consumption facilitated by flow of aid has undesirable implications for future growth not only because investment is less than it otherwise would have been if aid were entirely devoted to investment but also because a high level of consumption sustained by foreign aid may bring about changes in consumption habits. In other words, the upward shift of the consumption function may not be reversible when foreign aid, depending upon the exigencies of the policies followed by the donor countries, falls. Moreover, it is conceivable that the slope of the consumption function changes as a result or aid-financed additional consumption so that the marginal saving rate falls.

The moral as well as the rationale of the preceding discussion is that savings are a function not only of income but also of total resources at the disposal of a country and these consist of domestic product and foreign aid. It is not suggested that saving (redefined as income plus foreign aid minus consumption) declines consequent on the aid inflow. It may decline in the unlikely event that consumption increases by more than the amount of foreign aid. Similarly, propensity to save as a proportion of total resources (GNP and foreign aid) may fall if the propensity to save out of the additional resources of foreign aid is lower than the propensity to save out of income in the pre-aid situation. Thus, we may conclude that while economic growth is a complex phenomenon and a function of a large number of factors and while the statistical correlation between growth and aid is not self-evident, partly due to the difficulties inherent in a statistical exercise of the type mentioned earlier, the contribution of aid to growth is recognized. The extent of such contribution in any actual case depends upon the circumstances of each country, i.e., how it uses additional resources put at its disposal. The crucial factor is whether the country receiving aid pursues policies conducive to generating forces of growth. Availability of additional resources through aid is a necessary but not sufficient condition for generating growth. The fact that a certain amount of aid is consumed may, even have a beneficial effect on development if aid is spent on human resource development or on straightforward consumption of the poorer sections of the people, which either provides incentives for greater productive effort or increases sheer physical ability for harder work under circumstances where consumption and nutritional standards are very low (OECD, 1969).

The Commission recognizes that employment and distribution of income are important targets of development during the decade of the 1970s. A high rate of growth generates greater resources for distribution and provides greater flexibility for meeting demands of social justice. But a high growth rate is not a sufficient condition. Positive policy measures are necessary, within the context of a higher growth rate, to expand employment and to achieve a more equitable distribution of income. It is now increasingly suggested that conflicts between growth and distributive equity have been exaggerated. In a developing world it is necessary, in the interest of growth, to open up maximum avenues and sources of initiative and enterprise. Even when it is true that the rich save a larger proportion of income that the poor, it is not at all clear that the degree of inequality currently prevailing in many developing countries is a necessary price to pay or that a wider diffusion of investment opportunities and of productive employment will not, in the aggregate, increase the total savings of the economy. Furthermore, employment is an objective by itself. It satisfies human need for useful and creative activity and for participation in the tasks and benefits of development.

Development targets in the 1970s should include the objectives of employment and income distribution in specific, quantitative terms. For example, there is a growing consensus that an improvement in the standard of living of the lowest 25 per cent of the population should be a specific development target; similarly, specific sectors and projects can and should be identified for the purpose of providing employment. A greater emphasis on distributive justice as a necessary goal of development in the next decade will provide a partial, if not complete, answer to the critics of foreign aid who are on the left of the political spectrum.

How much aid should the poor countries, such as Pakistan, seek? The Commission suggests that if the developing world succeeds in growing at the minimum rate of 6 per cent per annum in the years to come, say until the turn of the century, they would be able to achieve self-sustaining growth. What does self-sustaining growth in this context mean? There are various ways of looking at it. One, the developing countries will receive no net inflow of foreign resources; there will be gross capital inflow which would just offset the outflow in terms of repayments and interest or dividend payments. This gross inflow will be at ordinary commercial terms at which foreign funds, equity or loan capital can be borrowed in the international capital market. Second, while the developing countries will not receive any concessional public aid, they will receive a net inflow, i.e., the gross inflow received on commercial terms will exceed the gross outflow, but then this net inflow should enable the developing world to grow at more than 6 per cent, say, after 2000 AD. In other words, self-sustaining growth in both cases implies that the developing world will be able to grow at 6 per cent without any net inflow of resources from abroad at concessional terms. They should be able, in the Commission's view, to attain a marginal rate of saving of 22–25 per cent to increase exports at a rate faster than income and to achieve a rate of investment of about 20 per cent by the end of the century which, thereafter, will be sustained with domestic savings. In both cases, moreover, they would continue to receive capital inflows at commercial terms, which would necessarily be mostly private foreign investment, in order to service the outstanding debt at the end of the century. The larger the outstanding debt, the larger is the need for private foreign capital to meet the debt-service payments and to grow at the same time at 6 per cent per annum. In order to keep the debt-service burden manageable for the developing world, the Commission recommends 40-year loans at 2 per cent interest with seven to 10 years' grace period. These are terms which are substantially more concessional than those prevailing today.

From the foregoing discussion, it appears that the amount of aid which a country should seek in the next few decades is closely linked with the amount of non-concessional capital inflow and private investment which she considers acceptable on sociopolitical grounds at the turn of the century or whatever is the length of the perspective-planning period. There is an underlying presumption that the greater the dependence on foreign capital, private and public, the greater is the sacrifice involved in terms of political and economic independence. International aid, as is explained later, is unlikely to be divorced entirely from a certain loss of political manoeuvrability or freedom in the internal and/or external affairs of a country even though no overt strings are attached. Similarly, it is also felt by the developing countries that domestic economic policies suffer from a loss of flexibility if a very significant proportion of domestic capital stock is owned by foreign direct investment.

Moreover, there are good reasons to suggest that the dependence on foreign aid for economic development should not be so heavy that if, for whatever reason, aid is cut off or its flow interrupted, the recipient country would be unable to sustain a minimum rate of growth on her own resources; this holds true during the intervening period, i.e., the period during which a poor country has not attained a self-sustaining growth in the sense in which the Commission defines it. The minimum desirable rate of growth is to be decided by each society by its own choice; it will probably in all cases have to be higher than the rate of growth of population. The rate and pattern of domestic resource mobilization, as well as of import substitution and export expansion, should be such that this critical minimum rate of growth must be sustained, if necessary, without aid. Even if a very poor country, to begin with, may not be in this position, it should be the aim of development policy to reach this stage within the shortest possible time (Rahman, 1968b). It is not that a developing country should, in fact, grow at this minimum rate and plan to repay all debts at the end of the perspective-plan period but that she should plan to repay the outstanding debt at the end of the period to the extent which is consistent with the target of politically acceptable flow of non-concessional aid and private foreign investment. She can indeed decide to develop with a large inflow of commercial capital and generate a higher growth rate, before or after the perspective-plan period, provided she stands ready to dispense with foreign aid of any kind, at any moment, and still continue to grow at the minimum acceptable rate. Such a policy would minimize the dislocation in the domestic economy in the event of the interruption of aid flow, which is subject to external forces beyond the control of the recipient countries.

Rationale of Aid - Developed Country's Point of View

The crisis which confronts international aid today has its source not so much in the developing countries as in the 'depressed' aid climate in the donor countries. The Commission strives hard to allay pessimism about the growth prospects of the developing world. That the developing world has achieved in the last decade a growth rate of 5 per cent is no mean achievement; more so in view of the fact that they had relied mainly on their own resources, foreign aid supplying only 15 to 20 per cent of the total investment resources. Even if aid helps the growth of poor nations, the critical question asked today is: why must the rich nations aid the poor in their development effort? That poverty anywhere is a threat to prosperity everywhere, especially in a shrinking world linked by instant communications, does not appear so real and crucial to the immediate self-interest of the rich nations. Admittedly, the growth of the poor nations helps widen the market for the rich and exploits resources for use by all, both the rich and the poor nations. That violence and instability are positively correlated with poverty and, therefore, that aid contributes to international peace is not convincing; revolutions often catch on when positive improvements in the levels of living take place and expectations and hopes are raised, since grinding poverty more often than not generates apathy and inertia rather than instability.

The ultimate sanction of foreign aid in the opinion of the Commission is the moral obligation of the rich to help the poor and for the strong to aid the weak. Whether this motive can sustain continued aid for many years and for many countries is doubtful. It appears on the basis of any realistic appraisal that the rationale of foreign aid in the coming years will be derived from a mixed bag or political, moral and economic considerations. It is not necessary to prove that aid can buy allies or stave off potential enemies or that it really ensures peace. If the politicians and peacemakers in the rich nations come to believe in the political role of aid then aid will flow at least partly from considerations of political objectives. The poor countries have to live with this rationale in the foreseeable future.

The Commission recommends a target for the supply of aid by the rich nations, viz., 1 per cent of GNP by 1975, including both public aid (0.7) and private investment (0.3). There is no particular economic argument for 1 per cent of GNP, which has no reference to the per capita incomes or the state of balance of payments of the individual donor countries and which may, therefore, involve inequity in international sharing of the burden of foreign aid. However, the absolute magnitude of the burden is so small that the Commission neglects considerations of equity in burden sharing. It is attainable in spite of the competing claims on resources in the domestic economies of the developed countries. The Commission considers this target of aid consistent with the various estimates of aid requirements of the developing countries, based on 6 per cent growth rate. The fixation of a target and its acceptance by the donor countries is expected to provide a focal point for mobilizing public opinion and political support for aid in the developed countries by serving as a symbol of the moral pressure of the world community on the rich and the affluent nations.

How to Allocate Aid?

The studies undertaken by OECD Development Assistance Committee (1969) show that in the past there has been a marked tendency for each recipient country to receive a mininium amount regardless of its size, plus a certain amount related to the size of its population. Aid as a percentage of GNP is not related to per capita income. However, aid receipts as a share of imports of the recipient countries appear to be strongly related to a combination of per capita income and some index of import needs (ibid., pp. 157–87). More illuminating is a detailed study of the allocation of foreign aid by different international organizations and individual countries (Levitt, 1968). The distribution of aid by the UN agencies seems to have been governed by random factors – not particularly related to relative degrees of poverty or considerations of absorptive capacity or the rates of growth. However, this does not seem very irrational, given the fact that the largest portion of the UN aid is technical assistance and pre-investment surveys, etc., and given the diverse pressures exercised on a world organization for a widest possible distribution of aid among the member countries. The inter-country distribution of IBRD loan seems to be positively related with population and per capita income; very small and poor countries seem to receive no aid from the IBRD. The per capita income, as a significant explanatory variable, derives its rationale from the fact that a higher-income country has a greater ability effectively to use as well as repay and service the debt. The US bilateral loans seem to follow the same pattern, i.e., positive correlation with GNP and population, excepting that two additional factors emerge as important explanatory variables, viz., non-economic relationship with the recipient countries, such as strategic considerations as exemplified by the amount of military assistance received by the recipient country, as well as the amount of foreign-exchange reserves. The later phenomenon indicates the importance of non-project aid for balance of payments support which often figures in bilateral aid operations. The most significant variable explaining the inter-country allocation of US government grants, as against loans, is the strategic consideration as indicated by the amount of military assistance received by the country concerned.

The Commission recommends an increase in the proportion of multilateral aid from the present figure of 10 per cent to 20 per cent of total aid. The suggestion that multilateral assistance involves no political considerations is not entirely true to the extent that the international organizations are dominated by powerful rich member countries. One can argue that a poor country receiving aid on a bilateral basis from a number of rich countries can partly neutralize the political influence of one country against that of the other.

Multilateralization of aid through the international agencies, on which the big donors are not able to exercise a dominating influence, may not be highly attractive to them so that with an increase in the share of the multilateral aid, total aid may suffer; the pressure for political control of the international agencies by the big donors may also increase correspondingly. So long as the use of foreign aid for non-economic purposes is not entirely eschewed, an unlikely possibility at the present state of international relations, the dilemma is not easy to resolve.

The Commission recognizes the problem and urgently pleads for partnership between the donor and the recipient countries in a multilateral institutional framework like consortium and consultative groups, etc., in which they would mutually monitor each others' policies and performance, formulate the criteria of performance and evaluate the performance of recipient countries on the basis of periodic reviews. In this multilateral framework of mutual consultation, while the recipient countries will consult the donor countries on the issues of economic policy and undertake to meet agreed standards of economic performance and efficient use of aid, the donor countries, in turn, will undertake to ensure an adequate and steady flow of aid and allocate aid according to the criteria of economic performance. It is necessary to remember that the aid relationship between the poor and the rich nations is basically unequal; to expect anything else is unrealistic. The equal opportunities of participation in the multilateral forums for aid negotiations do not alter the basic fact of equality of bargaining power; the mutual monitoring of policies of the debtor and the creditor countries, however desirable, can seldom extend to a critical analysis of the policies of the donor countries with the same effectiveness and ease as is usual in the case of the poor, debtor countries.

The criteria of performance which have been suggested by the Commission for judging successful economic performance are improvements in two specific ratios, i.e., the savings/income ratio and the ratio of exports to income. This is on the assumption that a rise in the marginal saving rates or in the export income ratios will help achieve self-sustaining growth. The reliance on these two criteria can only be partial, partly because of the statistical difficulties of measurement and partly because of their vulnerability to the influence of external factors. This implies that necessarily one has to go beyond these critical ratios to the particular economic circumstances, internal and external, confronting the recipient countries as well as to an evaluation of the whole gamut of economic policies.

One can think of at least four sets of considerations which should ideally govern the allocation of aid. They are: a) needs of developing countries; b) domestic development efforts including social, economic and institutional reforms and systematic effort in the implementation of necessary measures; c) performance in making productive use of external assistance, i.e., effective application in areas where gains are the greatest; and d) resources and potentialities of the countries themselves, so that resources are used where opportunities for exploiting known resources exist. The criterion d) is more relevant to the flow of private international investment; the criterion a) really is based on the principle of international solidarity and equity rather than the criterion of efficiency. An appropriate combination of the criteria of efficiency and equity is a matter of value judgment. No foolproof mechanism in terms of simple criteria, which the Commission started out to devise, is available.

Increasing Effectiveness of Aid

There are certain aspects of the current aid flow which vitally affect the developmental impact of aid as well as the real value of net aid which poor countries receive. Three issues in this general area are of central importance; composition of aid in terms of programme and project aid, tying of aid to sources of supply in the donor countries and terms of aid, i.e., rate of interest, maturity and grace period, etc., which govern the debt-service payments.

The Pearson Commission strongly endorses the demand, persistently made by the debtor countries, that for enabling them to absorb more aid effectively and to overcome the crucial bottlenecks on the domestic front, the donor countries should increase the flow of programme loans and move away from the overwhelming reliance on project loans as at present. This is particularly relevant where, as in Pakistan, the relatively higher proportion of project aid in the past has resulted in considerable production capacity, a large part of which remains unutilized for lack of imported raw materials and spare parts. A greater availability of programme loans or commodity assistance will accelerate growth with little or no addition to capital stock. The financing of the local costs of development programme, where necessary, is also recommended. If aid is to fill the gap in domestic savings, the financing of local expenditures, as is found by experience in many African countries, is an essential form of capital assistance. Moreover, the financing of local expenditure assumes critical importance with the increasing emphasis on investment in population control, education, research, health and housing, etc., all of which have a relatively small foreign-exchange component. This implies that foreign resource inflow takes place in terms of wage goods and/or raw materials for the production of wage goods.

Insofar as the fungibility of aid and domestic resources is now widely admitted and insofar as the efficient use of aid is to be basically a function of overall economic policies and of the pattern of resource use, the insistence on project aid loses its strongest rationale. While, on the one hand, project aid is intended to stimulate self-help for meeting the local expenditure of the aided project, on the other hand, it may, as well, direct domestic resources away from the more productive uses, just because foreign aid is available for a project which may be of low priority. One would have expected the Commission to provide some empirical evidence of how the efficient use of aid as well as the process of development is affected by the present aid policy of not financing local costs and insistence on project assistance.

On the question of tied aid, the Commission makes no radical suggestions. It recommends institutional arrangements for ensuring greater competition in the developed countries when bidding for contracts; it suggests the untying of aid funds in respect of their expenditure in the developing countries. The predominance of tied aid is an important instance of inequality in the aid relationship, not only because tied aid is to be repaid in untied foreign-exchange earnings of the debtor country, but also because the balance of payments difficulties, which is the major justification put up by the donor countries for tying aid, arise not from any actions or policies of the poor countries but from those of the rich countries in their mutual trade relations. The rich countries follow restrictive practices in their mutual economic relationships; the surplus, developed countries neither inflate nor adjust the exchange rate and liberalize trade but instead prefer to accumulate foreign-exchange reserves.

One of the critical problems facing the recipient countries is the mounting debt-service burden for many of them. In countries like Pakistan, the debt-service charges constitute about 20 per cent of the foreign-exchange earnings. It is often suggested that the debtor countries should eschew further borrowing, when the debt-service charges impose such a heavy and inflexible burden on its foreign-exchange resources. Whether a particular ratio or debt-service is critical or not depends partly on the scope for export expansion and import substitution, on the one hand, and on the prospects of aid in the future, on the other.

Two questions regarding the debt-service burden of the developing countries need to be re-examined. Firstly, what is the rationale of repayment of loans by the rich to the poor nations? After all, US assistance to the Marshall Plan countries (the second richest and developed group of countries in the world) after the Second World War was in the form of grants to the extent of 80 per cent of the total aid. Viewed in this light, the case for grants to the poorest countries of the world today is much stronger. Admittedly, the development assistance to the Western European countries was a one-time, short-term operation, lasting for about four years only. In the case of poor countries, it has turned out to be a long-term continuing commitment spreading over several decades. Undoubtedly, there is a grant element, though of a declining magnitude, in the loans which are extended today. But it is nonetheless not clear why, if the rich countries are to transfer resources for the development of the poor, the resources should flow back (at least in the foreseeable future); why the poor should not be encouraged instead to reinvest the entire resources generated out of the investment of the external aid.

One economic argument in favour of repayment and interest charges is that it ensures an efficient use of external aid for economic development. The obligation for debt-service payments does not necessarily ensure efficient use of aid; the efficient use of loan funds can be associated with an overall inefficiency in the use of the domestic resources which anyway finance the bulk of total investment. However, one may argue that if the entire aid is to be given in the form of grants, it may affect the quantum and flexibility of total resource flow to the developing world as a whole. For one thing, this may reduce the aggregate amount of capital flow to the developing countries. There is the usual resistance of the parliaments and Congress to 'give away'. Moreover, with a given total quantum of aid which is limited in relation to the needs of the competing and increasing number of claimants for foreign assistance, the repayment of loans by a given recipient country enables the donor countries to redistribute resources over time among the other poor nations in response to their needs and the productivity of investment, which change over time as between the different countries. Repayments create, so to speak, a revolving fund and enable a periodic reassessment and consequent reallocation, if necessary, of the external resources to the individual recipient countries.

This line of reasoning indicates that it is not efficient to extend all the economic assistance in the form of grants, especially since economic aid is expected to be a long-term phenomenon. However, it does emphasize the need for a much larger share of outright grants than at present, in the total economic assistance, especially for the poorer ones among the developing countries. Secondly, it is necessary to recognize that so long as a developing country requires a net inflow of resources, new loans have to be contracted in order to repay the old debts. This has been the historical experience with many of the developed countries in the past. Faced with heavy debt-service charges many debtor countries have requested the rescheduling of debt by the donor countries. The system of rescheduling, which needs to be repeated every few years, is inefficient and creates uncertainty; it further brings a bad name to the debtor country. The opportunity which the rescheduling operations offer for a mutual consultation on and reappraisal of economic policies of the debtor countries is only of marginal value and should be available anyway under the system of periodic reviews recommended by the Commission. What is necessary is to ensure that the need for debt rescheduling does not arise in the normal course of events – a possibility which can only be ensured if development assistance is put on a stable long-run basis not only in terms of its maturity, but also in terms of the commitment on the part of the donor countries to assist the poor countries on a long-term basis.

The foreign exchange difficulties which the debt-service burden gives rise to are partly due to the fact that, while the new loans are available mostly in terms of tied aid, which increase the cost of repayments, the repayments and interest payments are required in terms of multilaterally convertible foreign exchange. One way of overcoming the asymmetry between the form of aid and the method of its repayment is for the donor countries to accept repayment in terms of the local currencies of the debtor countries; this method of repayment is familiar in the case of aid granted by the socialist countries. Alternatively, debt-service payments of a particular recipient country may be available for expenditure in the other developing countries but not in the developed countries. This will have the additional advantage of promoting trade between the developing countries as a whole. The tying of repayments from the individual debtor countries to the exports of the developing world as a whole undoubtedly violates the principle of comparative advantage in the same way that tying of aid does in the first place. This is admittedly a suboptimal, second-best solution. Certainly, the untying of aid from the developed countries is the optimum solution from the point of view of the efficient use of world resources.

Private Foreign Investment

The above discussion relates to the quantum and allocation of capital inflow channelled through the governments of the developed countries. Private foreign investment plays a critical role in the Commission's recommendations on the ways and means of increasing international financial resources for development. Private investment is expected to contribute 0.3 per cent of the GNP of the rich countries by 1975, or even more than that if the public-aid target of 0.7 per cent of GNP is not reached by that time. Private foreign investment, the Commission hopes, not only helps fill the capital or savings gap but also brings along with it scarce foreign exchange as well as certain essential missing factors of production like technology, results of research and development, management, entrepreneurship and skill.

Do they in fact do so? The Commission admits that they did not undertake any detailed empirical analysis but they believe that their positive judgment on the substantial contribution of foreign direct investment to the development of poor countries is right. The Commission recommends very strongly the positive role of private investment for yet another reason, i.e., 'those who risk their own money may be expected to be particularly interested in its efficient use', implying that private investment is more productively used than aid administered and granted by the government agencies in the developed countries.

The contribution of private foreign investment can be judged with reference to its impact, firstly, on the growth of income and output and, secondly, on the balance of payments. That the contribution of foreign investment is often a subject of considerable controversy is partly due to the fact that while its immediate impact is easily identified, there has seldom been a systematic analysis of its indirect costs and benefits. The indirect effects on the balance of payments and growth, for example, involve, among other things, two sets of questions: a) what is the alternative use of the domestic resources which are employed in foreign enterprises and what is the impact on the balance of payments of the alternative use of such resources; and b) does foreign investment make a net addition to the total pool of investible resources in a country or does it replace or discourage domestic saving or investment? Does the foreign enterprise adopt different technology, in terms of relative inputs of capital and foreign exchange, than domestic enterprise? Is technology employed by foreign enterprises appropriate for the recipient countries, given its relative factor endowments? What effects does it have on the evolution of technology in the domestically-owned sector of the economy? Questions have been raised as to whether the multinational corporations, with their worldwide production facilities and inter-firm trade in final output and components, may not, in fact, restrict exports from or encourage imports of components and raw materials into one or more developing countries.

The increasing interest in foreign direct investment in recent years is partly a response to the relative inadequacy of public aid for development. If that is so, the high cost of foreign capital and entrepreneurship is a necessary supply price against which the recipient country is to set its own demand price for such capital and entrepreneurship. The difficulty which the developing countries face does not relate only to the availability of capital but also to the availability of technology from alternative sources. The knowledge of the technical processes of production is increasingly the exclusive preserve of the giant corporations of the developed countries which spend large amounts on research and development. In this case, the high price of foreign direct investment represents partly the scarcity price for the supply of technology.

The attempt by the poor countries to develop their own technology involves costs which, in many cases, are very high and may indeed be infinite, considering the underdeveloped state of science and technology in the majority of poor countries.

It is in this context that the Commission's recommendation for an increase in the share of public expenditure in the developed countries on research and technology which is directed towards meeting the needs of the poor countries assumes critical importance. The international agencies, it suggests, should develop research centres in the developing countries, both on a regional and on a national basis for the development of science and technology suited to the needs of poor countries. The international corporations seldom find it worthwhile to devote their research and development expenditure to evolve appropriate technology for the poor countries since their operations in the labour-abundant poor countries often constitute a small fraction of their total operations all the world over.

The developing countries faced with a situation in which the total gross inflow of the aggregate equity capital falls short of the gross outflow of foreign repatriations and remittances, frequently threaten to nationalize the foreign enterprises or restrict their outward remittance. There are two issues here. Firstly, there is the question whether the costs of a particular foreign direct investment are considered inordinately high, given the alternatives available for obtaining capital and technology from abroad. This has partly been discussed above. Moreover, profits may be very high, owing to the monopolistic conditions or the high rates of effective protection in the recipient country. In the latter case, the adverse effects may be sought to be mitigated by the taxation of monopoly profits or by a reduction of the effective rates of protection; alternatively, the recipient country may prevent foreign direct investment from these specific lines of activity, a recommendation which the Pearson Commission endorses. Apart from this, the mere fact that in a particular line of investment at a point of time, out-payments exceed the inflows of capital does not necessarily imply a high cost of foreign investment. For any borrower of equity or loan capital, the repayment or repatriation necessarily implies a net outflow, unless fresh borrowing is undertaken. However, there are specific instances where the outflow on account of a particular enterprise is high because there is no reinvestment of earnings in the host country due to the policies pursued either by the parent company abroad or by the host country.

Secondly, a problem which seems to worry many developing countries is that the total gross inflow of foreign investments at a particular point of time or over a period of time fails short or the total outflow on account of the dividends and the repatriation of capital. In a poor country, short of both savings and foreign exchange, the excess of the total outflows over the total inflows is obviously a cause for concern. Only a mature debtor country at a high level of income and foreign-exchange earnings is able to sustain such an outflow. In addition to the various factors discussed earlier, which increase the rate of return and consequently the magnitude of foreign remittances, this situation is partly due to the fact that the rate of growth of foreign-owned capital stock is lower than the rate of remittances. It is only an acceleration in the gross flow of foreign capital which can reverse the direction of the net flow. The threat of expropriation or restriction on remittances has often the unintended result of raising the expected rate of return necessary to attract the current and future inflow by way of an assurance against such risks. This further aggravates the problem of net outflow by raising the burden of remittances.

If there occurs an acceleration in the current gross inflow in order to reverse the net outflow, as the Pearson Commission strongly recommends, the proportion of the national capital stock owned by foreigners increases correspondingly, a situation which, for political and psycho-sociological reasons, is not often favoured by the poor countries. This is the familiar problem of 'alienation' of the capital stock of the host country; the resultant increase in the degree of dependence of a country on the interests and policies of foreign corporations may not be consistent with the social, economic and political objectives of the recipient country. The developing countries are indeed in a dilemma in this respect. The Commission notes with approval the restrictions placed by the host country on the amount of local borrowing undertaken by the foreign enterprises since this results in an extension of their control over the capital stock of the host country without the compensating benefits of a flow of capital from abroad. This measure will not contribute much to the resolution of the basic dilemma.

As one surveys the worldwide operations of the international corporations, one finds that a large number of poor countries are often faced with a small number of multinational corporations. The political implications of this unequal confrontation are aggravated because of the insistence by the foreign corporations on the rights of extraterritoriality. On the one hand, the developed countries insist on exercising their rights of control and regulation over their own citizens wherever they are, even if they operate within the boundaries of foreign sovereign states. On the other hand, the latter insist on exercising their sovereign rights on individuals and institutions operating within their frontiers, irrespective of their nationality. The confrontation between the multinational corporations and the recipient, poor nations often tends to turn into confrontations between the nation states themselves. Faced with this dilemma, the Commission hopes and pleads for restraint on the part of the developed countries. They 'should as far as possible keep aid policy and disputes concerning foreign investment separate'; an appropriate solution 'may be renunciation of the principle of extraterritoriality by all the member states of the United Nations and, for the long term, the eventual creation of a system of international incorporation of companies doing business in more than one country' (Commission on International Development, 1969, p. 107).

H.G. Johnson harbours more radical, one may say, utopian, hopes that since the concept of the nation states is not entirely consistent with the concept of the multinational corporations, the world community should eventually settle for a world government. In this case, however, the problems of poverty of the developing countries may lend itself to an easier solution through direct fiscal transfers, as within a national economy, combined with the politically neutral and economically beneficial operations of the giant corporations. To quote, this is

> ... an issue which should be seen in terms of a more general and fundamental problem of reconciling the traditional and anachronistic conception of an 'isolated state' with the economic and political facts of a rapidly economically integrating world economy in which there is an increasing mobility of goods, capital, labour (at least educated labour) and knowledge. In the long run, we shall have to become one world, politically as well as economically. (Johnson, 1970)

However, given the complex political and economic implications of private foreign investment in the context of a developing economy, it appears that a substantial reliance by developing economics, especially those which do not feel politically strong, on private investment as an engine of growth in the immediate future is unlikely. Historical experience indicates that it is often easier to benefit from the developmental impact of private foreign investment after: a) public investment, domestic and foreign, develops an adequate level of social and infrastructure; and b) after domestic private enterprise attains a reasonable size and competence, if necessary, with technical and managerial assistance from abroad. A viable and strong domestic private enterprise is better able not only to negotiate joint collaborative enterprises but also to withstand the competition of the foreign corporations in the domestic economy. It is worth serious consideration whether the international agencies cannot play a more active role in helping poor nations in bargaining and negotiating investment agreements with the multinational corporations.

Conclusion

The crisis of confidence in the role of foreign aid which led to the establishment of the Pearson Commission can be traced to a host of factors in both the developed and the developing countries. Undoubtedly, too much was expected of foreign aid by the donor countries and the political limits to economic policy-making in both the developed and developing countries were not clearly recognized. From the developed countries' point of view, the intermingling of political, moral and economic motives behind foreign aid can only be partially resolved in the years to come. Because of political considerations, aid often does not seem to promote socioeconomic reforms necessary for development. It is readily seen that aid cannot avoid strengthening the hands of those who are in power at the time of aid inflow; but if it does not discriminate between particular regimes in individual countries, in terms of quantum and terms of aid flow, then at least the forces of change within each country do not have to contend with the likelihood that pressure for a change within a country would bring in the aid donors on the side of the status quo. Basically, the donor countries must eschew, implicitly and explicitly, any preference for or against the sociopolitical forces within each country. To maintain such a position in practice is extremely difficult, especially if one remembers that aid has to be approved by the political processes of the parliament or the Congress in the donor countries; to meet this difficulty with honesty and ingenuity is a great challenge. It is, however, essential to separate the administration and allocation of economic aid, as far as possible, from that of aid for non-economic purposes, through strengthening multilateral institutions.

Experience of the last development decade has hopefully brought home the lesson that growth has to be combined with social justice and that targets of development must include not only rate of growth of GNP but also employment and equitable distribution of income. It will be a major task for the donors and recipients of aid alike in the next decade to devise ways and means of using objectives. If the aid-givers are to recognize and act upon the moral obligation of the rich to help the poor nations, then the privileged groups and regions in the aid-recipient countries must undertake, by the same token, to spread the benefits of development widely within the recipient country. This should be a part of the mutual obligation of the donors and recipients of aid.

To revive and indeed to strengthen the declining interest in aid in the donor countries would require systematic and concerted efforts on the part of the leaders of public opinion in these countries. For this, one needs more than a Pearson Commission; it is necessary to undertake a more direct persuasion of the political processes or forces in the rich countries. While some of the advanced countries have responded favourably to the Commission's target for aid, many others, including the United States, remain lukewarm to the concept of and commitment to a fixed target. As far as the allocation of aid between the developing countries is concerned, they will have to live with an unequal sharing of the benefits of aid as between themselves. Moreover, it should be acceptable to the developing world as a whole that a country with a per capita income, let its say, of US$ 300 or more, should not seek concessional aid and that a much larger proportion of grant, if it is available, should go to the poorest among the developing countries. One should probably add that a united approach, for that matter combined bargaining, on the part of all the developing countries is necessary not only for seeking concessions in trade from the developed countries, as is attempted to be done under the auspices of the UNCTAD, but also in the field of international assistance.

The Commission's discussion of the role of foreign private investment lacks critical analysis. A systematic analysis of the costs and benefits, direct and indirect, of foreign private investment from the recipient countries' point of view is urgently called for. Given the implications of inequality between the poor nations and the multinational corporations, as well as lack of adequate social and physical infrastructure in many countries, a major role for private foreign enterprise in economic development is unlikely in the near future. Indeed, one of the major tasks in the 1970s is to devise ways and means of ensuring a net gain from foreign private investment and of balancing it with an acceptable degree of non-economic or political cost.

As one scans the future of partnership in development, one is left with an uneasy feeling that in the fields of public aid, private investment and trade, the poor nations will remain unequal partners in the years to come; as domestic efforts for development are intensified and as the poor countries grow and achieve diversification of their economies, the inequality will hopefully diminish over time.

Notes

1 The regression coefficient of percentage change in income on foreign aid as a proportion of income varies between 0.13 and 0.10, whereas that on change in exports as a proportion of income varies between 0.21 and 0.62 (Cohen, 1968, pp. 283–4).

2 The regression coefficient relating change in income to aid as a proportion of GNP is 0.115 or 0.065 whereas that relating to exports as a proportion of GNP is .003 or .004 (Hagen et al., 1969, pp. 89–97).

References

Arskoug, K., *External Public Borrowing: Its role in economic Development* (New York: Praeger Publishers, 1969).

Chenery, H.B. and Bruno, M., 'Development Alternatives in an Open Economy. The Case of Israel', *Economic Journal*, March 1962.

Chenery, H.B. and Strout, A., Foreign Assistance and Economic Development', *American Economic Review*, September 1966.

Cohen, B.I., 'Relative Effects of Foreign Capital and Larger Exports on Economic Development', *Review of Economics and Statistics*, May 1968.

Commission on International Development, *Partners in Development*, Report of the Commission on International Development (Chairman: L.B. Pearson) (New York: Praeger Publishers, 1969).

Hagen, E.E. and Hawylyshyn, O., 'Analysis of World Income Growth 1955–1965', *Economic Development and Cultural Change*, Vol. I. No. 1, Part II, October 1969.

Johnson, H.G., 'The Multi-national Corporation as an Agency of Economic Development: Some exploratory observations', a paper prepared for the Columbia University Conference on International Economic Development, 15–21 February 1970.

Levitt, M.S., 'The Allocation of Economic Aid in Practice', *The Manchester School*, June 1968.

Lewis, W.A., *The Development Process*, Executive Briefing Paper 2 (New York: UN Center for Economic and Social Information, 1970).

Organization for Economic Cooperation and Development, *Assistance: 1969 Review* (Paris: OECD, December, 1969).

Papanek, G.F. and Jakubiak, S.C., 'Aid and Development', paper presented at the Dubrovnik Conference of the Development Advisory Service of Harvard University, June 1970.

Rahman, M.A., 'Foreign Capital and Domestic Savings: A test of Haavelmo's hypothesis with cross-country data', *Review of Economics and Statistics*, February 1968a.

Rahman, M.A., 'Perspective Planning and Self-Assured Growth: An approach to foreign capital from recipient's Point of View', *Pakistan Development Review*, Vol. VIII, No. 1, Spring 1968b.

Chapter 28

Foreign Assistance and Economic Development: the Case of Pakistan*

Among the developing countries Pakistan has been a significant recipient of foreign economic assistance. Moreover, starting with a low level of assistance in the early 1950s, the rate of flow of gross assistance increased considerably during the subsequent years, especially in the 1960s. During the 1960s, the high rate of flow of economic assistance was associated with a high rate of economic growth. However, towards the end of the second decade, Pakistan's economy developed stresses and conflicts of serious proportions which threatened not only economic progress but also political stability – indeed Pakistan's survival as one political entity. It is hoped that analysis of Pakistan's experience may provide lessons for the other developing countries.

Quantum and Composition of Economic Assistance

Excluding dependencies, annual average per capita external economic aid to Pakistan was quite high among the developing countries and was only marginally lower than the per capita aid to the independent countries which had special links with the developed, aid-giving countries. This is seen in Table 28.1. Moreover, her dependence on aid, measured as the percentage of aid to the total imports of goods and services, has been greater than all other groups.

Pakistan's dependence on foreign aid fluctuated over the years, as is seen in Table 28.2.

There has been a consistent increase in the ratio of external assistance to GNP throughout the period up to 1964–65.[1] Between 1964–65 and 1969–70, there was a fall in the ratio of assistance to GNP – quite a significant decline, but a small decline in the ratio of foreign assistance to investment and a small increase in the ratio of assistance to foreign payments. During the late 1950s and 1960s, there continued to be a high and growing dependence on aid as a

* First published in *The Economic Journal*, Vol. 82, March 1972.

Table 28.1 Foreign economic assis
69) per capita and as pe

Groups of countries
Oil-producing countries
Consortium and consultative groups
Independent countries with special links
Dependencies
Other developing countries
Pakistan

Source: OECD, 1970, p. 203.

Table 28.2 External economic assis

	Percentage of GNP
1949–50	1.3
1954–55	2.5
1959–60	2.8
1964–65	6.6
1969–70	3.8

Sources: Stern and Falcon, 1970, p. 11; Plan
of ratio of foreign aid to imports are
Vol. II, Table 57.

means of financing foreign payments, i.e., as a source of foreign exchange, whereas as a source of funds for financing domestic investment, there was a moderate decline in the late 1960s. This was partly a reflection of a change in the composition of investment, especially in the Third Plan, towards less import-intensive activities, including agriculture.

The increase in the inflow of foreign economic assistance into Pakistan was associated with the inflow of foreign military assistance in the mid-1950s. At this time, Pakistan entered into a military alliance with the United States, an alliance which was professedly directed against the threat of Communist aggression emanating from China and the Soviet Union, but which led to a sharp deterioration in her diplomatic relations with neighbouring countries. Up to 1957/58, military assistance constituted more than 37 per cent of the total foreign assistance to Pakistan (Mohammed, 1958, pp. 158–75). The scale of military assistance to Pakistan was geared not only to the security needs of Pakistan but also was related to the strategic requirements of the regional defence organizations such as CENTO and SEATO of which she was a member. The circumstances of Pakistan's emergence as an independent state and her hostile relations with India led Pakistan to accord top priority to defence expenditure over all other expenditures, including development expenditures (ibid.).

The slowing down of foreign economic assistance during the late 1960s was partly due to the outbreak of Indo-Pakistan hostilities in 1965, which led to the cessation of foreign aid during 1965–66. This was partly due to the fact that Pakistan slowly shifted away from the close military alliances of the 1950s, if not legally, at least in terms of her active association in CENTO and SEATO. She sought to diversify the sources of external economic assistance and to improve her bilateral relationships with the Communist bloc countries. During 1960–65, 3.6 per cent of her total external aid was received from the Communist bloc countries, and during 1965–70 their percentage share went up to 13.4 per cent, an almost fourfold increase (Stern and Falcon, 1970). Within the Communist bloc, the countries participating in aid programmes to Pakistan increased from two to five. The decline in the share of the Western bloc countries was accompanied by a greater diversification within the group itself. First, there was a rise in the share of the multilateral agencies. Secondly, among the bilateral donors, while the share of the USA and Germany declined from about 46 per cent and 11 per cent to 27 per cent and 8 per cent respectively, the share of the other Western countries increased.

As of June 1969, Pakistan received $5,684 million in loans and grants, including PL 480 aid from the United States government, from various national and international agencies, but excluding military assistance. Roughly 57 per cent of the total foreign economic assistance was in the form of loans and credits, about 21 per cent in grants and 22 per cent was in the form of PL 480 aid, which was a mixture of loans and grants. Excluding PL 480 aid, the project assistance constituted 44 per cent and the non-project, i.e., commodity assistance, constituted 34 per cent of the total assistance (Planning Commission, op. cit.). The composition of total aid as it stood in June 1969, given above, has changed over the years. The relative importance of PL 480 has declined over the years; this was a natural consequence of the acceleration in the rate of agricultural growth, starting in the Second Five Year Plan period. Pakistan reached the high water mark of non-project commodity assistance during the Second Plan period (1960–65). While project assistance increased by more than 60 per cent during the Third Five Year Plan (1965–70) as compared with the Second Plan period (1960–65), non-project assistance increased by barely 20 per cent. The changes in the composition of foreign economic assistance over the years are shown in Table 28.3.

The impact of foreign economic assistance on Pakistan's economic development and its evaluation raises a large number of issues. The following discussion limits itself to an examination of a few aspects of inter-relationship between economic assistance and development. While foreign assistance flowed in increasing volume over the years, did Pakistan raise the rate of mobilization of domestic resources for development? How did the rates of savings, investment and income growth respond to flow of aid? Did foreign assistance affect the pattern of utilization of resources and the pattern of investments? What were the effects on inter-regional and interpersonal income distribution? Did aid make a difference to the economic policies and institutions for development for better or for worse? These and related questions are examined in the following sections.

Savings, Investment and Economic Growth

Pakistan achieved a significant increase in the rate of saving and investment during the early 1960s compared with the 1950s. Associated with this was an acceleration in the rate of growth in the early 1960s, after a decade of comparative stagnation in the 1950s. What has been the relationship between the rate of savings and investment, on the one hand, and the rate of capital inflow, on the other? The changes over time of savings, investment and inflow of foreign resources are indicated in the following two graphs, first in terms

Table 28.3 Forms of assistance (million

	1st Pla[illegible] (1955–6[illegible]
Gross assistance	5,070
a External assistance	4,645
Project assistance	
Indus basin	2,779
Non-project assistance	
PL 480 food aid	1,618
Technical assistance	248
b Private foreign assistance	425

* Figures for the last year of the Second Five Year

Sources: *The Second Five Year Plan*, pp. 85–6; *T*

of actual amounts of flows of the three series and secondly in terms of three-year moving averages, which are intended to smooth out the short-term fluctuations. The data for years prior to 1951/52 are not available. Between 1951/52 and 1954/55, i.e., years which preceded the inauguration of economic planning in Pakistan, there were considerable annual fluctuations in the rates of both capital inflows and domestic savings and the relationship was mainly inverse. The improvement in saving rates from 1952/53 onwards was associated with a decline in the flow of foreign resources from the same period, as is evident from both Figures 28.1 and 28.2. The increase in saving rates towards the end of this period was basically the result of institutional improvement in the mobilization of domestic saving and for the organization of investment activity. The rate of investment was quite low in this period, varying between 4 and 5 per cent, and growth in income was inconsequential. Thus the improvement in the rate of saving was more a function of institutional changes rather than of growth of income or the rate of growth of income. At independence in 1947, Pakistan inherited very little of the financial infrastructure such as banking and financial institutions or of the administrative apparatus such as a tax collecting machinery or an entrepreneurial class for organizing private investment. These shortcomings were slowly being remedied by the mid-1950s.

In the following period, i.e., the period of the First Five Year Plan (1955/56–1959/60), while the rate of capital inflow registered an increase, both the rates of saving and investment recorded a decline. The rate of private saving could not be maintained at the level reached in 1954/55, in view of the slowing down of the rate of growth, accompanied by an actual fall in per capita income in East Pakistan. The financial savings by households, which increased from 1.2 per cent of GNP in 1949–50 to 3.2 per cent in 1954–55, declined to 2.9 per cent in 1959–60. The lack of any financial discipline in the public sector, by a government which was highly unstable and changed frequently, was reflected in the inability to raise taxes and to contain increases in non-development expenditure. This resulted in a very nominal public saving of 0.5 per cent of GNP during 1955–60. A large part of foreign assistance was used to import food to meet the domestic shortage. While the saving rate fell, the proportion of investment financed by foreign assistance rose. There was a strong presumption, therefore, that foreign aid in this period not only contributed to the financing of consumption expenditure but also substituted for domestic savings. To complicate matters and depress the saving rate was the additional external factor, i.e., a deterioration in the external terms of trade during this period, mainly due to a rise in import prices.

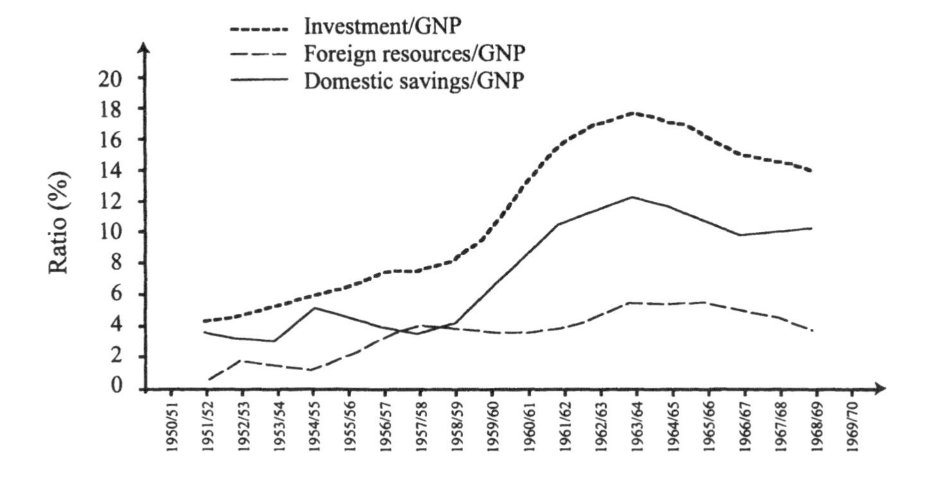

Figure 28.1

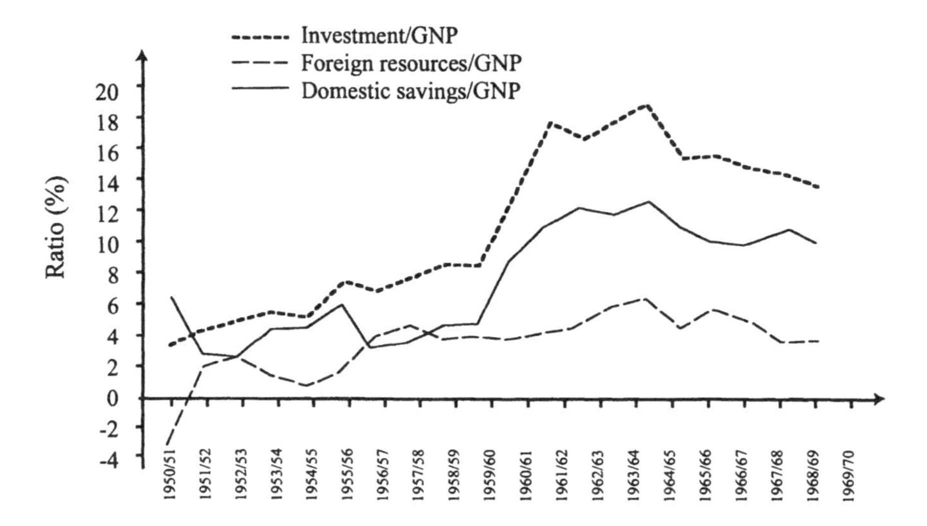

Figure 28.2

Thus it is difficult to establish any close one-way relationship during this period between the rates of domestic saving and of capital inflow. While during the first half of the 1950s there were years which witnessed an inverse relationship between the flow of savings and capital inflow, in the latter half of the 1950s the reverse was true. In spite of an acceleration in the rate of capital inflow in the late 1950s, compared to the early 1950s, the rate of domestic savings declined. The rate of increase in GNP and in per capita income was very small during this period. The rate of increase of GNP, 2.5 per cent, hardly matched the rate of increase in population. As noted, there were a number of independent factors which militated against a growth of saving, especially public saving. The absolute rate of capital inflow was quite small during the second half of the 1950s. But it was increasing at a constant rate at first, and then the rate of increase slowed down between 1956–57 and 1959/60. Though the slowing down of the rate of inflow slowed down the rate of investment, the crucial factor contributing to the stagnation in income was the absence of any growth in the agricultural sector. The latter was traceable in part to a set of policies, including controls on prices and distribution, which discouraged production.

During the five-year period starting 1960, Pakistan witnessed considerable acceleration in the rate of capital inflow as well as in the rate of domestic savings. Pakistan received the biggest 'push' in her history in terms of an increase in external economic assistance. The rate of capital inflow increased to 6.6 per cent of GNP in 1964/65 as compared with 2.8 per cent in 1959/60. Furthermore, there was an acceleration in the rate of increase in capital inflow starting 1961/62, as seen in Figure 28.2. It was not only that the rate of capital inflow during this period increased by about 50 per cent (Figure 28.2) – but that the absolute rate of inflow was very substantial in that it reached a very high figure of about 6.2 per cent of GNP in 1964–65 (Figure 28.1). The rate of investment increased by 70 per cent between 1959/60 and 1964/65; the rate of savings increased by about 90 per cent (Figure 28.2). The Second Five Year Plan also witnessed an acceleration in the rate of growth of income by more than twofold as compared with the First Plan period (1955–60). The rate of growth of income during this period was about 5.2 per cent. The increase in the rate of saving in this period was attributable to an increase in all the different forms of saving – more so in terms of an increase in corporate (including non-corporate non-financial saving) and public saving than in household financial saving. The household financial saving increased from 2.9 per cent to 3.7 per cent whereas the other components of saving combined together increased from .3.1 per cent to 8.0 per cent between 1959/60 and

1964/65. Public saving increased from 0.5 per cent during the First Plan period to 1.9 per cent during the Second Plan period. During the same period, Pakistan achieved an increase in tax income ratio. The ratio of tax revenues to GNP, which increased only marginally from 5.9 per cent in 1954–55 to 6.1 per cent in 1959–60, increased to 8.2 per cent in 1964–65 (*The Fourth Five Year Plan*, p. 47). This substantial increase in the rate of domestic saving has often been attributed to the acceleration in the rate of capital inflow (ibid., p. 9).

The increase in the rate of capital inflow is expected to increase the rate of domestic saving in two ways. First, it raises the rate of investment and hence the rate of increase in income, which in turn raises the marginal and average rates of saving. This holds true even if a part of foreign assistance goes to increase consumption, since the part of it which increases investment increases income and hence the rate of saving at a higher level of income. The net result in terms of a change in the rate of saving depends upon the relative pull of the saving-depressing effect of an increased consumption and the saving-raising effect of the increase in income. No less important are the policy measures designed to mobilize savings out of increased income. Secondly, an increase in capital inflow is expected to increase the rate of saving by providing the scarce foreign exchange component of investment. In a country like Pakistan, investment is critically dependent on imported inputs, i.e., imported machinery, so that domestic savings are frustrated if the foreign exchange component is not available. It is difficult to assess quantitatively how far savings would have been frustrated if foreign exchange through foreign capital inflow had not been available. Be that as it may, there are, however, two factors worth mentioning in this context. First, while the foreign exchange component of current production in a rural resource-scarce economy like Pakistan was important in increasing the rate of growth and hence of saving, the major reason for the accelerated growth during the 1950s was the breakthrough in agriculture, which was less import-intensive than the manufacturing and other modern sectors. There was at the same time considerable excess capacity in the industrial sector. The shortage of imported raw materials was a principal contributory factor to this phenomenon. The unbalanced composition of foreign assistance in favour of project assistance as against non-project commodity assistance contributed to the scarcity of raw materials in relation to the availability of capital equipment. Secondly, it needs to be mentioned that increase in foreign exchange availability during this period was partly due to a large increase in Pakistan's own foreign exchange earnings. Compared with the virtual stagnation in export earnings during the First Plan, Pakistan's exports grew at the rate of 7 per cent per annum during the Second Plan period. While an acceleration in the rate of growth of income during Second Plan provided the opportunity and resources for increasing the rate of saving and investment, there were important favourable changes in economic environment, institutions and policies not directly related to aid flow which increased the rate of saving in the early 1960s. While during the late 1950s there was considerable political instability and no commitment on the part of the government to economic development, the 1960s witnessed a stable, authoritarian government with a heavy bias towards economic development. The discipline of economic planning of public investment was enforced for the first time. Financial discipline was reflected in a large increase in public saving. Agriculture received considerable attention and was slowly losing its status of neglected sector. The bias against export promotion inherent in the import substitution policy was partially corrected through the introduction of the export bonus scheme, which provided the much needed stimulus to manufactured exports. One could even suggest that a rising export performance and improvement in the domestic saving rate encouraged a greater aid flow rather than the other way round. Pakistan provided a promising picture of a creditworthy nation to which aid flowed in increasing amounts. Along with an improvement in economic performance was associated enhanced ability to formulate aid-worthy projects as well as to negotiate aid, thanks to the strengthening of planning machinery as indicated later on (White, 1967). One could hypothesize on a two-way relationship between domestic saving and foreign capital inflow during this period. The accelerations in the rate of capital inflow provided the impetus to growth of income, savings and investment along with associated institutional and policy changes; the growth of income and savings in their turn had a beneficial cumulative effect on the rate of capital inflow, one factor, so to speak, reinforcing the other. The question might legitimately be asked whether during the Second Plan period, Pakistan could not have increased her rate of saving more than she in fact achieved, if she had been determined to reach her plan targets with a lower capital inflow. There are examples of other countries at her level of per capita income, which generated higher levels of saving. Her import bill had a wide variety of consumers' goods, which could have been cut off to generate more foreign exchange for development imports. As it was, the considerable acceleration in the rate of growth of income did provide a large increase in saving.

During the latter part of the 1960s, the rate of domestic saving, the rate of foreign capital inflow and the rate of investment all slowed down. Whether the positive relationship is a causative one is difficult to suggest since there were during this period independent factors in operation to reduce both the inflow of capital and the rate of domestic saving. The outbreak of the Indo-Pakistan hostilities resulted in the suspension of economic assistance, and its subsequent resumption took time to get under way again. The international climate for foreign aid worsened during the same period. In respect to domestic saving, the most immediate factor was the impact of the war on domestic public saving. There was a considerable rise in defence expenditure from 29.1 per cent of non-development expenditure and 2.4 per cent of GNP in 1964/65 to 48.4 per cent and 5.4 per cent of GNP in 1965/66. Even though its relative importance of non-development expenditure declined in subsequent years, it was, substantially higher than it was in 1964/65; it averaged about 40 per cent of non-development expenditure and 3.5 per cent of GNP during the rest of this period (Minister of Finance, 1969). The second factor tending to reduce the rate of saving was two consecutive years of bad harvests during 1965/66 and 1966/67. With a decline in the flow of PL 480 food grains, Pakistan had to spend a large part of her foreign exchange earnings in importing food. The fall in agricultural output adversely affected incomes and saving rates in the first two years of the Third Plan in the rural areas, whereas the rise in food prices increased the cost of living and depressed the saving rate in the urban areas. Thirdly, the political agitation against the Ayub government, associated with massive riots, demonstrations and industrial strikes all over the country during 1968 and 1969, created considerable uncertainty among investors about the future of the economic policy as well as that of the political system. This dampened the rate of investment and hence the rate of reinvestment of profit.

The rate of growth of income during the late 1960s was maintained at a slightly higher level than during the early 1960s, in spite of a fall in the rate of investment. This was primarily because of the maintenance of, or, indeed, an acceleration in, the rate of growth in agriculture in the latter part of the Third Plan. A number of factors were responsible for the acceleration of the growth rate in agriculture. In the first place, there was the technological breakthrough, consequent on the availability of the new variety of high-yielding seeds. Secondly, though the necessary increase in investment was minor, there was a considerable commitment towards the effective use of investment resources in ensuring abundant supplies, through imports, and efficient distribution of such critical inputs as fertilizer, seeds and insecticides. There was also greater administrative support for devising a flexible distribution system for the supply inputs and intensification of extension services, through a judicious combination of private and public agencies (Ministry of Commerce, p. 7, Table V).

The shortage of foreign exchange for development imports caused by the slowing down of aid and by the need to import food grains was met by a revision in the priorities of the Third Five Year Plan in the direction of those sectors which had lower import content and which were export-oriented. There was an increase in investment in the raw material processing industries as well as in those producing agricultural implements and fertilizers. There was also a shift of emphasis towards utilization of existing capacity as against the creation of new capacity (Ministry of Finance, 1966–67, pp. 179–92).

The fact that the domestic saving rate declined in spite of an acceleration of the rate of growth of income was partly due to the fact that contribution to the growth of income emanating from a rise in expenditure on administration and defence did not contribute to savings. An interesting aspect of the relationship between defence expenditure and domestic saving during this period is worth mentioning. The outbreak of war was met by the suspension of both military and economic assistance. Thus deprived of external resources, Pakistan did turn to internal sources and demonstrated the ability and willingness to increase the rate of mobilization of domestic resources to a degree which she might never have demonstrated otherwise. This experience may lend support to the hypothesis that a shortfall in aid could be accompanied by intensification of efforts for the mobilization of domestic resources, if social pressures for and imperatives of development were considered equally demanding. If the additional resources devoted to defence could have been instead diverted to public saving for development purposes, the rate of public saving during the Third Five-Year Plan period would have increased by 1.5 per cent to 2 per cent of GNP. Since the foreign exchange component of the defence expenditure was the same as that of expenditure, the diversion of resources away from defence to development would have encountered no bottleneck in terms of an increase in foreign exchange requirements (IBRD, 1966, p. 10).

The next most important reason why, in spite of the maintenance of a high rate of growth of income, the rate of domestic saving decreased, could be sought in two factors. First, the inability of the tax system to mobilize additional taxation from additional income; second, the inefficiency of the financial institutions and the capital markets in mobilizing private savings. The tax system has been singularly inelastic and regressive. Whatever increase in tax yields took place over the years was to a very large extent due to the imposition of new duties or the enhancement of the rates of existing duties. The existing system was heavily dependent on indirect taxes. The direct tax system was ridden with exemptions and rebates of various kinds, which reduced its elasticity in response to changes in income. The average ratios of all taxes to GNP increased from 7.4 per cent during the Second Plan period (1960–65) to 8.6 per cent during the Third Plan period (1965–70). The increase was entirely due to the increase in the ratio of indirect taxes to GNP (from 5.4 per cent to 6.6 per cent) while the ratio of direct taxes to GNP remained unchanged at 2 per cent between the two periods (Islam, 1970). The serious deficiency of the tax system was its inability to capture any part of the substantial increases in agricultural income which occurred during the late 1960s, especially among the medium and large farmers. Agricultural income, tax and land revenue constituted no more than 2.6 per cent of total direct taxes in 1965/66.

The flow of commodity aid and PL 480 provided an automatic mechanism by which the public sector could mobilize resources for its use. This was done in two ways: first, imports financed by commodity aid paid import duty so that the receipts from the imports went up as commodity aid went up; secondly, PL 480 food grains were in the first instance sold to private consumers; the sales proceeds were then re-lent to the private sector or directly invested by the public sector in the public works projects. The PL 480 grains thus provided a direct addition to public saving. If the food grains were directly given away as grants or loans to private individuals, this additional resource mobilization for development would not have taken place. With the slowing down of the flow of foreign aid, it was all the more necessary for the public sector to devise an elastic tax system to mobilize an increasing proportion of the rising agricultural output for financing development. The increase in agricultural output owing to the technological breakthrough was similar in its potential to the inflow of PL 480 food grains from abroad, for the purposes of mobilization of resources for development. Both were net additions to the domestic resources, which could be used for either consumption or investment or for a combination of both.

The combined result of the inelastic tax system and the rise in non-development expenditure during the Third Five Year Plan period was reflected in a fall in the public saving rate. The rate of public saving, having increased from the low figure of 0.5 per cent of GNP during the First Plan to 1.9 per cent during the Second Plan, declined to 1.1 per cent during the Third Plan (Haq, 1963, pp. 261–3; *The Fourth Five Year Plan*, pp. 45–6; IBRD, 1970, Table 2.11). The institutional framework and incentives for stimulating private savings were not sufficiently geared up during the period to compensate for all the fall in public saving. In fact there was a fall in household financial saving from 3.7 per cent during 1964/65 to 2.7 per cent during 1967/68, after having increased from 2.9 per cent in 1959/60 to 3.7 per cent in 1964/65.[2] There were two factors which might have inhibited the growth of private saving. First was the low rate of interest. Over the past decade, the maximum rate of interest on fixed long-term deposits rose from 3.4 per cent to 7 per cent whereas the returns from investment were as high as 15–25 per cent. The deliberate policy of keeping interest rate low was widely believed to have discouraged financial asset accumulation and contributed to the slow development of the market for bonds and fixed securities (Chowdhury, 1969). Second was the absence of the financial institutions to supplement the equity capital of small entrepreneurs with limited equity capital of their own. This inhibited the investment of small savings in productive investment and hence discouraged the growth of savings among small entrepreneurs.

In conclusion, there were two features of the relationship between domestic saving and foreign capital inflow which deserve special mention. First, over the years, once the annual fluctuations are smoothed out, there was a much slower rate of growth in the rate of capital inflow than either in the rate of domestic savings or in the rate of investment. The acceleration in the rate of savings and investment, when such an acceleration did occur, as in the early 1960s, was much steeper than in the case of foreign capital inflow. Secondly, the proportion of domestic investment which was financed by the inflow of capital declined over the years. The relative rates of domestic savings and foreign capital inflow were very much different during the 1960s as compared with the 1950s.

Pattern of Resource Allocation and Foreign Assistance

To what extent the pattern of economic growth and the priorities in the allocation of investment resources have been affected by foreign economic assistance is difficult to assess quantitatively in the absence of any access to detailed negotiations involved in specific aid programmes and the consultations which took place between the aid givers and Pakistan's policy makers. Two conclusions emerge from an analysis of the sectoral distribution of foreign loans. One, the distribution of foreign loans is much more skewed than total domestic investment, being mainly concentrated in industry, water and transport whereas domestic investment is more widely distributed. The high priority sectors in the total investment programme also received a larger proportion of total loans, but not in the same order of importance as indicated by the distribution of total investment expenditure. Secondly, two sectors – social sectors and agriculture, particularly the former – received a very small proportion of foreign loans. This, however, is an incomplete picture, since the distribution of foreign grants, as distinguished from loans, is not included above (Stern and Falcon, op. cit., p. 68, Table IV–I; *The Fourth Five Year Plan*, p. 28, Table 7; Ministry of Finance, 1965 and 1969–70).[3] The sectoral distribution of foreign grants from all sources is not available. However, the pattern of distribution of project grants (as against commodity grants) from the United States reveals a relatively greater allocation to social sectors than in the case of loans and credits, even though the bulk of it went to water, power, industry and irrigation (US Aid Mission to Pakistan, 1968, Table 9.5).

The above analysis does indicate the extent to which the intersectoral distribution of total investment expenditure was affected by the fact that foreign aid funds were available for certain sectors and not others and also by the fact that the aid donors had periodic consultations with the Pakistan planning authorities about the priorities of the development programme. This was specially so during the 1960s, when Pakistan's annual development programmes and five year plans were subject to scrutiny by the 'Consortium' countries under the leadership of the IBRD.[4] The most outstanding case was that of Pakistan revising the Second Year Plan after its initial presentation to the Consortium meeting. After the Plan was discussed and its foreign aid requirements were broadly agreed to, the Consortium persuaded Pakistan to re-examine the cost estimates as well as the targets of the Plan. While it is possibly reasonable to suggest that Pakistan's planning priorities among the very broad sectors such as industry, agriculture and transportation would have remained roughly the same with more or less foreign aid, it is quite likely that, as is evident from qualitative and fragmentary evidence, the availability of aid, including the discussion of the plan priorities with the aid giver, influenced the selection of or the relative emphasis on particular projects within these sectors. A comment on the United States–Pakistan aid relationship made in 1957 is worth quoting in this context.

> Aid is provided on the basis of specific requests by the recipient country. In practice, this means consultation with and agreement by USA. In the end, projects/purposes selected for aid, while desirable, might not be exactly what Pakistan would give priority to and some needs considered vital by Pakistan might be excluded. The contents of aid might also be different, e.g., more technicians might be included than originally considered necessary. Not only must all aid requirements be such as are acceptable to USA and fit into their ideas, but frequently they must be further examined by US technicians to certify their soundness. Further, when any equipment is requested for a project technicians must also be requested. From, the US viewpoint, this reflects their anxiety for a balanced programme and for sound projects. (Ministry of Economic Affairs, 1957, p. 3)

There was one particular sector of the development programme in Pakistan which owed not only its origin and initial impetus but also its subsequent extension to foreign economic assistance. This was the programme of rural public works, which was conceived in consultation with and with the help of the foreign aid-giving agencies in connection with the inflow of food grains and raw materials under US PL 480 aid programme. This was conceived as an employment and income generation programme, which was to help to absorb the imported food grains and raw materials without a depressing effect on domestic agricultural prices. During the Second and Third Plan periods respectively, the public works programme constituted 3 per cent to 3.4 per cent of the total investment expenditures (*The Fourth Five Year Plan*, op. cit., p. 28, Table 27). In East Pakistan, it consisted mainly of labour-intensive projects such as drainage, irrigation, embankments and road-building in the rural areas. In West Pakistan, the PL 480 aid financed the local expenditure on the large-scale water development programme (known as the Indus Replacement Works), as well as on the prevention of water-logging and salinity. The sales proceeds, in terms of local currency, of the PL 480 food grains sold to the consumer and provided the finance for investment (Gilbert, 1964; Planning Commission, 1961). The size of the programme varied in response to the changing availability of PL 480 funds. In the absence of the PL 480 aid, there was a reluctance to reallocate resources to the rural public works programme, at the cost of, let us say, large-scale industry. This programme subsequently became involved in the controversy relating to the rural political institutions established by the Ayub regime under the title of 'basic democracy'. Therefore, because of its unpopular political use, the programme shared the stigma attached to its associate and lost some of its developmental as well as distributional impact.

An important aspect of the pattern of investment in Pakistan has been the comparative neglect of investment in the social sectors. The investment in the social sectors constituting 20–21 per cent of total investment, consisted of physical planning and housing (mostly residential construction) to the extent of 15 per cent to 12 per cent; the rest, i.e., 6 per cent to 8 per cent, went to education, health and social welfare, etc. Education claimed barely 3.8 per cent of the investment resources during the Second Plan; its share went up to 5.1 per cent during the Third Plan. Moreover, the investment in the education and health service was basically urban-oriented and catered to the needs of the high income groups. In education only a third of the total expenditure was spent at the primary level. Between 1949 and 1969 the annual compound rate of growth of educational enrolment was 5.8 per cent at the primary level as against 7 per cent at the secondary level and 12.7 per cent at the level of college and university education (Maddison, 1970, p. 29). The educational system was scarcely designed for meeting the manpower requirements of development. About 4 per cent of the children at the secondary level received technical education, which, however, seldom prepared them for industrial employment, quality and content being inappropriate for the purposes of such employment. There were critical scarcities of medium-level technicians, on the one hand, and an abundance of liberal arts graduates or scientists on the other. The structure of wages and salaries was not responsive to relative supplies of and demand for different skills. The result was the migration of the high level manpower, which was partly offset by technical assistance from abroad, which brought in foreign technicians and experts. The health services are meagre and are concentrated in the urban areas, as was indicated by the fact that 80 per cent of the doctors in West Pakistan and 60 per cent in East Pakistan live in towns of over 25,000 population. Most of the hospitals were likewise in the urban areas (ibid., p. 41). Neither in the field of education nor in health or in other social sectors did foreign aid play any significant role. The neglect of these sectors by Pakistan's planning authorities was either accepted by the aid agencies without question or it coincided with their scheme of priorities, in which investment in education and the social sectors ranked quite low compared with the already productive sectors. There is no evidence so far that aid agencies showed any keen interest in or serious concern with the low level of investment in the social sectors, in the course of their dialogues over the years with Pakistan on the plan priorities and economic policies; they were more concerned with the exchange rate policy, the tax policy or the strategy of agricultural development or of industrialization. The neglect of the social sectors was attributable partly to the poor state of knowledge about the appropriate criteria for evaluating investment programmes in these sectors in a manner which was comparable to the cost-benefit analysis relating to the directly productive projects. It could also be that external aid for investment, for example, in the educational development, was more sensitive as it more closely affected cultural patterns and social mores.

There were two particular aspects of the pattern of economic development achieved with the assistance of substantial foreign aid which had brought the social and political fabric of Pakistan to the brink of a total collapse by the end of the 1960s. The first aspect was the growing inequality in income distribution which accompanied growth. The process of capital accumulation in Pakistan proceeded in the past, as in many other countries, through the transfer of resources from the agricultural sector, which as late as the mid-

1950s constituted more than 60 per cent of GNP. The transfer of resources from agriculture took place in the context of a stagnant agriculture; agricultural growth rate in the 1950s was lower than the rate of population growth. The majority of the agriculturists were poor and enjoyed little or no increase in income over the years. Therefore the transfer of resources from agriculture to the industrial sector was mainly in terms of a transfer from the rural poor to the rich in the industrial sector, thus accentuating inequality of income, not only between the urban and rural areas but also between rich and the poor in both the rural and urban areas. In the mid-1960s (1963–64) the rural per capita income was 65 per cent of the urban income; the highest 10 per cent of households received 28 per cent of the rural income and 30 per cent of the urban income, whereas the lowest 10 per cent of households received a 2.5 per cent share of the rural and 2.2 per cent of the urban income (Bergan, 1967). Over the years, the income of the rural poor, especially in East Pakistan, declined in the 1950s and did not rise significantly in the 1960s (Bose, 1968). In the urban areas, real wages probably declined slightly over the period from 1954 to 1963/64. During 1965–67, real wages in industry declined again, the rise in the cost of living having exceeded money wages. Workers received less than one third of value added in large-scale industry (Khan, 1967, 1969). As mentioned earlier, domestic savings, both public and private, declined during the second half of the decade, despite a growth in per capita GNP. The rise in private consumption took place mainly among the upper income groups, as demonstrated by increased sales of consumer durables and luxury items. At the same time, there was a decline in the per capita availability of several essential commodities between 1964/65 and 1968/69. This is seen in Table 28.4 below

Table 28.4 Percentage change in per capita availability

	East	West
Food grains	-3.2	+8.7
Pulses	-3.2	0.0
Edible oils	-34.8	+22.2
Vegetable ghee	-15.0	-14.8
Tea	-4.8	+10.3
Cotton cloth	-54.2	31.9

Source: Griffin and Khan, 1972.

It is important to note that the growth of inequality in Pakistan is not the unintended or unconscious by-product of Pakistan's development strategy. Pakistan's policy makers actively pursued policies which promoted inequality and initiative in view of scarce entrepreneurship and meagre capital formation. Not until the cumulative social unrest and dislocation of 1968–69 did they re-examine whether the degree of inequality had been carried too far and whether growth itself was not hampered by the limitation of opportunities or enterprise to a few large industrialists or to large farmers in the rural areas. The flow of foreign aid took place in the framework of this particular set of socioeconomic objectives, and helped to facilitate the implementation of the policy of growth through inequity without seriously questioning whether the conflict between growth and equality was not overdone. The aftermath of the Green Revolution fed through foreign-aided supplies of modern inputs and new technology brought in an aggravation of the inequalities. The aid-financed supply of tractors and agricultural machinery at an overvalued exchange rate and low tariff rates, and a liberal supply of credit on easy terms from the agriculture-financing institutions helped to polarize the agricultural population even further. Tenants were evicted, self-cultivating landlords annexed newer and larger areas under their own supervision and landless labourers faced fewer job opportunities as machines were readily available (Bose and Clark, 1969).

In the field of industry, the government-sponsored financial institutions through which the foreign loans and credits were channelled to private industry, directed a substantial part of their financing to the large industrialists. The Pakistan Industrial Credit and Investment Corporation (PICIC) which had received equity participation from IFC, an affiliate of the IBRD, and foreign private enterprise, as well as local financial resources from the government, granted 91 per cent of its total advances between 1957 and 1969 in terms of loans of the size of Rs 1 million and above, of which 71 per cent consisted of loans of over Rs 2.5 million each. The Industrial Development Bank of Pakistan (IDBP), the second most important industrial financing organization, which was specially established to promote small and medium enterprises and was in charge of allocating substantial amounts of foreign credits and loans from public sources, did a little better but fell considerably short of its mandate, in terms of size distribution of its loans. As of 1969, 71.5 per cent of its outstanding advances were in loans of size of Rs 1 million or more, out of which 58.1 per cent were loans of the size of Rs 2.5 millions or more (*Annual Reports* of IDBP and PICIC). The small-scale industry in Pakistan is defined as that which has fixed capital of Rs 500,000 or less. If the debt equity ratio was assumed to be 40:60, which was the maximum ratio allowed by these two financing institutions, the small-scale industrialists would mostly fall in the size category of loans of Rs 200,000 or less. Only 10 per cent of the total advances of the IDBP as of 1969 went to this class of entrepreneurs. The PICIC distributed only 3 per cent of its advances between 1957 and 1969 in terms of loans of the size of Rs 500,000 or less (ibid.). The medium scale industries may be defined as those receiving loans of the size between Rs 500,000 and Rs 2,500,000. No more than 25 per cent of the loans of both the organizations went to this sector.

Apart from the deterioration in the interpersonal distribution of income and a widening gap between the rich and the poor, Pakistan's economic growth was accompanied by a more serious imbalance which eventually contributed to a political tragedy, i.e., a considerable and a growing disparity in per capita income between East and West Pakistan. Between 1959–60 and 1969–70, per capita income increased by 17 per cent in East and 42 per cent in West Pakistan. West Pakistan's per capita income, which in 1959–60 was 32 per cent higher than that of East, was by 1969–70 61 per cent higher. The differences in the other indices of relative welfare, other than per capita income, such as per capita consumption of essentials as well as health, education and housing facilities, were even more striking. It was not only that East Pakistan was poorer than West Pakistan, but that she was absolutely very poor. The implicit model underlying the strategy of regional development was to concentrate investment in the more prosperous western region, better endowed with physical and social infrastructure, where it was hoped that returns would be higher and income would grow faster, so that at a later stage as total income grew to an adequate size resources would be transferred from the richer to the poorer region (Haq, 1963, chs 2 and 4). The disparity in development expenditure even as late as the Third Plan period was considerable and West Pakistan's share was 60 per cent higher than that of East Pakistan, as seen in Table 28.5.

The disparity in private investment was even worse. Private investment in the East seldom exceeded one-third of total investment. East Pakistan's share varied between 13 per cent and 15 per cent throughout the period. The relative allocation of development expenditure has to be seen in the context of a larger population in East than in West Pakistan. East had 55 per cent of the population of Pakistan. Per capita development expenditure in East, even as late as the Third Five Year Pan, was about 45 per cent of the total per capita development expenditure. The disproportionately higher levels of development and non-development expenditures in West Pakistan, supported by fiscal and commercial policy which led to a thriving private enterprise in West Pakistan,

Table 28.5 Regional allocation of resources (millions of rupees)

	East Pakistan		West Pakistan	
	Development expenditure	Revenue expenditure	Development expenditure	Revenue expenditure
1950/51–1954/55	1,000	1,710	4,000	7,200
1955/56–1959/60	2,700	2,540	7,570	8,980
1960/61–1964/65	9.250	4,340	18,400	12,840
1965/66–1969/70	16,560	6,480	26,100	22,230

Source: Planning Commission, 1970, p. 22.

contributed to the aggravation of disparity. East Pakistan, even at the end of two decades of development, did not have more than one-third of the industrial assets of Pakistan.

There were no spill-over effects on East from development in West Pakistan; Pakistan was a two-part land, separated by more than 1,000 miles of Indian territory. The means of transport were expensive and time-consuming and hence limited the mobility of goods. Mobility of labour from overpopulated East to relatively affluent and land-abundant West was nonexistent. These characteristics of the Pakistan economy implied that East and West were two separate, unintegrated economic entities. This distinguished the East-West Pakistan economic disparity from similar cases of regional disparities elsewhere. Secondly, it was the region where the majority of the people lived which received a smaller share of investment resources and had no share in the economic decision-making power. This created a politically explosive situation. The third disturbing feature was that the poorer region of East Pakistan transferred resources for West Pakistan's economic development. The transfer of resources took place via East Pakistan's net export surplus which the rest of the world, including West Pakistan, siphoned off for West's development through the mechanism of foreign exchange control exercised by the central government. Moreover, East Pakistan suffered in terms of unfavourable terms of trade. She exchanged her exports mostly at world market prices, while she purchased her imports at prices considerably higher than world market. It has been estimated that between 1948/49 and 1960/61, East transferred on an average Rs 850 million per year; between 1960/61 and 1968/69, the net outflow of resources declined to Rs 640 million a year. This estimate was made in terms of the scarcity price of foreign exchange and not at the official rate of exchange, which was overvalued to the extent of at least 100 per cent. The burden of this transfer on the East Pakistan economy can be gauged from the fact that the average outflow during the First Period was about 6 per cent of the current price regional gross domestic product of 1959/60. By 1964/65 the, average outflow did not constitute more than 3 per cent of regional GDP. More importantly, this burden fell particularly on the rural sector, the agricultural exporters with lower than average per capita income of East Pakistan (Planning Commission, *Reports of the Advisory Panels for the Fourth Five Year Plan*, p. 89). During this entire period, East Pakistan received about 34 per cent of the total foreign assistance accruing to Pakistan (see Table 28.6), most of it during the 1960s (ibid., Vol. 1, p. 279).

Table 28.6 Composition of East Pakistan's share of foreign economic assistance of different categories

Project loans and credits	35%
Non-project loans	36%
Project grant and technical assistance	14%
Commodity grant	33%
PL 480	35%

But then she transferred at the same time a part of her resources to West Pakistan. By the time in the 1960s when political pressures forced attention to the problem of regional disparity, disparity had already widened at a galloping pace. More fundamental during the whole period was the fact that there as never any sustained effort, in terms of commitment of resources and administrative support, to map out the existing and potential natural resources in East Pakistan, to identify her development potential and areas of comparative advantage.

The interesting question is what role foreign aid played in the development or amelioration of the East–West disparity. The aid-giving agencies, it appears from all evidence, accepted the rationale of regional development strategy in the sense that the built-in advantages of West Pakistan should be further exploited to speed up growth in West so that the latter would generate resources for development to be invested in East in subsequent years. They did not seriously call into question a strategy of development which concentrated on the exploitation of immediate, easily identifiable opportunities of investment in the richer region, which were profitable in the short-run, to the neglect of the systematic exploration of the development potential of the poorer region and hence of the long-run comparative advantage of both of the regions. It was expected that in due course efforts would be directed towards East Pakistan. The aid-giving agencies pushed forward with financing the Tarbela project in West Pakistan, the world's single biggest development project, costing in 1960 prices on a conservative estimate $1,047 million (Maddison, 1970, p. 17),[5] while at the same time neither financial resources nor technical assistance of sufficient magnitude could be obtained to provide the much-needed big push, financially and organizationally, to small-scale industry or agriculture in East Pakistan, the two key elements in East Pakistan's development efforts. That there was no automatic mechanism for the transfer of resources and shift of the focus of development efforts from West to East Pakistan, as West developed and accumulated resources, was not recognized. The timing and adequacy of such efforts were ignored or overlooked by the aid agencies. That in spite of East Pakistan's low human infrastructure, in terms of technical and managerial skill and trained manpower, the major part of the technical and financial assistance continued to flow on into the West, with the resultant accentuation in disparity in 'absorptive' capacity, lends support to the hypothesis.

The attitude of the aid-giving agencies was partly explained by the fact that during the 1960s they were very greatly concerned with making a success of aid efforts in at least a few countries of the world. The performance criteria for judging the success of aid efforts were the growth of GNP, the rates of saving and investment, growth of exports and the like. Income distribution, including regional disparities, was not among the accepted criteria. Judged by the accepted criteria of the time, Pakistan was a moderate success in terms of the growth of GNP, especially in agriculture, even though her performance in increasing the saving rates left much to be desired. Considering that only a few countries in the early 1960s did as well as Pakistan in these respects, and that at least a few success stores were essential for the future of international aid, the aid agencies chose to neglect the distributional problems of Pakistan. This was facilitated by the fact that the established political authority in Pakistan set rather narrow and strict limits within which these considerations were to be allowed to affect economic policies.

Foreign assistance has often been viewed by the major donor countries not only as a supplement to the domestic resources of a recipient country but also as an instrument for initiating changes in policies and institutions which contribute to growth as well as to great efficiency in resource allocation. To what extent, if at all, can important changes in policies and institutions in Pakistan be traced to the influence of aid-givers? Take the case of agriculture.

Beginning in 1960, controls on production, distribution and prices of food grains were relaxed substantially. The availability of PL 480 wheat contributed to the successful operations of the buffer stocks in food grains, which were to stabilize prices and to provide incentives for increased agricultural production. PL 480 provided, so to speak, the safety margin for the introduction of and experimentation with new and flexible policy measures, away from direct controls, to stimulate agricultural production. The rural public works programme, which was initiated by PL 480 food aid especially in East Pakistan, as already mentioned, was traceable to the influence of the aid givers.

There was, however, a risk involved in the opposite direction from the easy availability of food grains under PL 480. During the 1950s, PL food grains were available as grants or were to be repaid entirely in local currency. This could have slackened domestic efforts for agricultural development. This possibility was, however, forestalled by a decline in the supplies of surplus wheat in the United States as well as by a hardening of the terms of PL 480 assistance. By the 1960s Pakistan moved in the direction of an increased emphasis on agriculture. The policy makers were prepared to introduce necessary changes, and foreign aid provided the flexibility and cushions, if necessary, against the uncertainty involved in any change in economic policy. Foreign aid, especially foreign technical assistance, played a critical role in spreading knowledge about the new rice and wheat technology developed in Mexico and the Philippines, as well as in providing large quantities of the initial supplies of the new varieties of seed for experimentation under the new conditions of soil and climate.

The changes in import policy in the early 1960s have been traced to the influence of the aid givers. It is now widely acknowledged that the liberalization of import controls in 1963/64 was associated with a large amount of commodity assistance as a concomitant of Pakistan's adoption of a liberal import policy (Mason, 1966, pp. 44–5). The 'import liberalization' policy led to a more efficient allocation of resources for investment and a fuller utilization of already-installed productive equipment. It was not only that foreign aid was instrumental in the adoption of the liberalization programme, but also experience with the liberalization programme and the associated spurt in private economic assistance, economic activity brought in an additional flow of foreign economic assistance, especially non-project aid (White, 1967, pp. 59–84).[6]

At the same time, it is important to note that a few other specific desirable changes in economic policy did not have any relationship with foreign aid. For example, the most important measure of trade liberalization and export promotion, i.e., the Export Bonus scheme, which was introduced in 1959, was wholly unconnected with foreign aid or with any persuasion or suggestion from any aid-giving agency, either national or multinational. The scheme was adopted on the advice of a German expert, invited by Pakistan to examine the most effective method of promoting exports. The initiative was entirely on the side of Pakistan, worried as she was with stagnant export receipts, on the one hand, and a rising import bill, on the other. However, once this scheme was introduced, it was reviewed from time to time in consultation with aid-giving authorities who sought to prevent a proliferation of rates of exchange and to reduce the degree of discrimination between commodities and sectors (IBRD, various *Annual Mission Reports* on Pakistan, especially 1968–70).

Since the general gamut of fiscal and exchange rate policies in Pakistan contributed to the misallocation of resources via the underpricing of imports or of capital equipment, foreign assistance had to operate within this general framework and the additional resources provided by foreign aid, were subject to the same misallocative policies as the rest of the domestic resources. It might have augmented the misallocative effect of the inefficient policies. Pakistan maintained the rate of interest in the organized capital market at a level lower than was warranted by the scarcity of capital. Foreign aid temporarily augmented the supply of capital, buttressed the tendency towards underpricing and hence encouraged capital intensity, on the one hand, and wastage of capital through underutilization, on the other (Winston, 1971, 1970). Furthermore, preference of the aid givers for project aid over non-project aid contributed to underutilization of capacity because of the inadequate supply of imported raw materials. An industrialist's entitlement to licences for the imports of raw materials was linked up with the installed capacity on the basis of 'one shift' working. There was an incentive for the creation of new capacity even when a part of it was unutilized. Considerable excess demand under tight import restrictions yielded such high profits that the addition of new capacity, even when a part of it remained unutilized, was worthwhile. In spite of considerable excess capacity at the end of 1965, the increase in the flow of project aid during the subsequent five-year period was greater than that of non-project aid. Moreover, a relatively greater proportion of the imported capital goods was available at the official rate of exchange and under low tariff rates.

A large proportion of aid to Pakistan, both project and commodity aid, was tied to the sources of purchase in the aid-giving countries. It is not only that the imports under tied aid are more expensive – a fact which is widely known – but that tied aid contributes to inefficiency in the utilization of foreign resources. A sample of 20 tied project loans in 1967 revealed a weighted average cost of 51 per cent over international competitive bidding (Haq, 1967).[7] Eight important items of iron and steel procured under the United States commodity aid had cost in 1967 44 per cent to 111 per cent more than the cheapest international source (Planning Commission, 1968–69, p. 71). The high cost of the imported equipment as well as of the imported raw materials raised the cost of the domestic industries and reduced their competitiveness vis-à-vis industries abroad. Consequently, a larger degree of subsidy in the export market and a greater degree of protection in the domestic market were necessary for manufacturing industry. Tied aid also had seriously qualified the degree of liberalization of import controls which was introduced in 1963/64. Liberalization in Pakistan was mainly confined to the imports of important raw materials which were financed by tied commodity assistance. The amount up to which the letters of credit could be opened for the 'liberalized' imports had to be limited to the amount of tied aid which was available from a particular country. So long as there was tied aid, it could be disbursed only by means of licenses which were tied to countries and commodities, since commodity aid from a particular country could finance only a specified list of commodities. There could not be a unified foreign exchange market and the result was a multiplicity of exchange rates for the different categories of foreign exchange which was not interchangeable. The consequent compartmentalization of the foreign exchange market had no rationale from Pakistan's point of view. The large number of exchange rates for imports did detract from the efficient use of resources. By providing additional but temporary supplies of foreign exchange to match a part of the excess demand, foreign aid reduced the urgency of the need for a long-run policy for efficient exchange rate management. It was only very recently, in the late 1960s, that some attempts were made to rationalize the system, mainly in response to a slowing down of the inflow of foreign assistance. In the early 1960s, when aid inflow was on the increase, exchange rates proliferated at a rapid rate and changed quite frequently. Whether, in the absence of foreign aid, Pakistan would have devalued or persisted in maintaining the overvalued exchange rate by a further tightening of the exchange restrictions is difficult to say. However, judging from the experience of the late 1960s, when the slowing down of aid encouraged the move towards a rationalization, it was likely that a smaller inflow of aid in the 1950s would have increased the urgency for a more efficient exchange rate policy. Under a regime of differential exchange restrictions, the liberalization of the import of only a limited number of commodities, financed by foreign aid as it happened Pakistan in 1963–64, tended to aggravate the distortion in the allocation of resources. It reduced the effective protection for the sectors competing with 'liberalized' imports, in relation to the rest of the economy. It increased the degree of effective protection for those sectors which used the liberalized imports as intermediate inputs. The 'liberalized' imports in Pakistan were mainly industrial raw materials, which also had lower rates of import duty. The domestic intermediate goods sector, including industries in which Pakistan had potential comparative advantage, suffered a setback (Lewis, 1960, ch. 4). The readjustment of the structure of effective protection was not a part of a general realignment of the structure of protection, but was incidental and arbitrary.

Foreign aid significantly influenced the development of one particular development institution in Pakistan. Foreign aid helped the development or the planning process and that of planning machinery in several ways. It directly provided technical assistance to the planning machinery to improve the quality of planning. The assistance from abroad went hand in hand with the beginnings of a planning organization; they both grew and changed together and both the domestic staff and foreign counterparts 'learned by doing' the job of economic planning and development administration. Economic planning was also helped by external assistance in an indirect way. Towards the latter part of the 1950s, there was an increasing conviction on the part of the donor countries that aid was best administered on the basis of a consistent economic plan. This provided a considerable stimulus to the preparation of a plan in Pakistan; it increased the prestige of the planning machinery in the administrative set up and with the operating agencies by linking it up with the foreign aid negotiations (Waterston, 1963). Foreign aid compelled an improvement in the quality of project preparation as well as the examination of economic policy alternatives, especially as in the 1960s Pakistan's economic plans and projects came in for review and scrutiny in the annual meetings of the aid-giving countries under the sponsorship of the World Bank.

The relative role of public and private enterprise in Pakistan was often cited as an example of the predominance of the ideas of the aid-giving agencies which had a bias in favour of private enterprise. The United States' loans to industry, for example, were offered only to the private sector. It is difficult to test this hypothesis; the bias towards private enterprise in Pakistan predated the increase in the inflow of foreign aid and can be traced to the political and social composition of the early policy makers in Pakistan. The bureaucrats were in favour of controls over private enterprise rather than of substantial direct participation. However, foreign economic assistance, which is directly offered to the government of the recipient country or which indirectly augments resources in the public sector via taxes on aid-financed import or via sales proceeds of commodity aid such as PL 480 aid, tends to increase the role of the public sector in the economy.[8] Of the total outstanding foreign loans (both project and non-project) in 1965, 57 per cent was in the public sector; the proportion went up to 75 per cent by 1969. Much of this assistance was, however, concentrated in water, power and transport development of large projects with long gestation lags and large capital requirements, which did not attract private enterprise and were necessarily in the public sector (Government of Pakistan, 1969–70, pp. 53, 56).

Foreign aid also influenced the evolution of an appropriate institution for the distribution of new agricultural inputs (Gulick and Nelson, 1965, pp. 15–18). Although the preference of the Pakistani policy makers was for an exclusive or a very predominant role for public enterprise, it was increasingly realised in the course of dialogue with the aid-giving authorities, that a substantial increase in the inflow of fertilizers, for example, under commodity aid could not be handled and distributed efficiently if private enterprise was not allowed to share significantly in the trade, owing to the limited capacity of the public sector. There had been changing emphasis over time in the relative role accorded to the different types of agencies for the distribution of the agricultural inputs, such as the cooperatives, the private traders and the public sector government trading organizations. The shift of policy towards and encouragement of competition between different agencies and towards a larger role for private enterprise was partly due to the pressure or persuasion emanating from the aid-giving agencies.

Conclusion

Among the developing countries at similar stages of development, Pakistan received a significant amount of foreign economic assistance. Foreign assistance helped her achieve a higher rate of savings and investment than would otherwise have been possible. A part of foreign aid might have gone to increase consumption expenditure. However, the rate of mobilization of domestic resources for development declined in the late 1960s, even though both GNP and per capita income continued to increase. Pakistan's effort was low partly because the incidence of tax on agricultural income was low, in spite of an accelerated increase in such income consequent on the 'green revolution'. The impact of foreign aid on development strategy in terms of priorities between sectors was not recognized to be considerable, even though in the selection of particular projects and programmes it had significant impact. Foreign aid was an important influence in strengthening the machinery and process of planning in Pakistan, in initiating the policy of import liberalization in 1963–64 as well as in the relaxation of controls over food grain prices and distribution in the early 1960s. Pakistan's exchange rate and fiscal policy during this period had serious shortcomings, which contributed to the misallocation of resources, including resources provided under the foreign economic assistance.

What are the prospects of self-sustaining growth? Pakistan has accumulated a very large debt burden. Debt service payments in 1968 constituted 8.4 per cent of her total outstanding debt. In a sample of 81 countries, only 10 other countries had a higher ratio (OECD, 1970, p. 51). Most of Pakistan's loans were contracted in the 1960s and hence amortization payments have not started to accumulate at a rapid rate. While amortization payments constituted 5.2 per cent of the outstanding debt in 1968, the interest payments were 3.2 per cent of the debt. The corresponding figures for the sample of 81 countries were 5.2 per cent and 3.2 per cent respectively. Over the years the relative proportion of loans to grants increased considerably. While during the First Plan, loans constituted 41 per cent of total assistance, by the Third Plan they constituted 93 per cent of total assistance. Moreover, in the late 1960s, with a slowing down of the aid from the public sources, Pakistan relied increasingly on supplier's credits, which involved both shorter terms and higher rates of interest than the bilateral or multilateral public loans (Ministry of Finance, 1969–70, p. 199; Stern and Falcon, op. cit., p. 30). The seriousness of the debt service payments vis-à-vis her foreign exchange resources is indicated by the fact that the ratio of the former to the latter increased from 3.6 per cent in 1960–61 to 10 per cent in 1964–65 and 22.5 per cent in 1969–70 (ibid.). In 1970 Pakistan in fact requested from the major donors a rescheduling of her debt payments in view of the stringency of foreign exchange resources.

During the late 1960s the rate of savings recorded a decline. At the same time the ratio of exports to GNP, which was as high as 5.4 per cent during 1965–66, declined to 4.7 per cent by 1968/69 owing to a combination of circumstances, including dislocation in industrial production owing to industrial unrest and political agitation. Pakistan requires to achieve a very substantial increase in the rate of mobilization of domestic resources as well as in the rate of exports in order to be able to achieve a reasonable rate of growth while at the same time avoiding a moratorium on her debt repayments for some years to come.[9]

Notes

1 The statistical estimates of aid, foreign payments (which include both visible and invisible payments) and gross investment for the period prior to 1954–55 leave much to be desired and therefore should be treated with great caution.
2 There are no direct estimates of domestic saving by sectors and by years available in Pakistan. Savings are estimated as residual, i.e., investment minus foreign resource inflow. There are reasonable, direct estimates of household financial saving and of public saving for a number of years (Lewis and Khan, 1964; Chafur, 1969).
3 Loans do not include PL 480 loans.
4 The Consortium countries include all the principal aid donors in the Western world, who tend to coordinate their aid efforts to Pakistan.
5 It was an irrigation and power project which was not only enormous but was also expected to have a very delayed impact, benefits accruing in the late 1970s, which were furthermore contingent on subsequent supplementary investments in drainage and which in the absence of such supplementary investments could become negative.
6 The liberalization programme was cited as a successful example of aid relationship. 'The nature of that success is mediated by the fact that credit for getting the liberalization program started has at various times been separately claimed by officials in the government of Pakistan, in the USAID, and in the World Bank. There was simultaneous communication at several levels and in several directions. The participants responded to a situation.' There were three reasons for the success without causing strain in aid relationships. First, the liberalization programme was a limited operation with a specific, objective which was defined quantitatively and temporarily. The success could be easily measured. Secondly, economic policies of which the liberalization programme was a natural product, had already proved a moderate success, i.e., the active role and performance of the private sector was already evident in the early years of the Second Plan. The liberalization policy was to accelerate the rate of progress along the proven lines. Thirdly, the operation was one in which the separate functions of the participants were easily discernible. The link between the abolition of import controls and programme aid was simple.
7 If Pakistan could diversify the sources of aid and offer to each aid-giving country projects which were cheapest in the particular country and also if she could use her own foreign exchange judiciously to bring down bids in the aid-giving countries, she could bring down the cost of tied aid to about 12 per cent above world price.
8 The First Five-Year Plan of Pakistan cites in 1955 that one of the reasons why the public sector would come to play a bigger role was the practice of having a large amount of foreign aid channelled through the budgets and public expenditure.
9 This article was committed to and written before the recent events in Pakistan. The author was then Director, Pakistan Institute of Development. Economics, with which Austin Robinson was associated since its inception.

References

Bergan, A. 'Personal Income Distribution and Personal Savings in Pakistan', *The Pakistan Development Review*, Summer 1967.

Bose, S.R., 'Trend or Real Wages of Rural Poor in East Pakistan', *The Pakistan Development Review*, Autumn, 1968.
Bose, S.R. and Clark, E.H., 'Some Basic Considerations on Agricultural Mechanics in West Pakistan', *The Pakistan Development Review*, Autumn, 1969.
Chafur, A., 'Financial Asset Accumulation by the Non-corporate Private Sector in Pakistan, 1959/60–1965/66', *The Pakistan Development Review*, Vol. X, No. 1, Spring 1969.
Chowdhury, N., 'Some Reflections on Income Redistributive Intermediation in Pakistan', *The Pakistan Development Review*, Vol. IX, No. 2, Summer 1969.
Gilbert, R.V., *The PL 480 Programme for the Third Plan*, unpublished manuscript.
Griffin, K. and Khan, A.R. (eds), *Growth and Inequality in Pakistan: Essays and Commentaries* (London: Macmillan, 1972).
Gulick, C.S. and Nelson, J.M., 'Promoting Effective Development Policies: AID experience in the developing countries', AID Discussion Paper No. 9, September 1965.
Haq, M., *The Strategy of Economic Planning. A Case Study of Pakistan* (Oxford: Oxford University Press, 1963).
Haq, M., 'Tied Credits – A Quantitative Analysis', in J. Adler (ed.), *Capital Movements and Economic Development* (London: Macmillan, 1967).
IBRD, *Economic Development of Pakistan*, vols I and II (Washington, DC: World Bank, 1965).
IBRD, *Current Economic Position and Prospects of Pakistan* (Washington, DC: World Bank, 1966).
IBRD, *Annual Mission Reports on Pakistan* (Washington, DC: World Bank, various years).
Islam, N., *The Tax System of Pakistan* (New York: UN Division of Fiscal and Financial Analysis, 1970) unpublished manuscript.
Khan, A.R., 'What has been Happening to Real Wages in Pakistan', *The Pakistan Development Review*, Autumn, 1967.
Khan, A.R., 'Exercises in Minimum Wages and Wage Policy', mimeograph, Karachi, 1969.
Lewis Jr, S.R., *Pakistan: Industrialisation and trade policies* (Oxford: Oxford University Press, 1960).
Lewis, S.R. and Khan, M.I., 'Estimates of Non-corporate Private Saving in Pakistan, 1949–62', *The Pakistan Development Review*, Vol. V, No. 1, Spring 1964.
Maddison, A., 'Social Development in Pakistan, 1947–70', *Economic Development Report No. 169*, Centre for International Affairs, Harvard University, 1970.
Ministry of Commerce, *Report of the Working Group on Import Policy* (Karachi: Government of Pakistan, May 1970).
Ministry of Economic Affairs, *Critical Report on the Operations of Economic Aid to Pakistan from USA* (Karachi: Government of Pakistan, September 1957).
Ministry of Finance, *Pakistan Economic Survey 1966–67* (Karachi: Government of Pakistan, 1967).

Ministry of Finance, *Budget in Brief* (Karachi: Government of Pakistan, 1969).
Ministry of Finance, *Economic Survey* (Karachi: Government of Pakistan, 1969–70).
Ministry of Industries, *Priority List of Industry of Comprehensive Industrial Investment Schedule (1965–70)* (Karachi: Government of Pakistan, June 1968).
Mohammed, A.F., 'Some Aspects of Economic Impact of Foreign Aid on an Undeveloped Country: The case of Pakistan', PhD dissertation, George Washington University, June (date obscured).
OECD, *Development Assistance, Efforts and Policies of the Members of the Development Assitance Commitee 1970 Review* (Paris, 1970).
Planning Commission, *Memorandum for the Pakistan Consortium* (Islamabad: Government of Pakistan, 1968–69).
Planning Commission, *Reports of the Advisory Panels for the Fourth Five Year Plan* (Islamabad: Government of Pakistan, 1970).
Planning Commission, *The First Five Year Plan, The Second Five Year Plan, The Third Five Year Plan, The Fourth Five Year Plan*, Government of Pakistan, various years.
Sobhan, R., *Basic Democracies, Works Programme, and Rural Development in East Pakistan* (Dacca: Bureau of Economic Research, Dacca University, 1968).
Stern, J.J. and Falcon, W.P., *Growth and Development in Pakistan 1955–59*, Center for International Affairs, Harvard University, April 1970.
US Aid Mission to Pakistan, *Statistical Fact Book* (Karachi: Office of Assistant Director for Development Planning, 1968).
Waterston, A., *Planning in Pakistan* (Baltimore: The Johns Hopkins University Press, 1963).
White, J., *Pledged to Development* (London: The Overseas Development Institute Ltd, 1967).
Winston, G., 'Capital Utilisation in Economic Development', *Economic Journal*, March, 1971.
Winston, G., 'Over-invoicing, Underutilization and Distorted Industrial Growth', *The Pakistan Development Review*, Winter, 1970.

Chapter 29

Economic Policy Reforms and the IMF: Bangladesh Experience in the Early 1970s*

In the early 1970s, the 'newfangled' terminology of structural adjustment was not in vogue. However, it was known that in response to adverse changes in external environment, falling terms of trade, stagnation in export prospects/ earnings or a rise in import prices, it might be necessary to undertake measures to reallocate resources between export or import competing sectors as well as in each of these sectors, between commodities or groups of commodities. The concept of tradeables and non-tradeables was not popular in those days, even though the distinction between domestic goods or 'home goods' and the goods which competed with imports or were exported was well understood.

At that time, there was no structural adjustment loans from the Bank or the Fund, but there were loans, both bilateral and multilateral, which were tied to domestic policy reforms. For example, commodity or sector loans from the Bank, on the one hand, and drawings on IMF credit tranches (including Compensatory Financing Facilities for Export Shortfalls) were linked to policy reforms. These policy reforms included exchange rate devaluation, fiscal and monetary reforms, as well as pricing policies or subsidies bearing not only on agricultural outputs/inputs but also consumer goods such as edible oils and sugar or intermediate inputs like fuel.

In the years between 1972 and 1975, Bangladesh undertook negotiations for borrowing from the IMF and the Bank as well as from the bilateral donors. It was in 1973 when Bangladesh had its first experience with policy dialogue with the IMF, when a Fund mission visited Bangladesh in March 1973 and raised doubts about the then prevailing external value of the Bangladesh taka. The Bangladesh currency was considered to be 'overvalued', and the following arguments were advanced in favour of this judgment by the Fund.

---

* First published in R. Sobhan (ed.), *Structural Adjustment Policies in the Third World: Design and Experience* (Dhaka: Dhaka University Press, 1991).

1) In 1971, the Pakistan rupee which was the same as the Bangladesh taka (since they were the same country) was already overvalued.
2) By 1973, Pakistan had devalued to the level of rupees 26 per pound sterling as against the equivalent of roughly rupees 12 to pound sterling in 1971.
3) Bangladesh, on the other hand, had adopted a new parity of approximately takas 19 pound to sterling and, therefore, could be said to have devalued by much less.
4) This degree of devaluation was, however, more apparent than real for certain types of export earnings, because before 1971 some of the non-traditional exports were entitled to a system of 'export bonus' which in fact meant a higher rate of exchange than the official rate of taka 12 to pound sterling. After 1971, the export bonus system was abolished and the effect of the new official rate of exchange implied very little for these exports.
5) A substantial rise in internal prices relative to those of Bangladesh's trading partners occurred since 1971 and Bangladesh was therefore expected to find it more difficult to seek new markets for products previously exported to Pakistan, to develop new types of exports and generally, to expand export earnings.

Bangladesh authorities, however, were not convinced about the need for devaluation in early 1973, and they advanced the following arguments:

1) So long as physical and institutional bottlenecks remained the decisive hindrance to efficient and growing export trade, it was neither opportune nor worthwhile to alter the exchange rate. Serious transport difficulties impaired both internal and external trade; roads and bridges destroyed by the War of Independence during 1971 were often inadequately repaired or replaced; Chittagong port, the most important port of Bangladesh, needed to be completely cleared of sunken vessels; salvage operations to clear the second port of Chalna from mines laid during the war had not even started; inland water transport was also disrupted.
2) The export demand for raw jute and jute goods was strong but could not be met because it was not possible to arrange timely and regular shipment. In other words, it was not price so much but the physical supply on a regular and timely basis that limited exports.
3) Government, however, recognized that the jute manufacturing industry was working at a loss and that devaluation might be expected to improve the financial position of the industry but the extent to which this would occur was uncertain. In the circumstances prevailing in 1973, the major

result of devaluation might be to create pockets of excess profits in the marketing chain rather than to improve the financial position of the industry.

4) After a series of bad harvests, the internal price of food had already risen; devaluation was likely to seriously aggravate this unless the subsidy on rationed foodgrains was increased. Ills would increase the fiscal deficit of the government.
5) The effects on the prices of less essential commodities and on their profits, resulting from devaluation would also cause economic hardships not so much on the poor and the vulnerable but on the vocal and the politically powerful.

Towards the latter part of 1973 there was a further deterioration in both the external payments situation as well as the internal fiscal and monetary situation; supplies of essential imports were very tight; foreign exchange reserves were low and were expected to fall further. Furthermore, setbacks in agricultural output had increased import requirements. Disbursement of foreign assistance was slow and import prices were rising and money supply had greatly expanded.

In the face of these developments, the need for a 'standby arrangement' with the Fund was felt to serve as a cushion against further deterioration in the balance of payments and in foreign exchange reserves. It is in this context that the discussions were initiated with the Fund with a view to drawing on the first and the second tranches of its quota.

To facilitate the negotiations with the Fund, the government of Bangladesh, however, decided to take the initiative and to suggest and adopt a number of measures on its own, first, for the overall stabilization of prices and second, for the mobilization of domestic resources. The measures which were undertaken were as follows:

1) Limit on credit expansion was imposed, along with the imposition of sectoral limits in major industrial sectors, public enterprises and agricultural sector; advances against stocks of essential goods and scarce supplies were prohibited.
2) Import duties on tobacco, cement, and iron sheets were raised and fees on the issue of import license were increased as a step towards mopping up scarcity margins.
3) A number of excise duties, i.e., those on vehicles, trucks and public carriers were raised and also telephone rates were increased.
4) Subsidies were reduced by an increase in prices of rationed goods, i.e., 33 per cent for rice, 89 per cent for sugar, and over 50 per cent for edible oil and 13 per cent for kerosene oil.

The government argued that these actions were a sufficient evidence of the government's intention to correct financial imbalances which were crucial to the restoration of the external balance as well. The Fund, however, disagreed. The Fund, in fact, suggested that devaluation would provide a unique opportunity to mobilize domestic resources by a simple device of export taxes on jute and jute manufactures which constituted 90 per cent of exports rather than by trying to raise revenues through import duties which were vulnerable to pressures for exemptions and reductions.

The discussion on devaluation was deadlocked between the government and the Fund until early 1974. The government was willing, however, to give undertaking to direct its financial policy so as to reduce inflation by improvements in budgetary and public sector performance, to raise interest rates, to restrict credit, tax imports, and to remove subsidies on food and agricultural inputs. But it was not willing to consider devaluation.

By the middle of 1974, the balance of payments position of Bangladesh had further deteriorated substantially and discussions were again started between government and the Fund. The Fund mission argued that the devaluation of an agreed magnitude and with a uniform exchange rate, backed by appropriate fiscal and credit policies, was the only way open for Bangladesh to have access to new credit tranches. At one stage in the course of the negotiations, the subject of a dual exchange rate came up, i.e., a moderately depreciated rate for limited number of transactions such as import of essential items and a more sharply depreciated rate for the bulk of the transactions. The Fund's response was that such an arrangement could at best provide access only to one credit tranche.

By the middle of 1974, there was general agreement inside the government that devaluation would be inescapable at some point but there were differing views on its magnitude, as well as on the nature and the extent of accompanying measures and thirdly on its timing.

What would have been considered as an appropriate degree of devaluation in early 1973 by the Fund was no longer acceptable by 1974 since, in the meanwhile, there had been a progressive rise in costs and prices within Bangladesh. This had further impaired her competitive position in the world markets. In addition the Fund suggested that Bangladesh should devalue before the Consortia Meeting in October 1974. This was considered an appropriate timing since this would create a more congenial atmosphere for dialogue with the donors and facilitate a larger commitment of assistance at that time.

It was, however, not clear to the government of Bangladesh whether the devaluation, if undertaken before the Consortia meeting, would actually result in an increase of the overall aid flow. Neither the Fund nor the Bank could offer an opinion on this point or hold out firm hopes in this regard, excepting in very vague and general terms. During this period in 1974, as these negotiations were underway, Bangladesh was progressively entering a period of famine for lack of food aid as well as for lack of foreign exchange resources to buy food in time. Would Bangladesh be provided with more food aid by food exporting countries if she devalued? It was not clear at all to Bangladesh that if she devalued before the Consortia meeting, there would be a large flow of food assistance.

In the event Bangladesh did not, in fact, devalue during October 1974 at the time of the Consortia meeting. And at the Consortia meeting some pledges of assistance were indeed made even in the absence of devaluation. Following the Consortia meeting, she did in fact receive substantial commitments of assistance including drawings on a newly-created oil facility of the Fund. Further drawings on this facility, however, could not be made without meeting the Fund's requirements on devaluation.

Between the Consortia meeting of October 1974 and early March 1975, international pressure continued to mount on Bangladesh for devaluation. By March 1975, the government had eventually reached an agreement with the Fund to devalue the exchange rate by 58 per cent. Bangladesh obtained an assurance from the Bank and the Fund that they were ready to do their utmost to mobilize greater external resources. The Fund agreed to make available under its own arrangement a credit of $75 million from two credit tranches, i.e., second and the third credit tranches; first tranche was already withdrawn. In addition, Bangladesh obtained an access to oil facility for an additional amount of $20 million.

The measures which Bangladesh were to undertake to accompany the devaluation were as follows:

1) strict scrutiny of request for additional credit for public enterprises as well as an increase in interest rates;
2) various limits to credit expansion to be placed at specific dates during 1975–86;
3) reduction of the rate of inflation from 75 per cent in 1974/75 to 10 per cent in 1975/76. This was based on the assumption that food production would increase satisfactorily since financial policies could do very little to prevent a rise in prices when there was a harvest failure;
4) constraint on wages and salaries and restraint on credit expansion;

5) gradual reduction of subsidy on food grains and eventual termination of subsidies on agricultural inputs except when necessary to ensure the use of new technology;
6) export taxes on raw jute and jute manufactures which would be flexibly administered. This was a method of distinguishing between traditional and non-traditional exports, and recognizing the relative inelasticity of demand for jute exports. In a small way, it amounted to introducing a system of multiple exchange rates;
7) liberalization of import regulation systems, which were to be more streamlined. Import licensing procedures were to be made more liberal and more automatic. Efforts were to be made to place most essential items on open general license as soon as the foreign exchange position eased.

In light of the above, several questions may be raised regarding the decision to devalue in 1975. First, was the timing of devaluation correct? Opinions differed on this. The need for agreement with the Fund was manifest by June 1974. There were important indicators as described below. Reserves had plummeted to the level covering only 12 per cent of annual imports which recovered to only 17 per cent by June 1975. Inflation was in the range of 40 per cent per annum. Terms of trade deteriorated by 40 per cent between 1972–73 mostly due to a rise of 80 per cent in import prices. Total liquidity grew by 30 per cent over 72–73; domestic credit expanded by 59 per cent. Deficit financing exceeded 1 per cent of GDP for every year during 1972–73 and 1973–74. This, coupled with shortages of key imports and food, contributed significantly to domestic inflationary pressures, which in turn eroded the competitive advantage of Bangladesh's exports. Domestic production cost and export prices went seriously out of line with the prices in the world market.

However, even before the devaluation, an informal understanding with the Fund was reached when the government made a drawing on the first credit tranche in 1974–75. This led to the adoption of a number of restrictive monetary policies and to demonetization in April 1975. Total liquidity growth decelerated to 58 per cent between June 1974 and June 1975; the total domestic credit grew by no more than 13 per cent while net credit provided to the government increased by only 12 per cent i.e., by less than the rate of growth in total domestic credit. This reduced the extent of monetary overhang that existed prior to June 1975 when devaluation was undertaken.

If Bangladesh were to devalue in August 1984 she would have certainly obtained immediate access to $75 million from the Fund. In view of the great shortages of food and finance at that time, this would undoubtedly have helped significantly. On the other hand, it was doubtful whether the bilateral donors would or could have provided more than what they in fact provided in late 1974.

On the most important question of food supply in late 1974 , the decisions by the major food exporting or donor countries could hardly have been affected by the delay in carrying out devaluation.

However, devaluation in 1974 would have affected the relative price of rice and jute and that also in time to influence the cultivation of jute for 1975 season; sowing was in early 1975 and the relative price in late 1974 determined the acreage devoted to the growing of jute in early 1975.

The entire controversy between the Bangladesh government and the external donors, especially the Fund and the Bank, centred around the relative importance of devaluation vis-à-vis other measures of financial and economic stabilization as well as measures for improvement in economic efficiency. To many in Bangladesh the emphasis by the donors on devaluation as the 'acid test' of economic management and an appropriate development policy was found to be excessive.

From the perspective of Bangladesh, the mobilization of domestic resources, including the elimination of subsidies, reduction in deficit financing and a generation of a surplus by the public enterprises, were as if not more important. These measures would have contributed to the improvement of the internal balance as well as pre-empting an overvaluation of the exchange rate. At the same time there was a gradual need for an improvement in organization and economic management, especially in the field of agriculture and nationalized industry, if the relative price incentives provided by devaluation were to lead to an improvement in exports.

Devaluation represented one policy reform in the context of numerous policy measures and could be implemented by a stroke of pen. But its consequences always took time to work themselves out. The measures on the domestic front needed a greater commitment and a more concerted action. A dialogue on these measures should have preceded devaluation; the timing of devaluation itself as well as the timing of any 'pressure' by the Fund and the donors with regard to devaluation should have been determined in the light of progress made in achieving a consensus on the range of measures that are needed to make devaluation effective.

In the event, real devaluation was less than the nominal devaluation in view of rising wages and prices as well as accompanying increase in domestic credit creation, which followed devaluation (Tables A29.1 and A29.2 in the appendix).

Even after devaluation in 1975, export incentive schemes of various kinds were continued for individual non-traditional exports both agricultural and industrial. For example, rebates on taxes, both on export and imports, were continued. Import duties on import components of exports were exempted; export duties as well as excise duties on exports were waived. For a number of minor items, there were special import licensing arrangements which allowed the exporter to have access to foreign exchange amounting to a certain percentage of export earnings. This was in addition to the import quota to which the exporters were entitled in the same way as any other domestic producer.

In spite of various types of export incentives, the net export subsidy arising from all these incentive schemes did not exceed 10 per cent of gross export value. At the same time, however, in many cases the rate of nominal protection reached as high a figure as 100 (Matin, 1981; Parkinson, 1981).

It was true, however, that devaluation was not accompanied by import liberalization in the sense of eliminating all import quota or restrictions and allowing all imports to be determined by market supply and demand. For a number of items larger import quotas were allowed. Secondly, import quotas, instead of being granted for each individual item, were now administered for a broader range of commodities so that the choice between individual items within the broad range was left to the forces of supply and demand. Thirdly, a large part of imports were continued to be carried out by either government ministries or by the Trading Corporation of Bangladesh, which had a monopoly of imports of certain consumer items as well as intermediate inputs. Furthermore, to the extent there was a fall in the volume of the government imports, it was basically due to a shortfall in government's development expenditure caused by resource constraint. Furthermore, the fall in imports, especially in the private sector as well as in the nationalized sector, was due partly to the increase in credit constraint which was enforced prior to the devaluation.

However, even in a regime of quantitative restrictions on trade, especially imports, devaluation contributed to the increase in the supply of foreign exchange and to the reduction of the excess demand for imports and consequently reduced the pressure on the import control regime for allocating foreign exchange. Devaluation also partly reduced the anti-export bias, since the local currency price of exports rose by the full amount of devaluation, whereas the price of import substitutes (insulated from import competition) tended to rise only to the extent that devaluation increased their raw material and other costs and that the elasticity of demand for their products permitted.

Following the devaluation, there was an increase in the annual percentage change in domestic credit creation. In fact, there was a slowdown in the rate of increase in domestic credit creation between 1973/74 and 1974/75, which was the year preceding devaluation. The rate of increase in credit creation was accelerated next year and rose to 17 per cent during 1975–76 compared to 13 per cent in 1974/75. The fall in the rate of inflation following devaluation, in spite of the increase in credit creation (deficit financing), was principally due to very good harvests caused by good weather unconnected with fiscal and monetary policies.

During the year following devaluation, there was a fall in the value of both traditional and non-traditional exports in dollar terms; they recovered subsequently; the value of non-traditional exports reached the pre-devaluation level, whereas the traditional exports exceeded the pre-devaluation level (Table A29.1, appendix).

The nature of disagreement between the Fund and Bangladesh regarding the magnitude and the timing of devaluation and the implementation of associated policies seemed to follow a historically fixed pattern in respect of the negotiations of the Fund with a member country. In fact, in matters such as these a great deal depends on judgment. Getting the 'right' real exchange rate is a difficult problem. Moreover, it is not easy to determine what is the size of devaluation which is needed to achieve a 'target' real exchange rate and what are the effects of a change in real exchange rate.

Whether the real exchange rate was overvalued or not depended on the evaluation of the *equilibrium* real rate of exchange. The *equilibrium* real rate of exchange was the price of tradeables relative to non-tradeables which, for a given set of 'equilibrium or sustainable' values of other factors including taxes on trade, world prices and capital flows, etc. leads to a simultaneous determination of external and internal equilibrium. This rate must also at the same time be one which would facilitate the long term growth. This definition of equilibrium real rate of exchange refers only to real variables; nominal variables determine the actual rate of exchange.

There were several implications of this approach. First, if there were changes in any of the other variables that affected the internal and external equilibrium for example, terms of trade, then the equilibrium the real rate of exchange would be different. Any change in import tariff, except tax of interest rates, etc. would also affect the equilibrium rate. Second, while analysing the Interaction between the long-term determinants such as changes in demand and technology affecting exports and imports, on the one hand and the equilibrium real rate of exchange, on the other, the permanent and temporary changes in these variables need to be distinguished.

Third, the sustainable equilibrium real rate of exchange could be affected by changes in the long run, determining variables to such an extent that even when a current real rate of exchange was greatly depreciated in relation to the past, still it could be overvalued and the disequilibrium could persist.

A general equilibrium framework is ideally needed specifying the relationship between exchange rate, exports, imports, capital flows, and inflation, etc. to determine the 'target' real exchange rate. In the absence of such a model, resort is made to a rough and ready procedure which adjusts for changes in relative prices in a given country vis-à-vis her trading partners as compared to the year when the exchange rate is assumed to be right or appropriate. This index is liable to an error and can very easily be a subject of debate or disagreement between the Fund and a member country; in view of recognized limitations of this Index, it should be used with caution and with as much additional information as trade flows, capital movements as well as exchange rates in the possible parallel market as possible.

The extent to which a nominal devaluation affects real exchange rate and for how long depends, among other things, on the substitution elasticities between tradeable and non-tradeable goods in production and consumption as well as on the associated policies which are undertaken including the monetary, fiscal and wage policies. Therefore, the size of recognized devaluation cannot be determined in isolation.

While it is generally accepted, though adequate information is lacking, that devaluation works in the direction of improving competitiveness of exports and the balance of payments, it is not easy to foresee or predetermine how long improvements last. This depends on other policies.

What is the relative importance of fiscal, monetary and wage policy, depends on circumstances. There is no recognized theory which provides, guidance as to the appropriate sequence in which various policy reforms should be implemented. The judgment of the policy makers/shapers, is critical in this respect. If fiscal or monetary discipline can not be retained or enforced to reduce inflation and keep wages from rising, the real devaluation following a nominal devaluation would be inadequate; the rate of exchange will soon revert itself to the earlier level. Given the fiscal deficit or monetary indiscipline, the successive doses of nominal devaluations may be needed to reach a target real exchange rate. This may result in a situation of runaway inflation and unmitigated balance of payment deficit rather than an improvement in the internal and external balance.

If the appropriate, fiscal and monetary policies were not politically feasible, the only way was to rely on exchange rate adjustment, if necessary, through successive devaluation. In those years, i.e., early 1970s, a floating exchange rate was not as yet a popular idea among the international monetary Institutions.

The two roles of exchange rate sometimes clash, the stability of the nominal exchange rate serves as a monetary anchor as well as an anti-inflammatory factor. With a fixed exchange rate, the quantity of money becomes endogenous; governments afraid of inflation tend to keep exchange rate fixed or to devalue very slowly so that overvaluation is the result. This subverts the second role of the exchange rate, i.e., incentive to exports. Exports are discouraged in the presence of overvaluation, resulting in a deficit in current account. When inflation Is high and variable, fiscal deficits and inflation rates must be brought under control before the trade policy or exchange rate reforms can be efficiently implemented. Macroeconomic stability should, in fact, precede trade reforms.

The donors might have thought in the Bangladesh case that the devaluation must be the first step, and would act as a lever to force the authorities to undertake the supportive monetary and fiscal policies. This was, however, not proved to be the case, In fact, prior to the devaluation, fiscal deficit was cut, demonetization was undertaken and the rate of inflation was reduced.

References

Matin, K.M., *Bangladesh and the IMF: An exploratory study* (Dhaka: Bangladesh Institute of Development Studies, 1986).

Parkinson, J., 'Bangladesh and the Fund', in J. Faaland (ed.), *Aid and Influence* (London: Macmillan, 1981).

Appendix

Table A29.1 Devaluation and expor

	Official exchange rate
1973/74	8.0
1974/75	8.0
1975/76	14.50
1976/77	14.50

Table A29.2

Rate of change (%)

Rate of change (%)

Cost of living of industrial workers
Cost of living of rural workers

Sources: Matin, 1986, pp. 127–31; BIDS, 1986

PART X

TRADE AND AID: WORLD TRADE ISSUES

Chapter 30

Progress of the GATT Negotiations on Agriculture and Developing Countries: Options and Strategies*[1]

Introduction

The current GATT negotiations known as the Uruguay Round have entered their last phase; they are expected to be completed by the end of this year, i.e., 1990.[2] Regarding agricultural trade, attention to date has been focused on the major disagreements between the EC and the USA about the speed, content and techniques of trade liberalization. It is feared that unless they are resolved, the Uruguay Round as a whole may be jeopardized. In the event the implications for developing countries of the various proposals for reforms under consideration at GATT, as advanced by major trading nations, have not received as intensive an examination as they deserve in the light of the great importance of agriculture in developing countries.

This chapter is an attempt to highlight the progress of negotiations on the liberalization of agricultural trade, with a bearing on the main issues of interest to developing countries; furthermore, it explores the various options they face in achieving a resolution of these issues. It examines a few selected aspects which are high on the agenda of developing countries such as 'separate and differential treatment' of developing countries, non-traditional exports of developing countries, impact of liberalization on the net food importing countries, price instability and compensating measures for developing countries. This chapter ends upon with a summary containing an enumeration of the main policy options for developing countries.

The GATT contracting parties agreed at the midterm review in April 1989 that there was to be substantial progressive reduction in agricultural protectionism, sustained over a period of time, designed to correct distortions. It was also decided that agricultural reforms would be dealt with in a comprehensive manner covering market access and export subsidies as well

---

* First published in *Quarterly Journal of International Agriculture*, 1990.

as domestic support programmes, including sanitary and phytosanitary regulations (GATT, 1989a).

Regarding the issue of market access, the tariffication approach seems to be gaining ground, whereas in respect of the domestic support measures, there is some agreement on the use of an aggregate measure of support to agriculture (International Policy Council on Agricultural Trade, 1990; Zietz and Valdes, 1968).[3] Tariffication is the conversion of all non-tariff barriers including variable levies, minimum import price and import quotas into bound tariffs. In the transition period, such an immediate conversion may not be feasible. The use of the 'price wedge' for tariffication purposes may result in very high tariffs for commodities for which quota restrictions are very severe, In these cases, the import quota can be gradually increased so that the gap between the domestic and world price goes down so as not to exceed the ceiling imposed by the level of the bound tariff already agreed upon in advance (Tangerman and Miner, 1990).

Tariffication is not an alternative to the use of an aggregate measure of support; it is complementary to assessing various domestic measures of support, which have trade-distorting effects and are to be included in the quantification of the aggregate measure of support. The aggregate measure of support (AMS) has a number of versions. In its comprehensive version, it includes not only a wide variety of domestic measures of support to agriculture but also border measures which restrict trade, i.e., tariff and non-tariff barriers. In a more restrictive version, it includes only domestic measures which affect trade directly or indirectly (OECD, 1987). Three categories of domestic support measures have been identified: a) measures which push domestic prices above world market prices, and income support policies linked to growth in production; b) income support policies which are not tied to production, such as marketing, environment and conservation programmes and bona fide disaster assistance; c) all other policies that do not meet the criteria for the previous two categories. It has been suggested that measures in the a) category should be phased out, while those in the b) category should be permitted and those in the c) category should be disciplined and reduced through negotiations (UNCTAD, 1990). The use of AMS is not without difficulty. There are difficulties in estimating AMS for all commodities in all countries. Also, AMS being an aggregate measure can be reduced by increasing some restrictions and reducing others.

It is, however, most likely that the use of the AMS would be mainly a device or mechanism for evaluating or monitoring the progress in agricultural liberalization, whereas the commitments by the countries would be made in terms of specific policy instruments such as domestic support measures as well as tariffs and export subsidies, etc.

There are a few unresolved issues raised by the EC and Japan which stand in the way of the ready acceptance of tariffication. The EC wants to be able to vary tariffs within limits to stabilize domestic prices rather than undertaking domestic compensatory measures to stabilize income or to offset effects on those who suffer from price instability. Further, the EC seeks to undertake tariff harmonization as for example, to raise tariff on oilseeds, which currently have low tariffs, to bring them in harmony with other more highly protected commodities (Zietz, 1989; Goldin and Knudsen, 1990). Japan would like exemption for rice from the process of tariffication as a way of eliminating trade restrictions on rice; rice, she claims, deserves excess tariffs in view of the need for food security in Japan.

Special and Differential Treatment of Developing Countries

How do the developing countries stand in relation to the various proposals for the reform of the GATT rules and for the liberalization of trade in agriculture? In the past the implications of the 'special and differential treatment' of the developing countries were as follows: developing countries were not required to reciprocate the trade concessions which were granted by developed countries; in other words, concessions granted to each other by the developed countries as a result of reciprocal bargaining were extended to the developing countries under the MFN treatment (most favoured nation treatment) provided under the GATT rules. Developing countries were permitted to resort to quantitative controls for balance of payment reasons; they were also allowed preferential access to developed country markets (known as the generalized scheme of preferences (GSP)) as well as to enter into preferential trading arrangements among themselves.

The preferential access to the developed country markets under the GSP was frequently granted at the discretion of the developed countries and in respect of a very narrow range of commodities, mostly non-agricultural, which did not pose a serious threat or competition to the domestic production. A very limited number of countries benefited from the scheme; many others could not take much advantage of the preferential schemes because of their limited production and export capacity. Also, the developed countries frequently attempted to use the GSP as a means for extracting concessions from developing countries in such areas as trade in services, foreign private investment and intellectual property rights, etc. Increasingly, the developed countries implemented a process of graduation, by reducing the size and extent of concessions that were granted to high or middle income developing countries (Brown, 1986; Karsenty and Laird, 1987).

A growing diversity of interest among developing countries has clearly emerged in the current round of trade negotiations. First, a group of major exporters such as Thailand and Argentina, etc., has combined in the so-called CAIRNS group with a number of developed countries such as Canada, Australia and New Zealand, to press for more liberal access in export markets. Second, the food importing countries, such as Jamaica and Egypt, which are likely to suffer from a rise in food prices consequent on trade liberalization, have been pressing for compensatory measures. Third, a group of countries which are neither net exporters nor importers but are more or less self-sufficient, such as India, Indonesia, and Pakistan, etc., are mainly interested in freedom to pursue domestic policies for promoting their agricultural development.

This distinction among the developing countries, however, is not a rigid one nor are their interests totally separate. The net food importers at the same time are also often agricultural exporters. Similarly, the major cereal exporters also import some other agricultural commodities, even though not cereals. Today's self-sufficient countries can become tomorrow's importers or exporters. Moreover, with the growing diversification of the export structure of developing countries they have developed a common interest in promoting a larger market access for their non-traditional exports.

The developing countries are loath to abandon their case for 'special and differential treatment' at this stage of the negotiations. It is something which they won through hard struggle over the past decades; they can be persuaded to modify them only in return for some offsetting concessions in other areas of trade negotiations. While the 'special and differential' treatment may not be for them the most important negotiating objective, it is certainly considered an important tool in their negotiating strategy.

The nature and content of the 'special and differential' treatment which may be sought in the future would be different from what was pursued in the last decades (Whalley, 1990). For one thing, it seems to be now accepted that developing countries, especially the middle and high income developing countries, would be required to make some reciprocal concessions in order to gain access in the markets of developed countries for commodities of interest to them. In recent years, however, a number of the developing countries, irrespective of their stage of development, have already liberalized trade either unilaterally or, in some cases, as a part of their structural adjustment programmes, under the auspices of the World Bank or International Monetary Fund. They desire that the liberalization of trade restrictions already undertaken by them should be considered as a part of 'reciprocity' on their part in exchange for trade concessions to be obtained under the GATT.

There seems to be a growing consensus that the 'special and differential' treatment to be accorded to developing countries in the future could be along the following lines. First, the developing countries should participate in the policy of 'tariffication' but would be allowed temporarily to have higher rates of tariffs, which may be reduced over a longer timeframe. For some commodities they may end up with a higher average level of tariffs than in the case of developed countries. Second, the unrestricted use of quantitative controls as in the past for an unlimited period, irrespective of the stages of their development, on the plea of balance of payments deficit, is to be eschewed under the new GATT rules. However, the right to use quantitative controls for balance of payments reasons may be permitted only for a temporary time period and also that under international surveillance by GATT (US Treasure Acting Secretary, 1988; Anjaria, 1989). Third, at least some developing countries, such as the low income or the least developed amongst them, may gain access to the markets of developed countries at a very much lower rate of restrictions or tariffs over a longer period than the rest of the developing countries. Fourth, the domestic measures for the promotion of agricultural development should be permitted under the discipline of GATT. But, there may be a number of development policy measures which are in the grey area, i.e., neither prohibited nor freely permitted under the rules to be set up under the GATT, but which may be allowed for a transition period in developing countries to promote their agricultural and rural development. Fifth, the developing countries may be allowed to undertake a few measures which interfere with trade liberalization such as the use of trade controls or variable levies in order to stabilize domestic food prices. For countries with very low per capita income or with a high percentage of their domestic consumption expenditures on food, the objective of domestic food price stability is frequently a political imperative. Sixth, food importing countries may be provided compensation to meet the adverse impact of a possible rise in food prices in the world market as a result of trade liberalization.

Non-traditional Agricultural Exports

The developing countries have considerable interest in ensuring that the trade liberalization measures and new GATT rules cover the widest possible range of agricultural commodities extending beyond the traditional commodities like cereals, tropical beverages and agricultural raw materials. They should cover such non-traditional exports as horticultural and floricultural commodities, which have bright prospects in world trade. This is due partly to the high income elasticity of demand for them and partly to an increasing diversification of the pattern of food consumption in the developed and middle income developing countries. The developing countries are likely to have comparative advantage in the exports of such products because they are labour intensive and are often capable of being produced efficiently by the small farmers (Islam, 1990).[4]

The average rates of tariffs in the industrialized countries on the imports of horticultural products, both raw and processed, exceed the average rates of tariff on other agricultural commodities. They are, in addition, subject to high non-tariff barriers including, in particular, sanitary and phytosanitary regulations.

The GATT negotiations on agricultural commodities are carried out in separate committees, i.e., the Committee on Agriculture, which excludes horticultural countries, and the Committee on Tropical Products which includes amongst non-traditional products only tropical fruits and nuts and flowers and plants. There is a great deal of uncertainty as to the appropriate mechanism or committee for the negotiations on such products. It appears very important that the developing countries should seek their inclusion in many committees including those dealing with tariffs and non-tariff barriers.

In view of the very large number and heterogeneous nature of non-traditional exports such as horticultural products, it is difficult to negotiate trade concessions on a product by product basis, specifically since not much data or analysis for this set of commodities is so far available. For example, it is very difficult to produce an estimate of aggregate measure of support (AMS) for such a heterogeneous set of commodities, particularly if the impact *of* sanitary measures in trade flows is to be quantified as well.

The experience with the working of the Committee on Tropical Commodities indicates not only that the list of commodities covered by this committee is limited but also that within this list the concessions so far offered by the individual developed countries cover only a few selected items. Moreover, in many cases, concessions granted are conditional upon reciprocal concessions to be made by developing countries.

At the start of the Uruguay Round it was agreed that for tropical products, among which are included many non-traditional agricultural commodities, the developed countries will be more forthcoming and will make the speediest and biggest reductions in trade restrictions (GATT, 1989b).[5] This was considered both desirable and politically feasible, because while, on the one hand, this sector is important in a very large number of developing economies, on the other, trade liberalization in this sector is likely to meet less stiff or organised resistance in the developed countries, as the domestic sector affected by import competition is relatively of minor importance (ibid.).

What is most important for the future growth of horticultural exports of developing countries is a liberal trade regime with stability and certainty over time, since many of these products are new or have been introduced in world trade only in recent years. They require investment in export infrastructure, including marketing and distribution facilities. Furthermore, prospects of future market development are likely to be brighter for the processed rather than for the fresh products; in this regard, the adverse effect of the tariff escalation by the degree of processing is important. Without a substantial liberalization of trade in the processed horticultural products, the future expansion of exports will be limited.

The sanitary and phytosanitary regulations affect very significantly the exports of horticultural products more than any other group of agricultural products. This is an area in which agreement amongst the trading partners seems to be very advanced (GATT, 1989a, 1990). The principal elements of an agreement are as follows. Firstly, it has been agreed that such regulations should have minimum negative effects on trade. Secondly, they should be harmonized based upon internationally agreed standards, some of which are promoted by UN organizations or other international organizations. Thirdly, it is necessary to provide a scientific basis for national standards. Safety standards for the use of pesticides or fertilizer or their residues can be universal. But the standards for diseases control of plants and animals will vary from country to country because country situations widely differ. It is agreed that these standards should not cover quality or consumer preference but should be mainly concerned with safety, inspection techniques and procedures for testing.

There are two considerations to be kept in mind in the course of implementation of these regulations once they are agreed upon. First, the developing countries need an assurance that these regulations have the widest possible commodity coverage, and are not restricted to commodities which are currently under negotiation in the Committee on Agriculture, i.e. commodities of primary interest to developed countries. Second, when the standards are changed from time to time it is necessary to ensure that they do not unduly disrupt the domestic producers. Third, frequent changes in the standards set by developed countries make it difficult for the exporting developing countries to adjust easily or promptly; hence a period of transition is needed in order to allow adjustment in developing countries' export capacity.

There is a pressing need for transparency in the sanitary and phytosanitary standards, as well as for an effective procedure for the notification of national regulations or bilateral agreements. The process for the settlement of disputes under the GATT auspices should be strengthened; scientific evidence and expertise should bear on the decision-making process of the GATT dispute settlement mechanism.

The developing countries require technical and financial assistance, for the establishment of testing laboratories as well as the training of scientific personnel to enforce such standards. The appropriate international organizations, with adequate technical and financial resources, should be entrusted with this responsibility.

The Impact of Liberalization on the Net Food-importing Countries

There seems to be an emerging consensus that a rise in world food prices is most probable after the liberalization of trade, even though uncertainty persists regarding its extent. The uncertainty relates, amongst other things, not only to the extent of trade liberalization, including the commodity coverage, but also to the price elasticity of supply and demand in trading countries. There is an uncertainty regarding the extent of price rise due to liberalization as well as to the extent of gains from export earnings due to liberalization.

The extent of price rise is likely to be less if both developing and developed countries participate in the liberalization of trade. This will also be the case if liberalization is partial and not uniform across all commodities, which is probably the most, likely outcome. Again, with low international commodity stocks, prices can move significantly with a small change in production. In the short term, prices could rise, due to speculative reasons.

The impact of trade liberalization on the import price of food has two sets of consequences for developing countries. One is the adverse effect on the balance of payments insofar as their food import bill goes up in the short run. The rise in food import will be moderated to the extent that domestic production increases in response to higher price incentive. Second is the adverse impact on the low income consumers. The net food-importing developing countries request offsetting measures or compensation which could relieve their balance of payments problems as well as help them to cushion the adverse impact on the poor consumers.

Recently, for a few important net food-importing countries, an estimate has been made of the impact on their food import bill of a rise in price following the liberalization of trade. It has been found that the net import cost of basic foods for the years 1984–86, consisting of meal and meat preparations, dairy products and eggs, cereals and cereal preparations, oilseeds, fats and oils and miscellaneous foods, is likely to go up by 24 per cent for Mexico, 28 per cent for Morocco, 29 per cent for Egypt, 30 per cent for Jamaica and 33 per cent for Peru (IATRC, 1988; GATT, 1989c).

The question of compensation raises some practical problems of implementation. How to measure the rise in prices which is due to liberalization measures, as distinguished from other influences working on prices at the same time? What will be the product coverage for estimating the impact on the balance of payments? To what extent the gains in their export earnings due to trade liberalization offset the adverse impact on their food import bills? Should all countries, irrespective of income level, be eligible for compensatory payments? Should the compensation be provided mainly to the low income countries and not to the middle income countries? What will be the time limit for the termination of compensation payments? There is also the additional question as to whether compensation should be paid only for a rise in food prices and not for losses which a few developing countries may incur as a result of the termination of their preferential access to the markets of developed countries, which they currently enjoy, as for example, under the Lomé Convention of the EC and the Caribbean Basin Initiative of the US. The emerging consensus seems to be that compensation should be paid for a rise in food prices (the composition of food to be defined later), above the trend level, which otherwise might have occurred in the absence of liberalization.

On the question of compensation two additional issues are relevant. First, should the compensation take place within the framework of GATT as a part of the various trade concessions which are being negotiated? Second, should the compensation be provided outside the GATT mechanism but in coordination with and at the same time as the GATT negotiations are completed? In the latter case there is a need for coordination between the GATT, on the one hand, and other international agencies which might be involved with the compensatory arrangements, on the other. If compensation is provided within the context of GATT, the food-importing countries could probably be provided, for a definite period of time, with a higher level of trade concessions on their exports, i.e., a lower rate of tariffs or other restrictions on their exports.

The possibilities outside the GATT framework are: a) access to the IMF Compensatory and Contingency Financing facilities; and b) food aid. The recent changes in the IMF facility have made access to its resources more difficult or less automatic, specially in view of the various conditionalities regarding economic policies in the borrowing countries. The eligibility of a country to draw upon the facility depends on the IMF being satisfied that the members' balance of payments difficulties are not due to serious deficiencies in policies. Furthermore, the facility provides a loan, not a grant, and the interest rate on the loan is not concessional. Also, the facility seeks to compensate for prices of cereals and not of all food grains.

There is a more serious limitation on the IMF compensatory financing facility in that the increase in the import cost is offset against any rise in export earnings and it is the net increase in the import cost that is compensated. This loads to an anomaly in the use of the cereal financing facility that makes the developing countries reluctant to use it for this purpose. When the compensatory facility is used to counterbalance shortfalls only in export earnings, changes in imports are not used to determine the extent of their eligibility. It is only in the case of the cereal component of the compensatory facility that a netting of exports and imports comes into play. This asymmetrical inter-relatedness of the two components has made the developing countries reluctant to use the cereal facility. They prefer the compensatory mechanism in the IMF facility for exports which has less strict criteria (Ezekiel, 1985, 1989; Kaibini, 1988).

As far as the role of food aid as a compensatory mechanism is concerned it raises a number of questions. It is generally agreed that food aid, provided it not used as a hidden export subsidy to promote commercial exports, is permissible under the new proposed GATT regulations. However, food aid, if it is to be used as a compensatory mechanism, needs to be in addition to what the developing countries might be receiving in any case. To prevent food aid from being used as an export subsidy, the members of the GATT will have to adopt and enforce a test of legitimacy. The multilateral food aid – given as a grant – is unlikely to be used as an export subsidy. Moreover, contributions made in cash and used by the multilateral agencies to purchase food from the cheapest source, can seldom be used as an export promotion device. The contribution of cash aid to a multilateral agency to buy food from the cheapest source would be a very significant departure from the current practice. Unless pressure can be brought on the donors for providing high cash donations, total food aid may in fact be reduced under such a multilateral framework; a compromise may be struck by having a combination of cash and food donations. The existing arrangements under the FAO Surplus Disposal Committee which seek to monitor and ensure that food aid does not replace the commercial exports of the competing countries, or promote the commercial exports of the donor country need to be strengthened (Hopkins, 1983).

However, food aid availability, as a result of liberalization, might be reduced: in the future due to a decline firstly in food surplus, and secondly in the publicly-held food stocks. Food aid may be further discouraged by the more stringent conditions, as explained above, which are to be put on the grant of food aid under the new GATT rules. To serve as an appropriate compensatory mechanism, the agreement on food aid should be reached simultaneously as the GATT negotiations are finalized. Otherwise, the net food importing countries would have no assurance that additional food aid would be provided as a compensation. The current Food Aid Convention, which is a part of the International Wheat Agreement negotiated in the Kennedy Round of the GATT, is administered by the Food Aid Committee, i.e., a committee of food aid donors. This could also be used as a forum to reach an agreement on the compensatory use of food aid. Thus, it appears that both food aid and additional concessions in terms of trade liberalization can be used in some combination, to compensate the net food importing developing countries.

Trade Liberalization and Instability of Prices

As discussed earlier, global trade liberalization has the potential of raising agricultural prices at least in the short to medium run, and also of reducing the instability of world prices (Blantford, 1983; Tyers and Anderson, 1988).

There are at least three ways in which a country can contribute to the instability of world prices. One way is to prevent world price instability from being transmitted to the domestic economy; second, the domestic instability in its turn is transmitted fully or partially to the world market; third, trade or other policy measures which drive a wedge between the world and domestic prices are changed frequently. The contributions of the instability in policies to the fluctuations in prices has seldom been emphasized. The US policy, for example, as contrasted with the EC policy, is more unstable and thus contributes to the overall instability. This has important implications for the stock behaviour of the traders and the producers.

Even a partial trade liberalization, in the sense of liberalization by a group of countries as against worldwide liberalization, can contribute to the reduction in instability as well as a rise in the absolute level of prices. For example, a recent study shows that if only the OECD countries liberalize, the average degree of instability of food prices in the world market is reduced from 34 per cent (year to year fluctuation) to 23 per cent. However, if the developing countries also liberalize at the same time the variability is reduced from 23 per cent to 12 per cent (Tyers, 1990).[6] On the other hand, if the developing countries alone reduce trade restrictions, the reduction in price instability would not be significant. In fact, they will suffer from the price instability created by developed countries in the world market. Developing countries, therefore, cannot gain unless all the countries, including developed countries, liberalize.

The degree of instability of world prices in the future will depend not only on the extent of trade liberalization but also on the behaviour of stocks, held by the main exporters and importers, i.e., whether stocks respond to price changes in a normal fashion, including the release of stocks when the prices rise and their accumulation when prices fall. The reduction in the size of stocks of cereals held by the US during the late 1980s has partly contributed to the instability of world prices. In the coming years there is likely to be a reduction in the stocks–output ratio due to a decline in public stocks; since the behaviour of private stocks is uncertain, there might be a tendency towards increased instability. In order to achieve price stability in the world markets, a set of internationally agreed rules which govern the degree of insulation of the domestic markets is preferable to reliance on the uncertain stock behaviour of the loading cereal traders such as the USA and the EC.

Even if there is some reduction in the degree of instability in world prices, in the post trade liberalization period, the developing countries might still want to intervene to stabilize prices within a narrower range. Moreover, a higher level of output in the future may be associated with a higher degree of instability, as experienced in the past during the course of the Green Revolution. Furthermore, the developing countries will be reluctant to face the risks of wild price fluctuations, as took place during 1973–74, especially in view of the severe short-run impact on the poor (Ahmed and Bernard, 1989; Timmer, 1989).

The intervention in the domestic market to achieve price stability is also justified on the ground that the developing countries lack a sophisticated and well-functioning credit market with easy access for the agricultural producers and the food consumers, particularly the poorer ones among them to offset the impact of price fluctuations. Neither is it financially and administratively feasible for them to undertake direct income payments to those who are adversely affected by price instability (Siamwalla, 1990).

In order to reduce fluctuations in domestic prices, the developing countries would need, and therefore should seek exemption from the GATT rules so that they are able to use variable taxes and subsidies in border trade and/or to undertake through the state marketing agencies the purchases and sales in the domestic market. Even though a system of tax/subsidy on trade is more cost-effective than the establishment of large domestic stocks and open market operations, variable levies in their turn introduce instability in the budgetary resources of the country; they do not in any case alleviate the impact of variations in the cost of agricultural or food imports on the balance of payments.

It is necessary, however, that the domestic measures which the developing countries may take for the purposes of stabilization of prices should be justified, be transparent and subject to supervision and surveillance by the GATT. The purpose of the GATT supervision is to ensure that this does not open the door to protectionism and the interest groups in the particular commodity markets under the guise of stabilization do not raise the price level above the long-run trend. Since there is no undisputed and definite trend level in world prices on which everyone agrees, there is always a scope for the interest groups to press for a higher price than justified by the long-term trend.

The price-stabilizing measures adopted by developing countries may, however, have the effect of exacerbating international price instability of commodities involved are those in which the developing countries play a major role in world trade. They will gain most if the commodities are those in which they collectively have a small share in the world market such as cereals.

While the instability of food prices is of great concern to the developing countries, the impact of instability in prices of export products, specially tropical products, which have important consequences for their income and balance of payments, has received no attention in the context of the GATT negotiations. Given the limited success of international commodity agreements, the search for effective measures for compensating or offsetting fluctuations in agricultural export earnings of developing countries should be continued through an improvement in the IMF Compensatory and Contingency Financing Facility or in such regional schemes as STABEX under the Lomé Convention.

Evaluation

The developing countries should ensure that: a) the trade liberalization has the widest possible commodity coverage including commodities of vital interest to them; and b) that the likely 'side effects' of liberalization in terms of a possible rise of food prices in the world market or a constraint on their freedom of action to deal with the instability of agricultural prices are mitigated.

This in turn has implications for the future nature and content of the much debated 'special and differential treatment' for the developing countries, which takes into account their financial and trade needs and special constraints on development. Past experience with differential treatment for developing countries has not been encouraging. Preferential schemes have been selective and of limited significance. A gradual process of differentiation and graduation amongst the developing countries has set in. Non-reciprocity has kept the developing countries outside the main stream of the trade liberalization process. They may improve their economic efficiency as well as bargaining power by giving up the 'waiver' to use quantitative restrictions for balance of payments reasons, except in the short run and in very well defined and highly selective situations under the surveillance of the GATT.

The developing countries, particularly the middle and high income countries amongst them, are unlikely to gain trade concessions without granting reciprocal concessions. However, their large domestic market to which the developed countries seek access endows them with bargaining power in their trade negotiations, especially if cross-sectoral concessions (i.e., agriculture, industry, services and intellectual property rights) are, exchanged towards the end of the negotiation process in the Uruguay Round. To the extent that in recent years they have already liberalized their trade regime either unilaterally or as part of an adjustment programme under the auspices of the World Bank or the Fund, they should seek appropriate 'credit' in the course of their current negotiations under the GATT.

At the same time the developing countries need differential treatment in the course of the implementation of the trade reforms in a number of aspects. They include a greater flexibility in the implementation of the trade liberalization measures, such as a longer time frame or transition period for the reduction of trade barriers, and a more generous interpretation of permissible, domestic support measures for agriculture and rural development.

The commodities of special interest to the developing countries, such as tropical products, should receive, as was originally agreed at the time of the inception of the Uruguay Round, the deepest cuts in trade restrictions as fast as possible. The progress so far in the liberalization of trade restrictions on tropical products has been very limited, in spite of high hopes raised earlier.

Special attention should be paid to the need for an enlarged market access for nontraditional agricultural exports, such as horticultural products which hold bright market prospects in the future. In this context two issues deserve serious attention. First, the present structure of the GATT negotiating committees does not clearly specify where the nontraditional commodities such as the horticultural products are to be dealt with. In view of the urgency of this subject, developing countries should, therefore, pursue negotiations on these commodities in all the relevant committees dealing not only with agricultural and tropical products, but also tariffs and non-tariff barriers which cut across commodities. Second, significant progress has been made in negotiations on internationally agreed scientific standards to be enforced under the international surveillance and guidance of the GATT. It is essential that developing countries strive to ensure that all commodities of special interest to them are covered by these regulations. Third, in view of a lack of trained scientific manpower and institutions to enforce and adopt such standards, the provision of technical and financial assistance by the relevant international organizations and others should be assured.

Two additional issues of particular importance to developing countries are: a) compensation for a possible rise in food prices in the world market following trade liberalization; and b) freedom to undertake measures, including border measures, to stabilize domestic food prices within acceptable limits. The compensation for a possible rise in food prices is particularly relevant for the low income developing countries which in the short run may not gain advantages from a rise either in the volume or in the price of their export earnings. Several compensatory measures either singly or in combination deserve consideration, i.e., enlarged market access for the exports of the food deficit not importing countries, at least for a transitional period; additional food aid under a renegotiated Food Aid Convention; and IMF compensatory and contingency financing facility. The first two alternatives seem to be more feasible than the last. Irrespective of the particular mechanism chosen it is necessary to make such a compensatory arrangement an integral part of the Uruguay Round.

To the extent price instability in the post liberalization period remains a problem, in order to stabilize food prices, the developing countries should be able to undertake border measures which may result in a partial insulation of the domestic market from the world price movements and, therefore, their use would require exemption from the GATT rules. However, in order to pre-empt these measures from being used for raising domestic prices in the long run and thus encouraging protectionism, measures intended for promoting price stability should be transparent, well specified and be monitored under the surveillance of the GATT. The need to intervene in trade in order to promote stability could be partly alleviated to some degree, if international coordination of national cereal stock policies, under an agreed set of guidelines, is undertaken.

Summary and Conclusions

The developing countries, both as substantial exporters and importers of food and agricultural commodities, stand to derive significant benefits from the liberalization of world agricultural trade currently under negotiations in the Uruguay Round. They, therefore, have a particular interest in bringing agriculture within the framework of the rules and disciplines of the GATT which can liberalize market access and import protection as well as reduce or eliminate export subsidy and domestic support programmes. The alternative is discriminatory protection and bilateralism under which the developing countries with limited economic strength will suffer most.

Notes

1 At the time of writing, the author was Senior Policy Advisor, International Food Policy Research Institute, Washington, DC.
2 The analyses in the paper are based not only on the examination of the various comments relating to ongoing negotiations but also on discussions with the delegations from the developing countries participating in the GATT negotiations.
3 The magnitude of net welfare gains as well as their distribution between producers and consumers, which result from the stabilization of food prices, has been much discussed. The broader considerations relating to the adverse impact of price instability on macroeconomic stability, on the one hand, and long-run investment and growth, on the other, are no less important, even though they have not been explored quantitatively.
4 With a growing tendency in developed countries to consume organically-produced horticultural products, a characteristic which increases the labour intensity of their production, the comparative advantage of developing countries in this sector is likely to be strengthened. Furthermore, the processing of horticulture products is highly water polluting. The developed countries, worried about pollution and environmental degradation, are liable to impose strict environmental standards on their domestic processing industries. This may promote the relocation of these industries to developing countries with less strict standards. This would add, therefore, to the export prospects of developing countries in processed horticultural products. However, there are two qualifications. Firstly, the developed countries may insist on the developing countries introducing equally strict environmental standards for the processing industries. Secondly, failing that, they may impose countervailing duties on products from developing countries which do not impose strict environmental standards.
5 Two provisions of the GATT negotiating mandate on tropical products agreed upon by the contracting parties, are relevant in this context. Firstly: 'Negotiations shall aim at the fastest liberalization of trade in tropical products, including processed and semi-processed forms, and shall cover both tariff and non-tariff measures affecting trade in these products.' Secondly, 'the contracting parties recognize the importance of trade in tropical products to a large number of less developed contracting parties and agree that negotiations in this area shall receive special attention, including the timing of negotiations and the implementation of the results'.

6 Food Prices refer to the combined price index relating to grains, livestock products and sugar.

References

Ahmed, R. and Bemard, A., 'Rice Price Fluctuation and an Approach to Price Stabilization in Bangladesh', IFPRI Research Report, February 1989.

Anjaria, S.J., 'Balance of Payments and Related Issues in the Uruguay Round of Trade Negotiations', *World Bank Economic Review*, Vol. 1. pp. 669–88, 1989.

Blantford, D., 'Instability in World Grain Markets', *Journal of Agricultural Economics*, 34, pp. 379–92, 1983.

Brown, D., 'Trade Preferences for the Developing Countries: A survey of results', Working Paper (New York, Council of Foreign Relations, 1986), in J. Malley (ed.), *The Uruguay Round and Beyond* (Ann Arbor, The University of Michigan Press, 1989).

Ezekiel, H., *The IMF Cereal Import Financing Scheme*, IFPRI, 1985.

Ezekiel, H., *Food Aid, Food Imports and the Food Consumption of the Poor*, IFPRI, 1989.

GATT, *Uruguay Round – Mid-term Agreement*, Geneva, April 1989a.

GATT, Multilateral Trade Negotiations. Uruguay Round, Tropical Products, 30 May 1989b.

GATT, *Multilateral Trade Negotiations: The Uruguay Round. The Ways to Take Account of the Negative Effect of the Agriculture Reform Process on Net Food Importing Developing Countries*, MTN.GNG/NG5/W/119, 2 November 1989c.

GATT, *Multilateral Trade Negotiations, Uruguay Round, Synoptic Table of Proposals Relating to Key Concepts*, 29 May 1990.

GATT, *Focus*, Newsletter (Uruguay Round Special Issue), May.

GATT, *Tropical Products: Background Material for Negotiations*, January 1988.

Goldin, J. and Knudson, O., *Agricultural Trade Liberalization: Implications for developing Countries* (OECD/World Bank, Paris, 1990).

Hopkins, R.F., 'Food Aid and Development: Evaluation of regime principles and practices' in *Report on the World Food Program*, Government of the Netherlands Seminar on Food Aid, The Hague, pp. 73–92, October 1983.

International Policy Council on Agricultural Trade, *Agriculture in the Uruguay Round. Issues for the Final Bout*, January 1990.

International Agricultural Trade Research Consortium (IATRC), *Assessing the Benefits of Trade Liberalization*, August 1988.

Islam, N., 'Horticultural Exports of Developing Countries: Past performances, future prospects and policy issues', IFPRI Research Report 90, April 1990.

Kaibini, N.M., 'Financial Facilities of the IMF and the Food Deficit Countries', *Food Policy*, Vol. 13, No. 1, February 1988.

Karsenty, G. and Laird, S., 'The Generalized System of Preferences: A quantitative assessment of the direct trade effects and policy options', UNCTAD Discussion Paper 18, UNCTAD, 1987.

Siamwalla, A., 'Agricultural Trade Reform, Price Stability and Impact on Developing Countries', IFPRI Seminar, May 1990.

OECD, *National Policies and Agricultural Trade* (Paris, OECD, 1987).

Tangerman, S. and Miner, W. 'Negotiating Stronger GATT Rules for Agricultural Trade', Discussion Paper, Series 3, International Policy Council on Agricultural Trade, January 1990.

Timmor, P., Food Price Stabilization: Rationale, design and implementation (Cambridge, Mass., Harvard Institute for International Development, May 1989).

Tyers, R. 'Agricultural Trade Reform and the Stability of Domestic and International Food Prices', IFPRI Seminar, May 1990.

Tyers, R. and Anderson, K., 'Liberalizing OECD Agricultural Policies in the Uruguay Round: Effects on trade and welfare', *Journal of Agricultural Economics*, 39(2), pp. 192–216, 1988.

UNCTAD, *Review of Developments in the Uruguay Round* (Part V), January 1990.

UNCTAD, 'Review of Developments in the Uruguay Round (Part V), UNCTAD/ UNDP Interregional Project 'Support to Uruguay Round of Multilateral Trade Negotiations', January 1990.

US Treasury Acting Secretary, 'It is Time to Rethink the Role of GATT in Economic Development Strategies'. speech at the American Enterprise Institute, September 1988.

Whalley, J., 'Special and Differential Treatment for Developing Countries in the Uruguay Round', IFPRI Seminar, May 1990.

Zietz, J. and Valdes, A., 'Agriculture in the GATT: An analysis of alternative approaches to reform', IFPRI Research Report 70, Washington DC, 1988.

Zietz, J., 'Negotiations in GATT Reform and Political Incentives', *The World Economy*, 12 March, 1989, pp. 39–52.

Chapter 31

Is Agricultural Trade Liberalization Bad for Environment?*

Where trade and environment intersect, policy should be carefully designed to promote growth, alleviate poverty and protect the environment.

The effects of the liberalization of agricultural trade on the environment have been a subject of considerable discussion in recent years. The opening up of trade results in a reallocation of productive resources among agricultural sectors and subsectors. Trade leads to changes in relative output prices, and hence in the composition of output. Changes in the relative prices of inputs resulting from the opening of trade in turn lead to changes in technology, that is, the extent and intensity of use of inputs, including chemical and mechanical inputs, and so on.

Whether changes in the pattern of production and resource use will lead to environmentally adverse consequences cannot be determined *a priori*. The ultimate effect of changes in the pattern and intensity of use of natural resources on the environment depends on whether domestic policies exist to counteract adverse environmental consequences.

Case studies of environmental consequences resulting from changes in output patterns and use of inputs following the opening up of trade can be illuminating. In Egypt, for example, an export crop, gum arabic, has been found to combat desertification and land degradation. Therefore, a trade policy such as devaluation, which increases returns from such export crops, is environmentally friendly.

Specific instances of the environmental effects of changes in cropping patterns and associated changes in the pattern and intensity of input use cannot be generalized, of course. It all depends on the agro-ecological characteristics of a particular region, as well as the domestic policies affecting prices of inputs and outputs.

---

* First published in *ECODECISION, Environment and Policy Society*, Montreal, March 1993.

The Likely Consequences of Liberalization

What are the likely consequences of liberalization in world agricultural trade now under negotiation in the Uruguay Round? As high support prices for agriculture in the OECD countries are reduced, there will be less pressure or incentive to extend cultivation to marginal lands. In fact, marginal lands are likely to be withdrawn from cultivation, with beneficial environmental consequences.

Secondly, the use of chemical inputs such as fertilizers, pesticides and herbicides will be discouraged as the profitability of increased output declines. Resources will tend to move to service or industrial sectors. Service sectors are the least polluting, and the industrial sector is subject to increasingly strict pollution standards.

Overall, there is likely to be a decline in the level of environmental stress in OECD countries. At the same time, with the rise in world prices, the exporting developing countries would expand their agricultural production. They would therefore tend to use chemical inputs more intensively, and might also extend cultivation to marginal or low-potential lands. On the other hand, increased agricultural productivity and high farm incomes in developing countries may relieve pressure to extend cultivation to marginal lands.

However, it must also be stressed that the extend and intensity of use of agro-chemicals and irrigation water, with their possible adverse environmental consequences, are greatly affected by national input and output-pricing policies, including subsidies on inputs or water use in developing countries. The correction of price distortions introduced by government intervention is likely to promote their use in a more efficient and also more environment-friendly manner.

Liberalizing agricultural trade by improving agricultural and rural incomes is likely to slow migration to the overcrowded urban slums – and increasing source of environmental degradation in developing countries.

Price Stability

Trade liberalization in agriculture may give greater stability to world prices of agricultural products. This could induce developing countries to take a long-run view of returns on investment in the conservation of natural resources, that is, water and land. On the other hand, domestic prices in developed countries may be more unstable compared to price patterns under protectionist policies. This may make farmers in developed countries take a short-run view of agricultural investment.

For the world as a whole, it is likely that the decline in environmental stress in developed countries resulting from a reduction in agricultural support prices and subsidies will more than offset any possible increase in developing countries arising from the increased use of chemicals, irrigation, mechanization, and so on.

The question has been raised whether trade liberalization in OECD countries would necessarily reduce agricultural output and therefore relieve pressure on environmental and natural resources. Falling prices may persuade small farmers to consolidate into larger units or sell out to big farmers, thus securing economies of scale and increased efficiency. This would check the fall in input use which might otherwise have resulted from a fall in output.

Another possibility is that the technological innovations that are already underway as a result of current high support prices will come on stream in the near future and compensate for the disincentive effect of a fall in price. In the long run, lower prices will discourage technological innovation, but in the near future, the favourable impact of technological innovations which are not currently implemented will be felt.

In the long run, the world community faces the challenge of meeting increasing food demands from an expanding population, especially in developing countries, at high levels of income. Whether such an increase in food supply could be met without adverse environmental consequences will depend on several factors. These include the location of future increases in food production, that is, the agro-ecological potential of regions where future increases in production will take place; macro and sectoral policies which may either help or hinder an environmentally sound pattern of resource use; the internalization of environmental externalities; the availability of environment-friendly technology; and above all, a liberal international trade regime that allows a relatively free flow of food and agricultural commodities across countries and regions, based on comparative costs and environmental considerations.

Environmental Policy and Agricultural Trade

An important example of environmental policy that affects agricultural trade is provided by sanitary and phytosanitary regulations imposed by various countries to protect human, animal and plant health. Sanitary and phytosanitary standards that refer not only to products and product performance but also to process and production methods have widespread consumer support in all countries of the world.

Divergent standards among various countries increase the cost of doing business. In order to meet diverse standards of various importing countries, an exporting country has to invest in resources and technology. If there is a greater uniformity of standards, the cost of doing business will be correspondingly reduced. The objective is to seek ways of achieving a degree of harmonization of standard without offsetting or damaging any objective benefits that such standards confer.

To what extent sanitary and phytosanitary regulations in different countries do in fact act as a trade restrictive device is difficult to quantify. Shipments at points of entry in various countries are rejected or barred from entry on grounds of failing to meet the standards of sanitary and phytosanitary safety in importing countries. In fact, the exporting countries might be discouraged from undertaking production for export in view of the uncertainty created by regulations that may restrict or prohibit entry into the importing countries.

Consensus on Harmonization

In the ongoing negotiations of the Uruguay Round, it is broadly agreed that standards should be harmonized. In this context, harmonization means the 'establishment, recognition and application of common sanitary and phytosanitary measures'. It is further agreed among countries that 'equivalence' of standards should be pursued. Individual countries should strive to adopt internationally agreed standards. Secondly, international as well as national standards should be based on scientific evidence. In this context, international organizations such as FAO, WHO and the International Plant Protection Convention (IPPC) should provide norms for: a) acceptable levels of risk assessment; and b) methods of risk assessment. National risk assessment procedures should be scientifically-based and transparent to all countries.

Under these circumstances, a country's standards may exceed international standards. Harmonization of standards does not imply that the countries must have identical standards, but that standards could still be accepted as equivalent by the countries concerned in the light of information on assessment methods and acceptable risk levels.

Questions have been raised as to the feasibility of establishing internationally agreed-upon standards, based on scientific criteria, that cover a wide range of products and processes. First, it is argued that in view of the difficulties of reaching international agreements, international standards should be restricted to a very few 'high visibility' products or processes. Second, the scientific evidence is never absolute. There are always areas of uncertainly and margins of doubt about both assessment methods and acceptable risk levels. The latter is often a matter of value judgment and political consensus in a country.

Third, international negotiations setting such standards have been greatly influenced by the negotiating strength of various governments and industry groups engaged in these negotiations. In other words, standards are frequently politically negotiated rather than based strictly upon scientific evidence. Even when standards are nationally established, they are negotiated politically among various interest groups, including the relevant agencies or departments within a national government.

Fourth, the process of international negotiations on setting standards is not recognized by all concerned as sufficiently transparent. The various public interest groups, such as consumer groups and environmental NGOs, do not feel that they are full participants, or that their concerns are adequately reflected in the process. Thought may need to be given to the ways and manner in which a broader public participation and greater transparency can be ensured.

Assistance for Developing Countries

How far can developing countries adopt and implement internationally agreed or acceptable sanitary or phytosanitary standards? The developing countries require technical and financial assistance in order to develop the capacity, both administrative and technical, to formulate and implement such standards. Developing countries also need a 'longer time period' to adopt and implement the internationally agreed-upon standards. The level of technical or financial assistance available and the agencies or organizations that will provide such assistance need to be clearly spelled out. Unless the provision of such assistance is an integral part of the international agreement on standards, its availability in practice will remain highly uncertain.

Should trade policies be used to enforce compliance with environmental standards, for example through trade restrictions by an importing country on another's exports originating from sectors or commodities that do not conform to the environmental standards of the importing countries? This would inhibit efficient international specialization. Provided minimum scientific requirements are met, international agreement on environmental standards or their 'equivalence' or 'mutual recognition' would prevent or pre-empt the possibility of distortion of trade flows away from comparative cost correlations by arbitrary importation of trade restrictions on a unilateral basis. The use of trade restrictions to enforce environmental standards on trading partners opens up the possibility of abuse on the part of protectionist forces and makes strange bedfellows of baptists (environmentalists) and bootleggers (protectionists).

Trade policies should promote the efficient international division of labour and growth of trade based on comparative costs. However, trade policies or restrictions may have to be used as a last resort in the absence of feasible alternative measures to promote environmental standards. Care must be taken to ensure that they are used in the least trade-distorting way. In a number of international conventions relating to specific aspects of the environment, trade sanctions or restrictions have been incorporated as a way of enforcing compliance. For example, the Montreal Protocol on the Depletion of Ozone prohibits trade in products incorporating chlorofluorocarbons among members of the convention as well as between members and nonmembers.

Trade and environment intersect at various points. Trade affects the pattern and intensity of use of natural resources. Similarly, the national environmental policies that determine product and process standards affect trade. Both trade and environment policies need to be devised with a view to promoting growth, poverty alleviation and the protection of the environment.

For Product Safety Concerns and Information please contact our EU representative GPSR@taylorandfrancis.com Taylor & Francis Verlag GmbH, Kaufingerstraße 24, 80331 München, Germany

T - #0049 - 230425 - C0 - 219/152/31 - PB - 9781138741621 - Gloss Lamination